SILICONE
DISPERSIONS

SURFACTANT SCIENCE SERIES

FOUNDING EDITOR
MARTIN J. SCHICK
1918–1998

SERIES EDITOR
ARTHUR T. HUBBARD
Santa Barbara Science Project
Santa Barbara, California

SILICONE DISPERSIONS

Edited by
Yihan Liu

CRC Press
Taylor & Francis Group
Boca Raton London New York

CRC Press is an imprint of the
Taylor & Francis Group, an **informa** business

CRC Press
Taylor & Francis Group
6000 Broken Sound Parkway NW, Suite 300
Boca Raton, FL 33487-2742

First issued in paperback 2019

ISBN-13: 978-1-4987-1555-3 (hbk)
ISBN-13: 978-0-367-87256-4 (pbk)

Library of Congress Cataloging-in-Publication Data

Names: Liu, Yihan (Chemist)
Title: Silicone dispersions / [edited by] Yihan Liu.
Description: Boca Raton : Taylor & Francis, 2016. | Series: Surfactant
science | "A CRC title."
Identifiers: LCCN 2016028467| ISBN 9781498715553 (hardback : alk. paper) |
ISBN 9781498715560 (e-book)
Subjects: LCSH: Silicones. | Silicones--Industrial applications.
Classification: LCC QD383.S54 S558 2016 | DDC 547/.7--dc23
LC record available at https://lccn.loc.gov/2016028467

Visit the Taylor & Francis Web site at
http://www.taylorandfrancis.com

and the CRC Press Web site at
http://www.crcpress.com

Contents

Preface

Silicones have been sold commercially for more than 70 years and the market has dramatically grown. Silicones are used in a wide variety of applications, for example, as engine lubricants, construction sealants, molds, optical lenses, antifoams, electronic encapsulates, adhesives and release liners, and as additives in coating, cosmetic, personal care, and household products. The wide range of applications is attributed to the unique properties of silicones that are distinct from those of inorganic glasses and organic plastics. Silicones are manufactured and used in different forms: neat, solutions, or dispersions. As with all materials, the properties and utilities of silicones are inseparable from their forms of delivery. This book is on silicone dispersions. It is intended to fill a gap in the literature and provide in one collection a description of the preparation, properties, and applications of several commercially important categories of silicone dispersions. The book also provides references to scattered literature and patents.

The dispersion systems described in this book are of the following four types: liquid-in-liquid (emulsion), solid-in-liquid (suspension), air-in-liquid (foam), and solid-in-air (powder). The chapters in this book are each written by leading experts in the particular area. The book begins with an introduction to the basic properties of silicones, including interfacial properties and miscibility between silicones and other liquids. This provides the necessary background for understanding the many phenomena encountered in silicone dispersions that the subsequent chapters describe.

Chapter 2 provides a comprehensive review of the broad category of silicone emulsions and microemulsions. Aqueous emulsions of each of the specific type of silicones—fluid, gum, elastomer, resin, and volatile silicone and silane—are described with respect to their preparation and stability. The authors relate the stability of emulsion and various process features to the phase behavior of the system. This approach gives fundamental insights and allows the reader (and formulator) to appreciate simple rationales behind formulation and process design.

Emulsion polymerization is a cost-effective route to synthesize silicones for use in water-based applications. Chapter 3 provides an overview of the different routes to polymerize silicones in aqueous media. A detailed account is also given of the mechanistic and kinetic aspects of one specific route of silicone emulsion polymerization—that based on the ring-opening polymerization of cyclic siloxanes. Chapters 4 and 5 further demonstrate the use of emulsion polymerization to prepare elastomeric silicones, which are otherwise difficult to disperse. Chapter 4 addresses water-based silicone elastomers, and Chapter 5 addresses silicone elastomeric powders.

A theoretical treatment is presented in Chapter 6 where Janus emulsions involving water, silicone oil, and organic oil are used to elucidate a fundamental aspect of multiple emulsions. It is shown that the interfacial free energy dictates the structure of the multiple emulsion, a revelation that is somewhat surprising given that an emulsion is not an equilibrium system. The message delivered in this chapter has profound implications in the engineering of hybrid materials.

Chapter 7 describes aqueous suspensions of silica and silsesquioxane prepared from the sol-gel process. Progress in microcapsule technologies based on these sol-gel products is reviewed.

The next four chapters are devoted to four major, but by no means exclusive, application areas involving silicone dispersions. Chapter 8 deals with silicone antifoams. The essential elements of the complicated dispersion system composed of foam and antifoam compounds are discussed. New experimental evidence is reported that demonstrates a previously unrecognized spreading process and illustrates the dynamic interactions between foam and antifoam compounds. Chapter 9 covers the application of silicones in cosmetics and personal care. The author gives a historical account of how silicones entered the cosmetic and personal care market. The unique attributes that silicones bring to the various products are described. Some formulary guidance is also provided. Chapter 10 shows how silicones and silanes are used as water repellents to treat construction materials. The final chapter reviews the diverse applications of silicones in the exploration, well completion, and production parts of the oil and gas industry.

I hope the reader finds this book a useful reference and that it contributes to a better understanding of the material science of silicone dispersions. I thank most sincerely all the authors for their contribution to this book despite their busy schedules. Special thanks are due to Dr. Randy Hill and Professor Michael Brook for their valuable comments as reviewers to the original book proposal. My appreciation also goes to Barbara Glunn of Taylor & Francis Group for her guidance and patience in the preparation of this book.

Yihan Liu
Midland, Michigan
September 2016

Editor

Yihan Liu is a senior product development specialist at Dow Corning Corporation, Midland, Michigan, and was previously a senior scientist at Johnson & Johnson Consumer Companies, Inc., Skillman, New Jersey. With 20 years of industrial experience specialized in the research, product development, and applications of surfactants and emulsions, Dr. Liu has commercialized numerous products. She is the inventor of 10 issued US patents and 20 additional pending applications and is the author of 11 journal articles and book chapters in the field of colloid and interface science. She earned a BA in physics and mathematics at Saint Olaf College, an MS in physics at the University of Washington, and a PhD in materials science and engineering at the University of Minnesota.

Contributors

Steven P. Christiano
SiVance LLC
A subsidiary of Milliken & Co.
Spartanburg, South Carolina

Samuel F. Costanzo
Dow Corning Corporation
Midland, Michigan

Stig E. Friberg
Ugelstad Laboratory
NTNU
Trondheim, Norway

Ronald P. Gee
Dow Corning Corporation
Midland, Michigan

Randal M. Hill
Flotek Chemistry, LLC
Houston, Texas

Jean-Paul Lecomte
Dow Corning Europe SA
Seneffe, Belgium

Donald T. Liles
Dow Corning Corporation
Currently at: Si-Aqua Consulting, LLC
Midland, Michigan

Yihan Liu
Dow Corning Corporation
Midland, Michigan

Leon Marteaux
Dow Corning Europe SA
Seneffe, Belgium

Gianna Pietrangeli
Flotek Chemistry, LLC
Houston, Texas

Michael S. Starch
Wintermute Consulting Services
Midland, Michigan

Siwar Trabelsi
Flotek Chemistry, LLC
Houston, Texas

Isabelle Van Reeth
Dow Corning (China) Holding Co., Ltd.
Pudong, Shanghai
People's Republic of China

Mari Wakita
Dow Corning Toray Co., Ltd.
Chiba, Japan

Oliver Weichold
Institute of Building Materials Research
RWTH Aachen University
Aachen, Germany

1 Basic Structure and Properties of Silicones

Yihan Liu and Samuel F. Costanzo

CONTENTS

1.1 INTRODUCTION

Silicones were first prepared by Professor Frederic Stanley Kipping in 1904, who described the materials as "sticky messes of no particular use" [1]. In spite of his doubtfulness, Kipping and his coworkers continued the research for the next 40 years. Based on Kipping's work, industrial research on silicones began during the 1930s at Corning Glass Works and General Electric, which then led to the birth of the silicone industry in the early 1940s [2]. The first silicone product, Dow Corning® 4 Compound, was introduced in 1942. The product was used as a grease to coat the ignition wire harnesses of Allied airplanes during World War II. Following the war, the silicone industry shifted its focus to new applications and launched the first civilian product, a silicone-in-water emulsion, used as a tire release agent [2]. During the next 70 years, silicone materials blossomed into a global business. According to market report, worldwide silicone consumption reached 1.7 million tons in 2012 with an average annual growth of 6% during the 2007–2012 period [3]. The market is projected to continue to grow in the near future.

Today silicones are commercially available in thousands of product forms and used in a wide variety of applications ranging from construction, auto and aviation, medical, electronics, coatings, textiles, to household and personal care products [4]. Remarkably, the many benefits that silicones provide are the consequence of only a few basic properties. These properties are in turn attributed to the nature of the siloxane bond and the pendent organic groups. This chapter describes the nomenclature, structures, and basic properties of silicones. Since the literature has already adequately covered the chemistry and material science of silicones [5–7], this chapter is kept as brief as possible and aims to provide background information necessary for the understanding of the subsequent chapters. As this book is devoted to silicone dispersions, it is important to outline some of the most fundamental surface and interfacial characteristics of silicones. The reader is encouraged to read detailed discussions by Owen on silicone surface behaviors and their molecular origins [8,9]. Information regarding the compatibility of silicones with other liquids and—pertinent to aqueous dispersions of silicones—the mutual solubilities of silicones and water are provided as well; this may help explain certain behaviors in silicone dispersions. Silicone surfactants, which are employed in many dispersion systems and are a unique class of materials on their own, have been thoroughly described by Hill [10]; only a very brief introduction is given here. Several of the subsequent chapters will elaborate upon the role that silicone surfactants play in specific processes and applications.

1.2   STRUCTURE AND NOMENCLATURE OF SILICONES

The term "silicone" is used rather loosely in the literature. In general, it refers to the class of synthetic materials based on a polymeric siloxane backbone consisting of alternating silicon and oxygen atoms with organic pendent groups attached to the silicon atoms (Figure 1.1). The organic groups can be inert or reactive, apolar or polar. Methyl is the predominant organic group in most commercially available silicones. Silicones can be classified as fluids, elastomers, or resins depending on the extent of cross-linking. When no specific description is given to the structure, the phrase "silicone oil" typically refers to linear polydimethylsiloxane (PDMS), which is a fluid at room temperature and down to very low temperatures at all molecular weights. Silicone elastomers or silicone rubbers are cross-linked PDMS. Silicone resins are highly cross-linked network structures; they can be either liquid or solid at room temperature. "Functional silicones" typically refer to PDMS that have a portion of the methyl groups replaced by nonmethyl groups. Some representative examples of the various types of silicones are given in Table 1.1.

One convenient way to describe the structure of a silicone is by the MDTQ shorthand notation [11]. This notation recognizes the fact that all silicone molecules can be constructed from some combination of four basic building blocks. These are defined according to the number of oxygen atoms covalently bonded to the silicon, as illustrated in Figure 1.2. Thus, M, D, T, and Q each represents a siloxane unit having, respectively, one (mono), two (di), three (tri), or four (quad) oxygen atoms attached to the silicon. For example, the smallest siloxane, hexamethyldisiloxane, can be written as MM. The cyclic tetramer, octamethylcyclotetrasiloxane, can be written as D_4. A linear PDMS with n repeating dimethylsiloxane units and a trimethylsiloxy group at each end can be written as MD_nM. If, in addition, there are m siloxane units having functional groups

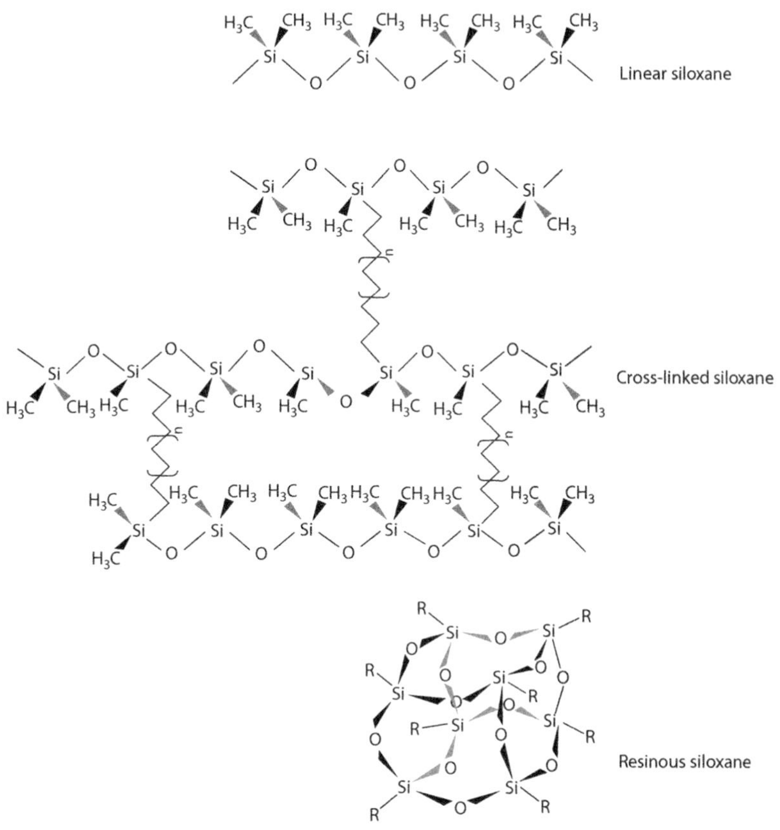

FIGURE 1.1 Molecular structure of polysiloxanes (silicones).

attached, the structure can be expressed as $MD_nD'_mM$. By this nomenclature, it is easy to recognize whether the molecule is linear (M and D units), cross-linked (mostly D with a small amount of T), or resinous (mostly T or Q). T-based resins, which may contain M and D units in addition to T units, are also called polysilsesquioxanes. Q-based resins are typically capped with M units and are therefore referred to as MQ resins. An uncapped Q structure is essentially silica. Silicones can be considered as organic–inorganic hybrid materials. As the mole ratio of silicon to carbon increases, the inorganic nature of the material becomes more pronounced. This is depicted in Figure 1.2.

Since each oxygen atom in the siloxane backbone is shared between two silicon atoms and each silicon is bonded to four other atoms, silicones can be represented by an average molecular formula of the form $R_nSiO_{(4-n)/2}$. It can be seen that M, D, T, and Q units correspond to $n = 3, 2, 1$, and 0, respectively. Since a silicone molecule can be composed of a variety of combinations of the four units, n can be any number between 0 and 3 and it can be a noninteger. This form of presentation is convenient and is often used in the patent literature.

TABLE 1.1

Some Representative Examples of Silicones

Name	Shorthand Notation	Structure
Polydimethylsiloxane (silicone oil)	MD_nM	$(CH_3)_3OSi\left[\begin{array}{c} CH_3 \\ \mid \\ OSi \\ \mid \\ CH_3 \end{array}\right]_n OSi(CH_3)_3$
Polydimethylsiloxane-*co*-polyoxyalkylene (silicone polyether)	$MD_nD'_m(EO_xPO_y)M$	$(CH_3)_3OSi\left[\begin{array}{c} CH_3 \\ \mid \\ OSi \\ \mid \\ CH_3 \end{array}\right]_n\left[\begin{array}{c} CH_3 \\ \mid \\ OSi \\ \mid \\ (CH_2)_3 \end{array}\right]_m OSi(CH_3)_3$ $(OCH_2CH_2)_x(OCHCH_2)_yOR$ $\mid$ CH_3
Poly(dimethyl methyl aminoethyl aminopropyl) siloxane (amino silicone)	$MD_nD'_m[(CH_2)_xNH_2]M$	$(CH_3)_3OSi\left[\begin{array}{c} CH_3 \\ \mid \\ OSi \\ \mid \\ CH_3 \end{array}\right]_n\left[\begin{array}{c} CH_3 \\ \mid \\ OSi \\ \mid \\ (CH_2)_3 \end{array}\right]_m OSi(CH_3)_3$ NH $\mid$ $(CH_2)_2$ $\mid$ NH_2
Poly(dimethyl methylalkyl) siloxane (silicone wax)	$MD_nD'_m(C_xH_{2x+1})M$	$(CH_3)_3OSi\left[\begin{array}{c} CH_3 \\ \mid \\ OSi \\ \mid \\ CH_3 \end{array}\right]_n\left[\begin{array}{c} CH_3 \\ \mid \\ OSi \\ \mid \\ CH_2 \end{array}\right]_m OSi(CH_3)_3$ $(CH_2)_{x-2}CH_3$
Methyl silsesquioxane (silicone resin)	T^{Me}	$\left[(CH_3)SiO_{3/2}\right]_n$

Besides silicones, there are other useful organosilicon materials such as silanes, silazanes, siliconate, silicate, and so on. Some examples are given in Table 1.2. Although this book focuses on dispersions of silicones, silanes will be included in certain discussions. Silanes are reactive monomeric molecules containing only one silicon atom. Besides being precursors in the synthesis of siloxanes [12], silanes are used primarily as coupling agents in the production of composite materials and as adhesion promoters [13]. Silanes are also used to treat various surfaces to provide hydrophobicity [13]. In this book, silanes will be discussed as cross-linking agents in the synthesis of water-based silicone elastomers, as monomers in the emulsion polymerization of silicone copolymers and resins, and as the main active ingredients in various dispersion systems for waterproofing applications. Polysilanes are yet a different class of materials having very different properties than silanes or silicones. The backbones of polysilanes are composed solely of silicon atoms. Polysilanes have unique electronic and optical properties and are used in electronic applications. Polysilanes will not be discussed in this book.

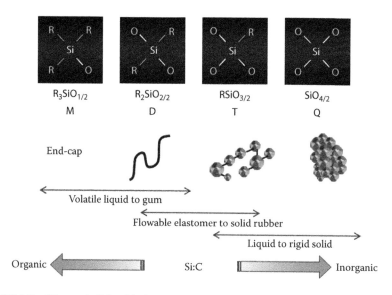

FIGURE 1.2 Siloxane building blocks, where R represents the same or different organic radical.

TABLE 1.2

Examples of Silanes and Other Silicon-Containing Materials

Name	Structure	Example	Main Uses
Silane	$R_3 -\!\!\underset{\underset{R_2}{\vert}}{\overset{\overset{R_1}{\vert}}{Si}}\!\!- R_4$	Dimethyldichlorosilane $(CH_3)_2SiCl_2$ Octyltriethoxysilane $C_8H_{17}Si(OCH_2CH_2)_3$	Precursor for making siloxanes; coupling agent; hydrophobing surfaces
Silazane	$-Si - N - Si -$	Hexamethyldisilazane $((CH_3)_3Si)_2NH$	Adhesion promoter; surface treatment
Siliconate	$R -\!\!\underset{\underset{O^-}{\vert}}{\overset{\overset{O^-}{\vert}}{Si}}\!\!- O^-$	Sodium methylsiliconate $CH_3Si(ONa)_3$	Water proofing
Silicate	$\cdots Si -\!\!\underset{\underset{O}{\vert}}{\overset{\overset{O^-M^+}{\vert}}{}}\!\!- O^-M^+$	Sodium metasilicate (water glass) Na_2SiO_3	Binder in paints and coatings
Polysilane	$\left[\begin{array}{c} R \\ \vert \\ Si \\ \vert \\ R \end{array}\right]_n$	Polydimethylsilylene, $[(CH_3)_2Si]_n$	Precursor to silicon carbide; lithography

1.3   THE SILOXANE BOND AND BASIC PROPERTIES OF SILICONES

Virtually all applications of silicones exploit one or another of the following properties:

- Chemical and thermal stability
- Low glass transition temperature
- Low viscosity and near-zero surface shear viscosity
- Low surface tension and moderate interfacial tensions
- High gas permeability

These properties of silicones are ultimately attributed to the nature of the siloxane bond (Si–O) and the organic group attached to the silicon. Table 1.3 lists the electronegativity of silicon, carbon, and oxygen and compares the characteristics of the siloxane bond with those of C–C and C–O bonds.

As can be seen, silicon is significantly more electropositive than carbon. This results in a considerable degree of ionic character in the Si–O bond. The bond energy of Si–O is relatively high, which accounts for the thermal stability of silicones. The Si–O–Si bond angle is rather wide and variable, and the bond lengths of both Si–O and Si–C are long in comparison to those of C–O and C–C. The reason for the wide and variable Si–O–Si bond angle lies in the silicon-heteroatom electronic configuration, the nature of which is still an ongoing scientific debate [14]. Nevertheless, the wide and variable bond angle, the long bond length, and the fact that the organic substituents are attached to only every other atom along the polysiloxane backbone, make the polysiloxane chain extremely flexible. This is especially true for PDMS due to the small size of the methyl group. The rotational energy barrier around the Si–O bond in PDMS is virtually zero. These features are best illustrated in Table 1.4, which compares the physical properties of PDMS to those of polyisobutylene (PIB), an all-hydrocarbon structural analog to PDMS. As can be seen, PDMS has a remarkably low glass transition temperature (T_g). This is a consequence of the extreme flexibility

---

TABLE 1.3

Electronegativity and Bond Characters[a]

	Si	C	O	
Electronegativity	1.8	2.5	3.5	
	Si–O	**Si–C**	**C–O**	**C–C**
Bond length (Å)	1.64	1.90	1.42	1.53
Bond energy (kJ/mol)	445	306	358	346
Ionic character (%)	50	12	22	0
Rotational energy (kJ/mol)	<0.8	6.7 (Si–CH$_3$)	11.3	15.1 (C–CH$_3$)
	Si–O–Si	**O–Si–O**	**C–O–C**	**C–C–C**
Bond angle (°)	143	110	112	112

[a] All values taken from References 6 and 9 which cite from various original sources referenced therein.

TABLE 1.4

Comparison of Physical Properties of PDMS and PIB (Measured at 25°C Unless Otherwise Indicated)

	PDMS	PIB	References
Structure	$\left[\!-O-\overset{\overset{\displaystyle CH_3}{\mid}}{\underset{\underset{\displaystyle CH_3}{\mid}}{Si}}-\!\right]_n$	$\left[\!-CH_2-\overset{\overset{\displaystyle CH_3}{\mid}}{\underset{\underset{\displaystyle CH_3}{\mid}}{CH}}-\!\right]_n$	
T_g (°C)	−123	−71	[17,18]
Viscosity (mPa·s)	4.6 (MW = 800)	3.7×10^3 (MW = 755, T = 30°C)	[19–21]
	9.7×10^2 (MW = 28,000)	9.59×10^6 (MW = 27,000)	
	9.7×10^4 (MW = 139,000)	1.39×10^9 (MW = 133,000)	
Permeability (Barrer)			[16]
O_2	600	1.4	
N_2	280	0.3	
Surface tension (mN/m²)	21	34	[8,22]
Interfacial tension against water (mN/m²)	43	60[a]	[23,24]

[a] Value calculated from the Young equation using water surface tension of 72.8 mJ/m², PIB surface tension of 34 mJ/m² [22], and water contact angle on PIB of 111° [24].

of the siloxane backbone in combination with the low intermolecular forces between methyl groups. The high flexibility and low intermolecular forces also account for the much lower viscosity of PDMS fluids as compared with their hydrocarbon counterparts. The contrast is particularly pronounced at high molecular weight. PDMS also has the lowest recorded surface shear viscosity among all polymers (<10⁻⁵ surface poise) [15]. Liquids and gases permeate more rapidly through silicone rubber than through other organic polymers. The very large permeability through silicone is primarily due to the large diffusivity; the latter is explained by the ease of segment rotation about the Si–O bond, which accommodates motion and results in a low activation energy for diffusion. Permeation and diffusion in silicone are also less sensitive to the size of the penetrating molecules than in other organic polymers [16].

1.4 SURFACE AND INTERFACIAL BEHAVIOR, COMPATIBILITY, AND SOLUBILITY OF SILICONES

1.4.1 SURFACE AND INTERFACIAL ENERGY

Owen has compared the surface energies of the homologues of n-alkanes, PDMS, n-fluoroalkanes, polyoxyhexafluoropropylene, and plotted the surface energies against boiling points of the oils [8]. Figure 1.3 is a reprint. As can be seen, fluorinated polymers

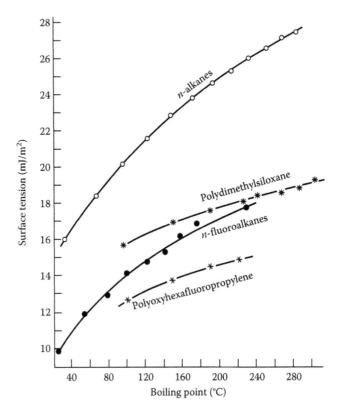

FIGURE 1.3 Surface tension at 20°C versus boiling point for *n*-alkanes, PDMS, *n*-fluoroalkanes and polyoxyhexafluoropropylene. (Reprinted with permission from Owen, M.J., *Ind. Eng. Chem. Prod. Res. Dev.*, 19, 97–103. Copyright 1980 American Chemical Society.)

as a whole have lower surface energies than PDMS, which in turn have lower surface energies than *n*-alkanes. The relative values reflect the ranking of group contribution to surface energy: $-CH_2->CH_3->-CF_2->-CF_3$ [25]. PDMS has a lower surface energy than a corresponding *n*-alkane because PDMS is permethylated along the entire chain, whereas an *n*-alkanes contains only terminal methyl groups, and the rest of the chain is made of methylene groups. By the same token, polyoxyhexafluoropropylene is perfluoromethylated along the entire chain, so its surface energy is lower than that of a corresponding *n*-fluoroalkane. The curves all have positive slopes; this is mostly due to the end group effect which, for PDMS and polyoxyhexafluoropropylene, is understandably less than for *n*-alkanes and *n*-fluoroalkanes.

Even though the Si–O bond is polar, its contribution to surface energy is minimized by the densely packed methyl surface that results from a highly flexible and freely rotatable PDMS chain. This element is best demonstrated by comparing the surface tension of PDMS, 21 mJ/m^2, with that of PIB, 34 mJ/m^2. For PIB, even though there are two pendent methyl groups on each repeating unit, a similar situation as in PDMS, its backbone is relatively rigid. Therefore, the PIB methyl groups cannot freely orient to achieve maximum surface exposure. This flexibility also explains why PDMS that have been grafted with polar functional groups display surface tensions that are very close to that of nonfunctional PDMS. The molecule can freely orient to hide the polar functional groups in the sublayer away from the surface.

While PDMS displays a lower surface tension than n-alkanes, its interfacial tension against water is, perhaps to one's surprise, lower than that of n-alkanes against water. Kanellopoulos and Owen showed an average interfacial tension of about 42.6 mJ/m^2 for PDMS against water. The value varies little with PDMS chain length except for the smallest siloxane, the dimmer MM, which has a slightly higher value, 44 mJ/m^2 [23]. With the presence of silanol groups (SiOH) on PDMS, the interfacial tension can be lower by several units. In contrast, the interfacial tension for n-alkanes against water ranges from 50 mJ/m^2 (hexane) to 53 mJ/m^2 (hexadecane) [23]. The comparatively low interfacial tension against water for silicone is again attributed to the flexibility of the siloxane backbone. When exposed to air, the silicone molecules orient to maximize the packing density of the methyl groups at the surface, whereas when exposed to water the molecules orient to allow interaction of the polar siloxane bond with water.

Interfacial tensions between PDMS and various common organic polymers have been reported in the literature and a compilation can be found in Reference 26. Values are typically in the range of several mJ/m^2, and the more polar the organic polymer, the higher the interfacial tension with PDMS. Interfacial tensions between PDMS and weakly polar hydrocarbon oils are also several mJ/m^2. For example, the interfacial tension between PDMS and sunflower oil has been reported to be 2.54 mJ/m^2 [27].

1.4.2 WETTING AND SPREADING

Silicones tend to spread over most surfaces. A common but not entirely correct explanation is that the phenomenon is due to the low surface tension of silicones. Sometimes silicones are described as surface active in reference to their tendency to spread. It is to be pointed out that spreading of a bulk material on a surface is not to be confused with it being surface active, or else water will be considered surface active as well since it spreads (and wets) over a clean driveway and it does that rather fast! Here we adhere to the concept that "surface active" refers to aggregation behavior of a compound in a homogeneous solution, such that the surface has an excess concentration of the compound thereby lowering the surface tension of the *solution*. Wetting and spreading, on the other hand, are dictated by three interfacial tensions at the contact line the liquid forms with the substrate. Figure 1.4 illustrates the concept. At equilibrium, the following relationship holds,

$$\gamma_{sv} = \gamma_{ls} + \gamma_{lv} \cos\theta \qquad (1.1)$$

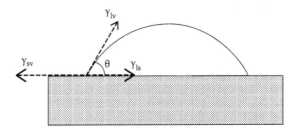

FIGURE 1.4 A liquid drop on a substrate at equilibrium.

where γ_{sv}, γ_{ls}, γ_{lv} are interfacial tensions between the substrate and the vapor phase, the liquid and the substrate, and the liquid and the vapor phase, respectively, and θ is the contact angle as drawn in the figure.

Typically wetting is viewed as when the equilibrium contact angle is less than 90°. A contact angle of 0° corresponds to complete wetting. Spreading of a liquid on a substrate, on the other hand, can occur merely under gravity if there is enough of the liquid, whether the liquid wets the substrate or not. But if the liquid completely wets the substrate and if in addition, the interfacial tensions are such that the spreading coefficient, S, defined as

$$S = \gamma_{sv} - (\gamma_{lv} + \gamma_{ls})$$ (1.2)

is positive, the liquid will spontaneously spread without gravity. Silicone wets many surfaces as a consequence of the low surface tension of silicone *as well as* a sufficiently low interfacial tension between silicone and the substrate. The low bulk and surface viscosity help the rate of spreading, though they do not determine the state of wetting. Spontaneous spreading occurs when the material over which silicone spreads has a relatively high surface tension but a low interfacial tension with silicone such that the spreading coefficient is positive. Silicone homopolymers such as PDMS are not surface active, but they may spontaneously spread over a high energy surface. An example is PDMS over water or over vegetable oil. In both cases the spreading is spontaneous.

Silicones *can* be surface active. Surface-active silicones, that is, silicone surfactants, which are described in more detail in Section 1.5, are amphiphilic silicone copolymers obtained by attaching functional groups or another polymer onto the siloxane backbone. The PDMS portion of the copolymer is hydrophobic but not lipophilic; it is sometimes described as siliphilic and the portion is referred to as a siliphile. An amphiphilic silicone copolymer may contain, for example, a siliphile and a hydrophile, or a siliphile and a lipophile, or all three. However, not all amphiphilic silicones are surface active. Only those that have a suitable amphiphilic balance (like the analogous concept of hydrophile–lipophile balance (HLB) for a hydrophile–lipophile surfactant) to render solubility *and* aggregation behavior are surface active. Like any amphiphilic polymer, the transition from complete molecular solubility to aggregation and thence to complete phase separation can occur not only with a change in the ratio between the two portions of the silicone copolymer, but also with an increase in the overall molecular weight even if the amphiphilic balance is kept the same. Many amphiphilic silicone copolymers are insoluble in a solvent of interest because of either an insufficient amount of the solvent-like portion or simply because the molecular weight of the copolymer is too high. Such insoluble amphiphilic silicone copolymers are not surface active in the particular solvent, but they may be readily dispersible in that solvent with an additional stabilizer. Examples are high molecular weight silicones having a few molar percent polar functional substitutions in the molecule. These silicone copolymers can be readily dispersible or even microemulsifiable in water using additional surfactant (see Section 2.5).

1.4.3 COMPATIBILITY AND SOLUBILITY

The above discussions around silicone surface and interfacial behaviors are with respect to the situation where the silicone phase is immiscible with the other phase in the system under consideration, be the other phase air (the problem concerns with silicone surface behavior) or another liquid or solid (the problem concerns with silicone interfacial behavior). When silicone is miscible or partially miscible with another phase in a system, the precise state of miscibility can be of essential consequence to the dispersion structure and behavior, so here we provide some information on silicone solubility and compatibility with other materials that are frequently encountered in silicone applications. We clarify that by "complete miscibility" we mean two components, when mixed, form a single phase solution at all proportions, and by "partial miscibility" we mean the two components form a single phase solution only at certain range of proportions, while outside of that range two phases coexist in the mixture.

Like organic (hydrocarbon-based) oils, nonfunctional silicones such as PDMS have very limited solubility in water, and for this reason nonfunctional silicones are commonly called oils too. However, silicone oils are not necessarily miscible with organic oils. PDMS is completely miscible at room temperature with paraffins smaller than C_{16}, regardless of PDMS molecular weight. Longer chain paraffins and high molecular weight PDMS are only partially miscible with one another. PDMS, with the exception of lower oligomers (viscosity less than 5 cSt or size smaller than MD_7M), is also immiscible with what are commonly called natural oils and fats which are composed mainly of fatty acids and their triglycerides. Solvents that are miscible in all proportions with PDMS of any molecular weight are those which are weakly polar or polarizable, such as benzene, toluene, xylene, carbon tetrachloride, chloroform, kerosene, and methyl ethyl ketone. More polar solvents such as acetone, butanol, isopropanol, and ethanol are miscible only with low molecular weight dimethylsiloxanes (viscosity below 10 cSt) but immiscible or partially miscible with higher molecular weight PDMS.

The mutual solubility of PDMS and water, albeit small, is fundamentally telling. At first glance, PDMS seems to be much less soluble in water than its hydrocarbon analogue. Figure 1.5 plots the solubilities of the homologues of PDMS and n-alkanes as a function of the number of repeating units in the molecule. A linear PDMS pentamer (MD_3M, or L_5) has an aqueous solubility (1.83×10^{-10} M) more than 5 orders of magnitude lower than that of n-pentane (5.28×10^{-4} M)! It is perhaps tempting

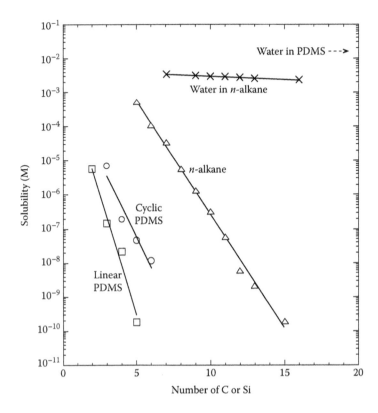

FIGURE 1.5 Solubility versus number of carbon or silicon atoms for: *n*-alkanes (triangle), linear PDMS (square), and cyclic PDMS (circle) in water; water in *n*-alkanes (cross); and water in PDMS (dashed arrow pointing to data located at number of Si ≈ 40). Data taken from Riddick et al. [28] (C_5 in water), McAuliffe [29] (C_6 in water), Yalkowsky et al. [30] (C_7 in water), Tolls et al. [31] ($C_8 - C_{15}$ in water), Schatzberg [32] (water in $C_7 - C_{16}$), Varaprath et al. [33] (linear and cyclic PDMS in water), Hakim et al. [34] (water in PDMS). Where needed, solubility values have been converted to the unit of mol/L (M) using density values from Reference 35.

to attribute the markedly lower solubility of PDMS to its high hydrophobicity. But, recognizing that the free energy of solubilization depends on the size of the solute, it is more meaningful to plot solubility against molar volume (Figure 1.6). Now one finds that PDMS is actually more soluble in water (also by orders of magnitude) than its *n*-alkane counterpart of the same molecular size. PDMS appears to be less soluble than alkane in Figure 1.5 only because the PDMS molecule is much larger. The smallest PDMS, MM (or L_2), is already at the size of *n*-undecane (n-$C_{11}H_{24}$); a linear trimer (L_3) is similar in size to *n*-hexadecane (n-$C_{16}H_{34}$). The higher aqueous solubility of PDMS (per molar volume) can be attributed to the polar nature of the Si–O bond and the ability of the PDMS backbone to flex to expose the polar bond. The fact that the slope of the exponentially decreasing solubility with increasing molecular size is less steep for PDMS than for *n*-alkanes (Figure 1.6) is also attributed to the flexibility of the PDMS backbone.

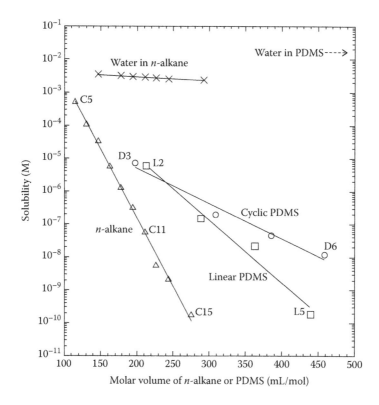

FIGURE 1.6 Solubility versus molar volume for: *n*-alkanes (triangle), linear PDMS (square) and cyclic PDMS (circle) in water; water in alkanes (cross); and water in PDMS (dashed arrow pointing to data located at number of Si $\approx$ 40). Data taken from Riddick et al. [28] (C_5 in water), McAuliffe [29] (C_6 in water), Yalkowsky et al. [30] (C_7 in water), Tolls et al. [31] ($C_8 - C_{15}$ in water), Schatzberg [32] (water in $C_7 - C_{16}$), Varaprath et al. [33] (linear and cyclic PDMS in water), Hakim et al. [34] (water in PDMS). Where needed, solubility values have been converted to the unit of mol/L (M) using density values from Reference 35.

The solubility of cyclic PDMS in water is greater than that of linear PDMS with the same number of siloxane units (Figure 1.5). This may be due, in part, to the smaller molecular size of cyclic siloxane as compared to its linear counterpart but additional factors must be at play. Note, for example, that D_4 is smaller in molar volume than L_4 but slightly larger than L_3, and yet solubility of D_4 is higher than solubilities of both L_3 and L_4 (Figure 1.6). D_6 is slightly larger than L_5 but its solubility is two orders of magnitude higher than that of L_5. The slope of solubility with increasing molecular size is less steep for cyclic PDMS than it is for linear PDMS. The same feature was also observed for cyclic and branched alkanes versus linear alkanes [29,31], and this has been ascribed to the higher vapor pressures of cyclic and branched alkanes as compared to linear alkanes [29]. The same reason may apply to the case of PDMS; the vapor pressures of cyclic dimethylsiloxanes are also found to be higher than those of their linear counterparts [36].

Interestingly, Hoffmann and Stümer had reported very similar features in the solubilization of PDMS and alkanes in surfactant aqueous micellar solutions [37]. The oil molecules are solubilized in the hydrocarbon core of the micelles and the solubility of oil per unit solution is generally much higher than the solubility of the same oil in water without surfactant. For example, in Hoffmann's system, about 0.3 parts of D_3 (or L_3 and C_{12}) was solubilized per part of surfactant in the micellar solution. Taking a surfactant concentration of 1 wt% in the micellar solution, this amounts to a solubility of 14 mM for D_3 in the micellar solution—about 3 orders of magnitude higher than its solubility in water. The difference is even larger for L_3 and C_{12}. Absolute values of solubility set aside, the solubility trends are impressively similar for micellar solutions and molecular solutions. In both cases, solubility of the oil decreases exponentially with the size of the oil molecule—which is expected—but the slope of the decay, interestingly, obeys the same descending order: n-alkane > linear PDMS > cyclic PDMS. In the aqueous surfactant micellar solution, solubilities are almost the same for PDMS and n-alkanes at the low end of the solute size [37], and in the aqueous molecular solution the difference is about 2 orders of magnitude (with solubility for PDMS being higher). As solute size increases, the difference in solubility between PDMS and n-alkanes increases and expands significantly.

In summary, the solubility of PDMS in water is quite low. Within the limited range of siloxanes that have been reliably measured, cyclic PDMS exhibits a higher aqueous solubility than linear PDMS, which in turn exhibits a higher solubility than n-alkane. The aqueous solubility of oil has profound effects on emulsion stability and emulsion polymerization, some of which will be demonstrated in Chapters 2 and 3.

The solubility of water in oil also has implications for emulsion stability. The solubility of water in PDMS or alkane is much higher than the solubility of either of the oils in water (Figures 1.5 and 1.6). This simply demonstrates the fact that when oil dissolves in water, the energy penalty for forming an entropy-reduced water structure around the oil molecule [38] outweighs the energy gain from dispersion interaction. Unlike the solubility of oil in water, the solubility of water in oil is relatively insensitive to the molecular size of the oil. It should be noted that the data for water solubility in PDMS (arrow in the figures) was taken from Hakim et al. who measured water content in a linear PDMS fluid of 50 cSt (corresponding to an average degree of polymerization around 40) as a function of relative humidity [34]. The authors extrapolated to a 270 ppm water content in the PDMS fluid at 100% relative humidity. In a previous publication by Vogel et al., the solubility of water in small PDMS (L_2, L_7, and D_4) was reported to be less than 1 ppm [39]. It is not clear where the discrepancy lies; one possibility is the purity of the silicone fluid. Hakim's measurement used a commercial grade, higher molecular weight PDMS fluid which may have contained a significant concentration of silanol groups, while Vogel's measurement used lower molecular weight oligomers that are typically purer. We expect the solubility of water in PDMS to be greater than the solubility of water in alkanes because of the polar nature of the siloxane bond. We also expect this value be insensitive to the molecular weight of the PDMS (since the size of the solute, here, water, is constant), and that silanol concentration may substantially change the value.

1.5   SILICONE SURFACTANTS

Silicone surfactants play an important role in many dispersion systems such as cosmetic products, coating formulations, agriculture adjuvants, foams, and anti-foams [10]. Most silicone surfactants are based on a linear PDMS backbone [40], while others are based on cross-linked structures [41,42]. Both block copolymer and block terpolymer structures are possible in silicone surfactants. As a side note, there are also silane surfactants [43,44] which, as the name implies, are based on silane instead of a siloxane backbone. The most widely used silicone surfactants are poly-siloxane-*co*-poly(oxyalkylene), also referred to as silicone polyethers (SPE) or sili-cone glycol copolymers. These copolymers are based on a PDMS backbone that is grafted with homopolymers of oxyethylene (EO), or copolymers of oxyethylene and oxypropylene (EO/PO). Figure 1.7 illustrates several possible structures for SPEs. Other hydrophilic groups such as sulfonate, carboxylate, phosphate, ammonium, betaines, saccharides, etc., and lipophilic groups such as alkyl or aryl have also been grafted onto the PDMS backbone to give a variety of silicone surfactants [40]. Reference 10 provides a review of the various aspects of silicone surfactants includ-ing their applications.

Silicone surfactants are best known for their tendency to lower the surface ten-sion of the liquids in which they are dispersed. This includes both aqueous and nonaqueous solutions [45]. The lowest solution surface tension that a silicone sur-factant can achieve is 21 mJ/m^2 [45,46], which is the same as surface tension of the bulk PDMS itself. This value is significantly lower than that achievable by the most effective hydrocarbon surfactants, which is around 26 mJ/m^2. The reason for this outstanding ability to lower a solution surface tension is, again, rooted in the extreme flexibility of the siloxane backbone. The flexible backbone allows the molecule to orient at the surface with the solvent-like functional groups toward the solution and the methyl groups toward the air, a situation that renders the surface completely methylated. This contrasts with a solution of hydrocarbon surfactants where methylene groups are exposed to the air. Indeed, silicone surfactants can lower the surface tension of nonaqueous media such as mineral oil and polyols

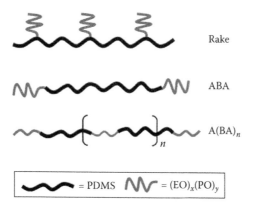

FIGURE 1.7   Common structures of silicone polyethers, rake (comb), ABA, and A(BA)$_n$.

while hydrocarbon surfactants cannot. For this reason, silicone surfactants were commercialized in the 1950s for the production of polyurethane foam [47].

Water-soluble or water-dispersible silicone surfactants self-associate in water in a similar manner as other common hydrocarbon-based surfactants. These silicone surfactants have a defined critical aggregation concentration (CAC), form liquid crystals, and show the same pattern of structure progression as other common surfactants with change in concentration, temperature, and molecular geometry [45,48–57]. Both globular and worm-like micelles have been identified [48,49,53,54,57]. Despite the fluidity of the siloxane chains, silicone surfactants show stable liquid crystal phases that can span wide ranges of concentration and temperature [49,53–55,57]. Even commercial grade silicone surfactants, which are polydispersed in structure and can contain an appreciable amount of impurities, form liquid crystals. (Polymeric SPEs can contain up to ca. 40% impurities, most of which are unreacted allyl ethers.) Silicone polyethers that contain both EO and PO tend to form micellar solution at all concentrations. It has been described that the high flexibility of the siloxane chain allows a silicone surfactant molecule to adapt its conformation to the packing constraints of various aggregation structures, including liquid crystals [54]. Data show that the siloxane chains in the aggregates are coiled and interdigitated [49,54]. The elastic bending modulus for lamellar bilayers formed from silicone surfactants and water has been estimated to be low [54], which is consistent with the observation that many of the silicone surfactants that form lamellar structure also form vesicles [52–56], and that the vesicles have varying sizes and shapes and display surface undulation [54].

As the surfactant self-associates in the bulk aqueous solution, adsorption at the solution surface or at an interface with another liquid reaches saturation. For an emulsion system, absorbed surfactant layer at the droplet surface (interface) stabilizes the droplet against coalescing with other droplets. It is natural to expect silicone surfactants to be used as emulsifiers for silicone emulsions. However, commercial silicone-in-water (Si/W) emulsions seldom use silicone surfactant as the main emulsifier. Water-based emulsions sometimes incorporate silicone emulsifiers to provide sensory benefits. Often in these cases either a water thickener or a co-emulsifier needs to be used to provide emulsion stability [58]. It was earlier speculated that because of the fluidity of the siloxane portion, a solid or liquid crystalline barrier would not form in the silicone surfactant boundary layer at the oil–water interface [40]. According to the Kabalnov–Wennerström theory [59], lack of rigidity in the surfactant boundary layer results in weak emulsion stability against coalescence. However, the hypothesized lack of rigidity in the adsorbed silicone surfactant layer and hence its inability to provide sufficient stabilization has not been directly proven. On the contrary, there is evidence that silicone surfactants stabilize O/W emulsions just as effectively as organic surfactants [60–62]. Cost and hydrolytic stability [40,63–65] may be other reasons to prevent popular use of silicone surfactants as emulsifiers for O/W emulsions. For water-in-silicone (W/Si) emulsions, on the other hand, high molecular weight, hydrophobic silicone surfactants are the preferred emulsifiers [66]. The mechanism with which high molecular weight silicone surfactants stabilize W/Si emulsions is different from the mechanism of stabilization in Si/W emulsions where low molecular weight surfactants are used (see discussions in Section 2.3).

For foams, silicone surfactants have shown unrivaled performance both as stabilizers and destabilizers depending on the structure of the silicone surfactant. The role of silicone surfactants in polyurethane foam formation is described elsewhere [47]. Chapter 8 describes the role of silicones and silicone surfactants in antifoams. Chapter 11 demonstrates the use of silicone and silicone surfactants in oil and gas production.

Silicone surfactants are also used as wetting and spreading agents in paints and coatings, textiles, and agricultural adjuvants. While high molecular weight silicone surfactants are effective in stabilizing emulsions and foams, low molecular weight silicone surfactants, especially those containing only 3–5 siloxane units and 5–12 EO units have demonstrated superior ability to promote wetting of aqueous solutions over low surface energy surfaces such as polyethylene [67], parafilm [68], and graphite [69]. As described earlier, spontaneous spreading of a liquid over a substrate is determined by the three interfacial tensions involved (Figure 1.4). Silicone surfactants promote wetting of their aqueous solutions over another liquid or solid owing to their ability to lower both the surface tension of the aqueous solution *and* the interfacial tension between the aqueous solution and the substrate. It has been shown that aqueous solutions of fluorocarbon surfactants do not spread on hydrophobic surfaces while aqueous solutions of silicone surfactants do, even though the former have lower surface tensions than the latter [67,70]. The "super-wetting" behavior known for silicone surfactants not only refers to the extent of wetting, but also to the speed of spreading; it involves mechanism beyond just equilibrium tensions. The phenomenon of super-wetting by silicone surfactants has been intensely studied since its discovery in the 1960s [71]. The subject, along with the general subject of surfactant enhanced spreading, remains as an ongoing topic of debate despite the many models established [70,72,73].

1.6 DELIVERY OF SILICONES

Silicones are sold commercially in the neat form or in solutions and dispersions. Even though silicones are less viscous than other organic polymers of comparable molecular weight, very high molecular weight silicones are produced which reach above tens of millions of centipoises in viscosity. Ultrahigh molecular weight linear silicones are known as silicone gums as they have a gum consistency and are not pourable. Cross-linked and resinous silicones exhibit viscoelastic properties. These high molecular weight silicones often provide excellent performance in applications but handling becomes a problem. As mentioned earlier, the most compatible solvents for silicones are the weakly polar or polarizable type, for example, aromatics and chloroform, most of which are harmful to the environment. Mineral spirits used to be a popular solvent to deliver silicones, but this is becoming costly. The growing awareness of human and environmental health drives the market toward replacing organic solvents with water whenever possible. In applications where the process or the final product is water-based, aqueous delivery of silicone becomes necessary. Aqueous emulsions of silicones have therefore been gaining steady market growth. In recent years, powder and granulated forms of silicone products have also seized market attention.

ACKNOWLEDGMENT

The authors thank Dr. Robert D. Kennedy for proofreading the manuscript.

REFERENCES

1. Thomas, N. R. *Silicon* 2010, *2(4)*, 187–193.
2. Warrick, E. L. *Forty Years of Firsts*. McGraw-Hill, New York, 1990.
3. Jebens, A. M., Kalin, T., Kishi, A. *IHS Chemical Economics Handbook, Silicones*, IHS Markit, London, New York, 2013. https://www.ihs.com/products/silicones-chemical-economics-handbook.html, accessed on October 1, 2015.
4. Andriot, M., Chao, S. H., Colas, A. et al. Silicones in industrial applications. In *Inorganic Polymers*, de Jaeger, R., Gleria, M. (eds.). Nova Sciences, New York, 2007, pp. 61–161.
5. Brook, M. A. *Silicon in Organic, Organometallic, and Polymer Chemistry*. John Wiley & Sons, New York, 2000.
6. Stark, F. O., Falender, J. R., Wright, A. P. Silicones. In *Comprehensive Organometallic Chemistry*, Vol. 2, Wilkinson, G., Stone, F. G. A., Abel, E. W. (eds.). Pergamon Press, Oxford, U.K., 1982, pp. 305–363.
7. Clarson, S. J., Semlyen, J. A. (eds.). *Siloxane Polymers*. Prentice Hall, Englewood Cliffs, NJ, 1993.
8. Owen, M. J. *Ind. Eng. Chem. Prod. Res. Dev.* 1980, *19*, 97–103.
9. Owen, M. J. Surface chemistry and applications. In *Siloxane Polymers*, Clarson, S. J., Semlyen, J. A. (eds.). Prentice Hall, Englewood Cliffs, NJ, 1993, pp. 309–372.
10. Hill, R. M. (ed.). *Siloxane Surfactants*. Marcel Dekker, New York, 1999.
11. Liebhafsky, H. A. *Silicones under the Monogram*. John Wiley & Sons, New York, 1978, pp. 99–100.
12. Brook, M. A. Silicones. In *Silicon in Organic, Organometallic, and Polymer Chemistry*. John Wiley & Sons, New York, 2000.
13. Plueddemann, E. P. *Silane Coupling Agents*, 2nd edn. Plenum Press, New York, 1991.
14. Brook, M. A. Atomic and molecular properties of silicon. In *Silicon in Organic, Organometallic, and Polymer Chemistry*. John Wiley & Sons, New York, 2000.
15. Jarvis, N. L. *J. Phys. Chem.* 1966, *70*, 3027–3033.
16. Robb, W. L. *Ann. N. Y. Acad. Sci.* 1968, *146*, 119–137.
17. Kuo, A. C. M. Poly(dimethylsiloxqane). In *Polymer Data Handbook*, Mark, J. E. (ed.). Oxford University Press, New York, 2009.
18. Ver Strate, G. W., Lohse, D. J. Poly(isobutylene), butyl rubber, halobutyl rubber. In *Polymer Data Handbook*, Mark, J. E. (ed.). Oxford University Press, New York, 2009.
19. Liu, Y. Unpublished data.
20. Fox, Jr., T. G., Flory, P. J. *J. Phys. Chem.* 1951, *55*, 221–234.
21. Fetters, L. J., Graessley, W. W., Kiss, A. D. *Macromolecules* 1991, *24*, 3136–3141.
22. Wu, S. *Polymer Interface and Adhesion*. Marcel Dekker, New York, 1982.
23. Kanellopoulos, A. G., Owen, M. J. *Trans. Faraday Soc.* 1971, *67*, 3127–3138.
24. Budziak, C. J., Varcha-Butler, E. I., Neumann, A. W. *J. Appl. Polym. Sci.*, 1991, *42*, 1959–1964.
25. Zisman, W. A. Relation of the equilibrium contact angle to liquid and solid constitution. In *Contact Angle, Wettability, and Adhesion*, Fowkes, F. (ed.), Advances in Chemistry. ACS, Washingon, DC, 1964, pp. 1–56.
26. Owen, M. J., Dvornic, P. R. General introduction to silicone surfaces. In *Silicone Surface Science*, Owen, M. J., Dvornic, P. R. (eds.). Springer, New York, 2012.
27. Ge, L., Shao, W., Lu, S., Guo, R. *Soft Matter* 2014, *10*, 4498–4505.

28. Riddick, J. A., Bunger, W. B., Sakano, T. K. *Techniques of Chemistry*, Vol. 2, Organic Solvents: Physical Properties and Methods of Purification, 4th edn. John Wiley & Sons, New York, 1985.
29. McAuliffe C. *J. Phys. Chem.* 1966, *70*, 1267–1275.
30. Yalkowsky, S. H., He, Y., Jain, P. *Handbook of Aqueous Solubility Data*, 2nd edn. CRC Press, Boca Raton, FL, 2010, p. 438.
31. Tolls, J., van Dijk, J., Verbruggen, E. J. M., Hermens, J. L. M., Loeprecht, B., Schüürmann, G. *J. Phys. Chem. A* 2002, *106*, 2760–2765.
32. Schatzberg, P. *J. Phys. Chem.* 1963, *67*, 776–779.
33. Varaprath, S., Frye, C., Hamelink, J. *Environ. Toxicol. Chem.* 1996, *15*, 1263–1265.
34. Hakim, R. M., Olivier, R. G., St-Onge, H. *IEEE Trans. Electr. Insul.* 1977, *EI-12(5)*, 360–370.
35. Open chemistry database from NIH/US National Library of medicine/National Center for Biotechnology Information, http://pubchem.ncbi.nlm.nih.gov/, accessed on October 1, 2015.
36. Flaningam, O. L. *J. Chem. Eng. Data*, 1986, *31*, 266–272.
37. Hoffmann, H., Stürmer, A. *Tenside, Surfactants, Deterg.* 1993, *30*, 335–341.
38. Evans, D. F., Wennerström, H. (eds.). *The Colloidal Domain*. VCH Publishers, New York, 1994, pp. 26–33.
39. Vogel, G. E., Stark, F. O. *J. Chem. Eng. Data* 1964, *9*, 599–601.
40. Grüning, B., Koerner, G. *Tenside, Surfactants, Deterg.* 1989, *26*, 312–317.
41. Fenton, W. N., Keil, J. W. US4125470, 1978.
42. Ekeland, R. A., Hill, R. M. US5958448, 1999.
43. Wagner, R., Strey, R. *Langmuir* 1999, *15*, 902–905.
44. Ferritto, M. S., Li, L. M. E., Petroff, L. J., Roidl, J. T., Surgenor, A. E. US2014/0302318, 2014.
45. Hoffmann, H., Ulbricht, W. Surface activity and aggregation behavior of siloxane surfactants. In *Silicone Surfactants*, Hill, R. M. (ed.). Marcel Dekker, New York, 1999.
46. Kanellopoulos, A. G., Owen, M. J. *J. Colloid Interface Sci.* 1971, *35*, 120–125.
47. Snow, S. A., Stevens, R. E. The science of silicone surfactant application in the formation of polyurethane foam, Chapter 5. In *Silicone Surfactants*, Hill, R. M. (ed.). Marcel Dekker, New York, 1999.
48. Kanner, B., Reid, W. G., Petersen. I. H. *Ind. Eng. Chem. Prod. Res. Dev.* 1967, *6*, 88–92.
49. Gradzielski, M., Hoffmann, H., Robisch, P., Ulbricht, W., Gruening, B. *Tenside, Surfactants, Deterg.* 1990, *27*, 366–379.
50. Yang, J., Wegner, G. *Macromolecules* 1992, *25*, 1786–1790.
51. Yang, J., Wegner, G. *Macromolecules* 1992, *25*, 1791–1795.
52. Lin, Z., He, M., Scriven, L. E., Davis, H. T., Snow, S. A. *J. Phys. Chem.* 1993, *97*, 3571–3578.
53. He, M., Hill, R. M., Lin, Z., Scriven, L. E., Davis, H. T. *J. Phys. Chem.* 1993, *97*, 8820–8834.
54. Hill, R. M., He, M., Lin, Z., Davis, H. T., Scriven, L. E. *Langmuir* 1993, *9*, 2789–2798.
55. Hill, R. M., He, M., Davis, H. T., Scriven, L. E. *Langmuir* 1994, *10*, 1724–1734.
56. He, M., Lin, Z., Scriven, L. E., Davis, H. T., Snow, S. A. *J. Phys. Chem.* 1994, *98*, 6148–6157.
57. Stürmer, A., Thunig, C., Hoffmann, H., Grüning, B. *Tenside Surfactants Deterg.* 1994, *31*, 90–98.
58. Meyer, J., Hartung, C., Unger, F. *SOFW J.* 2009, *135*, 62–70.
59. Kalbalnov, A., Wennerström, H. *Langmuir* 1996, *12*, 276–292.
60. Sela, Y., Magdassi, S., Garti, N. *Colloid Polym. Sci.* 1994, *272*, 684–691.
61. Feng, Q. J., Hickerson, R. S., Starch, M. S., Van Reeth, I. US2007/0190012, 2007.

62. Liu, Y., Costanzo, S. F. Emulsifiers for silicone emulsions. In *Silicone Dispersions*, Liu, Y., (ed.), CRC Press, Boca Raton, FL, 2016, 33–42.
63. Snow, S. A., Fenton, W. N., Owen, M. J. *Langmuir* 1990, *6*, 385–391.
64. Snow, S. A., Fenton, W. N., Owen, M. J. *Langmuir* 1991, *7*, 868–871.
65. Retter, U., Klinger, R., Philipp, R., Lohse, H., Schmaucks, G. *J. Colloid Interface Sci.* 1998, *202*, 269–277.
66. Anseth, J. W., Bialek, A., Hill, R. M., Fuller, G. G. *Langmuir* 2003, *19*, 6349–6356.
67. Ananthapadmanabhan, K. P., Goddard, E. D., Chandar, P. *Colloids Surf.* 1990, *44*, 281–297.
68. Zhu, S., Miller, W. G., Scriven, L. E., Davis, H. T. *Colloids Surf. A Physicochem. Eng. Asp.* 1994, *90*, 63–78.
69. Svitoa, T., Hill, R. M., Radke, C. J. *Colloids Surf. A Physicochem. Eng. Asp.* 2001, *183–185*, 607–620.
70. Stoebe, T., Hill, R. M., Ward, M. D., Scriven, L. E., Davis, H. T. Surfactant-enhanced spreading. In *Silicone Surfactants*, Hill, R. M. (ed.). Marcel Dekker, New York, 1999.
71. Bailey, D. L. US3299112, 1967.
72. Hill, R. M. *Curr. Opin. Colloid Interface Sci.* 1998, *3*, 247–254.
73. Kabalnov, A. *Langmuir* 2000, *16*, 2595–2603.

2 Silicone Emulsions and Microemulsions

Yihan Liu and Samuel F. Costanzo

CONTENTS

2.1 INTRODUCTION

Emulsions are heterogeneous systems in which one liquid is dispersed in another liquid. The dispersed phase exists as droplets that are kinetically stabilized by an emulsifier. The most common types of emulsions are oil-in-water (O/W) and water-in-oil (W/O). The word "water" and the symbol "W," when referring to an emulsion, may not necessarily mean pure water but rather an aqueous solution. In most cases, it is an aqueous solution of surfactant that contains micellar aggregates. The aqueous solution may also contain other water-soluble species, such as salt, thickener, and biocide. Similarly, "oil" and the symbol "O" refer to a water-immiscible liquid that is less polar than water. Emulsions can display a more complicated morphology than the simple O/W or W/O. Multiple emulsions such as water-in-oil-in-water (W/O/W) and oil-in-water-in-oil (O/W/O) have been long known. An emulsion may also contain more than two immiscible phases. Janus and Cerberus emulsions, for example, have been well demonstrated in recent years [1,2].

When properly prepared, an emulsion can be stable for years or decades, but as a heterogeneous system, it will eventually phase separate. Microemulsions, on the other hand, are single-phase solutions that contain two otherwise immiscible liquids that are solubilized with the aid of a surfactant. Microemulsions are equilibrium systems and as such are indefinitely stable and reversible. Emulsions and microemulsions are technologically useful as they deliver immiscible components in one macroscopically uniform system.

A silicone emulsion or microemulsion contains silicone as one of the components. Often in a silicone emulsion, the silicone component is completely immiscible with the other components or phases and therefore constitutes the silicone phase. Polydimethylsiloxane (PDMS) is the most common form of silicone. PDMS and many other types of silicones are water immiscible and for this reason are commonly referred to as oil or silicone oil. Most of the silicone emulsions sold on the market are of the silicone-in-water (Si/W) type. Silicone-in-water emulsions are a convenient vehicle for delivering silicones, especially when the silicone is too viscous or elastic to handle. In many applications, the emulsion form is necessary. For instance, if a water-immiscible silicone is to be added into a water-based formulation, a Si/W emulsion allows for a uniform incorporation of the silicone. The emulsion also provides a system to synthesize silicones. Aqueous emulsion polymerization is a particularly advantageous process to obtain aqueous emulsions of high molecular weight silicones.

Often it is more cost effective to emulsify a low molecular weight silicone precursor and subsequently allow the polymerization reaction to occur, rather than to conduct the polymerization reaction first and subsequently emulsify the polymer. Aside from silicone-in-water emulsions, water-in-silicone (W/Si) and nonaqueous silicone emulsions are also commercially available and used in various applications.

To comprehensively cover the subject of silicone emulsions and microemulsions in a single chapter is not possible. The content chosen here is undoubtedly influenced by the authors' interest, and the scope limited to the authors' knowledge. On the other hand, a comprehensive coverage may not be necessary since, as is obvious in the forthcoming discussions, the general principles of emulsion science apply to silicone emulsions, and there already exists a large volume of literature on the subject of emulsions. We aim to bring out the aspects of emulsion science that directly relate to the preparation and behavior of silicone emulsions and microemulsions.

This chapter is organized in the following way. First, a brief overview of the technology, development, and applications involving silicone emulsions and microemulsions is given in Section 2.2. The rest of the chapter is devoted to the principles behind the design and preparation of silicone emulsions (Sections 2.3 through 2.5) and microemulsions (Section 2.6). Since, for any emulsion, the essential concern is the stability of the emulsion, and emulsion stability is in turn governed by the type of emulsifier and, more importantly, by the phase behavior of the system water–oil–surfactant, the discussion of silicone emulsions begins with a description of emulsifier options (Section 2.3). This is then followed by a description of the general phase behavior of the water–silicone–surfactant (W–Si–S) system. With the fundamentals established, Section 2.5 then presents methods for the emulsification of various types of silicones. Finally, silicone microemulsions are described in Section 2.6.

2.2 A BRIEF OVERVIEW OF THE TECHNOLOGY DEVELOPMENT AND APPLICATIONS OF SILICONE EMULSIONS AND MICROEMULSIONS

Silicone was first used during World War II as a grease to coat the ignition wire harnesses of Allied airplanes [3]. After the war, the silicone industry began to seek civilian application of silicone materials, and the first commercial success came in 1945 with an aqueous emulsion of polydimethylsiloxane (PDMS). The product was used as a release agent for molds in tire production [3]. Compared with the organic wax, which had been the previous workhorse for mold release, silicones exhibited superior release performance and left much less buildup inside the mold. The application in tire mold release required dilution of the silicone, and because most solvents for silicones are highly flammable, delivering the silicone in an aqueous emulsion presented a perfect solution. The instant success in the tire industry promoted the use of silicone emulsions as mold release agents in other applications [4]. Meanwhile, emulsions of phenyl silicone resins were developed and used to provide electrical insulation to motors, generators, and transformers. This resulted in the first silicone emulsion patent, granted in 1948 to Westinghouse Electric Int. Co. [5]. The textile industry then discovered that emulsions of silicone resins provided water repellence to fabrics [6].

Development of silicone emulsions continued during the 1950s. Silicone emulsions during this era were typically prepared by mixing silicone with water and surfactant, followed by homogenization using the colloid mill. The surfactants used for silicone emulsions were those that had been commonly used to stabilize lipophilic oil-in-water emulsions, such as long chain alcohol ethoxylates, alkyl aryl sulfonates, and alkyl aryl polyether sulfates. The silicone emulsion products prepared in this manner showed excellent stability [7].

During the 1950s and 1960s, silicone emulsion polymerization was explored as an alternative route to produce silicone emulsions. Wehrly and Hyde, and later Weyenberg and Findlay, discovered that octamethylcyclotetrasiloxane (D_4) could be polymerized in an emulsion form using either a cationic or an anionic surfactant, and under acidic or basic conditions [8–10]. In the ensuing work, the process of silicone emulsion polymerization continued to be improved [11–18]. By copolymerizing cyclic siloxane and organo-functional silanes in emulsion polymerization, silicones that are composed of different combinations of MDTQ structural units and that contain different functional groups have been produced. (For a description of the siloxane and silane basic structures as well as MDTQ nomenclature, see Chapter 1.) The use of silanes in silicone emulsion polymerization also enabled the synthesis of silicone resins [19–21]. These resins tend to have different properties than silicone resins synthesized in organic solvents. The general mechanism of silicone emulsion polymerization involves the diffusion of the monomers—D_4 or silanes—into the aqueous phase where hydrolysis and condensation takes place. The resultant oligomers precipitate out of the aqueous solution as small droplets. The droplets continue to grow in size while the oligomers in the droplets further condense into polymers. A detailed account of the mechanism of silicone emulsion polymerization is provided in Chapter 3. Depending on the concentration of the surfactant and process conditions, emulsions that have droplet sizes as small as a few nanometers to as large as a micron can be attained. Viscosities of linear silicones as high as 60,000 cP can be reached by silicone emulsion polymerization. Solid resins can also be obtained. In particular, emulsion polymerization has been used to produce optically clear or translucent silicone emulsions that have droplet sizes below 100 nm [11,13,14,18,20]. These emulsions are suitable for products that require optical clarity, such as clear shampoos and shower gels. They are also used to treat fibers and fabrics. Often these emulsions are claimed or labeled as microemulsions due to the optical clarity. However, they are two-phase emulsions.

Another type of silicone emulsion polymerization, sometimes referred to in the silicone industry as suspension polymerization, was developed during the 1960s [9]. This type of emulsion polymerization employs reactive oligomeric or polymeric linear siloxanes as the starting "monomers." The starting "monomer" has no solubility in water. In silicone suspension polymerization, an emulsion of the "monomer" is first prepared, and then a catalyst is added to initiate polymerization. Polymerization occurs by either the condensation [9,22–25] or the addition [26–28] chemistry. Since the "monomer" is insoluble in the aqueous phase, polymerization proceeds inside the "monomer" droplet or at the droplet interface. Each emulsion droplet acts like a mini reactor, and the emulsion droplet size remains the same throughout the process. Very high molecular weight linear silicones with viscosities in excess of millions of centipoises can be reached using this type of emulsion polymerization [27,28]. With cross-link agents, the silicone can also be cured into an elastomer in the emulsion. This produces a silicone latex [22–26,29–32]. Silicone latex dries to a coherent and elastomeric film upon the evaporation of water. The film can be further strengthened by the addition of fillers in the emulsion. Silicone latexes were introduced to the market in the 1970s and have been used in coatings and sealants since [33,34]. An extension of silicone latex technology is the production of silicone elastomer powders [35,36]. Such powders are used in cosmetics, paints, and coatings. Silicone latexes and silicone elastomer powders are the subjects of Chapters 4 and 5, respectively.

Another technology using emulsion polymerization involves the synthesis of silicone–organic copolymers or hybrid latexes by combining silicone emulsion polymerization with free-radical emulsion polymerization of vinyl monomers. In these processes, organic vinyl monomers react and copolymerize with vinyl functional siloxanes. Alternatively, vinyl functional alkoxysilanes are used to link organic vinyl monomers with alkoxy- or silanol-containing siloxanes. Styrene and acrylates have been shown to polymerize in emulsions of vinyl functional siloxanes, which results in the grafting of styrene and acrylate onto silicones [37–39]. If no vinyl functional silane or siloxane is present, the organic component and the silicone component are not copolymerized but instead are polymerized in separate domains within the same droplet. This leads to a composite colloidal structure that contains organic and silicone domains. Mautner and Deubzer disclosed that emulsion polymerization with sequential additions of vinyl monomer, silane, and siloxane resulted in the formation of core–shell structured particles [40]. Hellstern et al. disclosed that the concurrent addition of a mixture of vinyl monomer, silane, and siloxane yielded a multi-lobed colloidal structure [41]. In the latter case, the structure was described as having an organic polymer lobe and an organopolysiloxane appendage. Kim et al. reported silicone–organic latex particles which the authors described as having an acorn structure. The acorn structure was revealed by electron microscopy to consist a silicone-rich domain partially engulfed by an acrylate-rich domain [42,43], and was formed through a seeded emulsion polymerization of vinyl monomers in the presence of a nonreactive silicone emulsion seed [44,45]. The vinyl monomers were believed to first swell the silicone droplets, polymerize, and then phase separate from the silicone phase as a result of increased molecular weight [43]. Evidently, the phase separation of the polymerized acrylate from the silicone does not result in isolated acrylate particles separated from the silicone particles, but rather the acrylate phase stays attached as an appendage to the silicone phase. It is meaningful to mention here that interestingly, Friberg et al. showed that Janus emulsions can be formed by simply mixing together water, silicone, and lipophilic oil [46]. The resultant Janus drop consists of a silicone lobe with an attached lipophilic oil appendage. The specific geometry is shown to be controlled by the relative phase volume ratio and interfacial energy. The detailed fundamentals are covered in Chapter 6. Silicone–organic emulsion copolymers and hybrid colloids combine the different properties from organic polymers and silicones into a single system and can provide a superior performance in applications. In paints and coatings, silicone–organic emulsion copolymer and hybrid latexes offer simultaneously the film-forming property of acrylates and the mar resistance, weather resistance, and water-proofing properties of silicones. These copolymers and hybrid latexes can also be flocculated and recovered in the form of free-flowing powders or agglomerated granules [41], which are useful as impact modifiers for thermoplastics.

Silicone microemulsions were introduced to the market in the 1980s. Historically, the term microemulsion had been used with confusion until research in the 1960s and 1970s revealed that what was called microemulsions were actually thermo-dynamically stable solutions instead of emulsions having small droplet size [47]. Unfortunately, the term microemulsion continues to be used with confusion in industry. Often an emulsion product is called a microemulsion when it appears optically clear, regardless of whether or not it is a single phase. In this chapter, we adhere to the definition that microemulsion is a single-phase solution. It is a thermodynamically stable, equilibrium system. Single-phase silicone microemulsions form when silicone is solubilized in a micellar solution or the surfactant phase. This occurs if the silicone molecular weight is sufficiently low or if the silicone contains certain amount of polar-functional groups grafted onto the siloxane backbone. Microemulsions of low molecular weight dimethylsiloxanes were discovered and extensively studied by Hill et al. in the 1990s and 2000s [48–56]. Microemulsions of polar-functional polymeric silicones entered the market in the 1980s; they are used for applications in textiles and personal care [57–59]. The formation and phase behavior of these latter microemulsions have not been well understood until recently. We will describe the topic in more details in Section 2.6.

Silicone emulsions and microemulsions are now used in a wide range of applications. The fabric and fashion industries use silicone emulsions to treat fabric and deliver benefits such as softening, ease of ironing, anti-wrinkle, and, depending on the functional groups grafted on the silicone molecule, water absorbency or repellency [60–63]. Emulsions and microemulsions of aminosilicones are particularly effective at delivering these benefits. Airbags and tents that are treated with emulsions of cross-linked silicones are found to exhibit greater mechanical strength and heat resistance [64,65]. Main parameters that affect the performance of the treated fabrics include the emulsion droplet size, the surfactant type, and the type and amount of functional group grafted onto the silicone [61,66,67]. Silicones are also used in top coats for leather and ink coatings [68–71]. Leathers are typically treated with acrylic and urethane binders to deliver a combination of softness, toughness, and durability. These binders are water-based. Aqueous emulsions of high molecular weight silicones are incorporated to provide abrasion and mar resistance, slip, soft hand feel, and to reduce squeak noise. In acrylic and polyurethane binders that are used to varnish ink surfaces, emulsions of silicones can be added to provide slip and anti-blocking.

Silicone is the most common material used in release liners for pressure sensitive adhesives, labels, food packagings, bakery sheets, etc. [72,73]. In paper release coatings, curable silicones are used in the neat form and in solvents, but increasingly in the aqueous emulsion form.

Many contemporary cosmetic and personal care products contain silicones. For hair, silicone provides ease of combing, styling, and damage repair [74,75].

Shampoos and conditioners are water-based; aqueous silicone emulsions are therefore readily incorporated into these formulations. The performance of silicone greatly depends on the amount of silicone deposited onto the hair fibers, and the amount of deposition in turn depends on the molecular weight of the silicone, type of functional grafting on the silicone, emulsion droplet size, and surfactant type [75–77]. For skin, silicone mainly provides sensory benefits [78]. Depending on the nature of the formulation, skin care and cosmetic products incorporate silicones in various forms including the neat form, Si/W emulsions, W/Si emulsions, and powder. The type of silicones used in cosmetic and skin care products varies greatly, ranging from low molecular weight volatile silicones to high molecular weight gums, elastomers, and resins. Each type is associated with specific performance characteristics. Most noticeable to the consumer is the sensory effect which dictates much of the silicone product design for this category of applications. More details are provided in Chapters 5 and 9. Silicone dispersions are also employed in healthcare products to provide sensory benefits [79]. In some instances, they are used as delivery vehicles for active ingredients [80–82].

Another major application of silicone is in antifoam. PDMS can provide certain level of antifoaming, but the more effective antifoams involve a dispersion of silica in silicone fluid. To control foam in aqueous formulations, aqueous emulsions of a mixture of silica and silicone fluid are used. The preparation of antifoam emulsions is challenging because the emulsion must remain stable until it is introduced to the target formulation, at which point the emulsion must be destabilized in order to release the antifoam active. Chapter 8 describes silicone antifoams and the dispersing process of silicone antifoam during application.

In construction, silanes and silicones are used to waterproof building materials. Water ingress causes corrosion and consequently deterioration of the load-bearing structures. Treatment using silanes and silicones can effectively prevent water from penetrating into the interior of the building materials [83]. Because most building materials are polar in nature, the delivery of silanes and silicones in the form of an aqueous emulsion helps the wetting and penetration of the product into the pores of the construction materials; thereby a more effective treatment is achieved [84,85]. Such emulsions can be mixed directly with building materials or applied as a post-treatment. Chapter 10 describes how silanes and siloxanes work in these applications.

2.3 EMULSIFIERS FOR SILICONE EMULSIONS

The word emulsifier is used here as a general term to include all the agents that aid in the emulsification process and stabilize the emulsion. An emulsifier can be a surface-active agent, that is, a surfactant, a non-surface-active agent that solubilizes in and thickens the continuous phase, or a component that is insoluble in either of the two liquid phases and therefore presents as a third phase in the emulsion. (In this chapter, we use the symbol "SP" to designate the third phase–the surfactant phase. The symbol "S" is reserved for the component surfactant.) Most discussions on emulsions deal with surfactants as the emulsifier, and therefore the word emulsifier is often used interchangeably with the word surfactant. As most of the systems discussed in this chapter involve surfactants as the emulsifier, we also use the two words interchangeably. In places where it is important to point out that non-surface-active or insoluble emulsifiers are used, it is specified. We categorize emulsifiers used in silicone emulsions into three main groups:

1. Hydrocarbon-based small molecule surfactants
2. Hydrocarbon-based polymeric stabilizers
3. Silicone emulsifiers

This categorization is arbitrary and solely for the purpose of organizing the subsequent discussions; there is obvious overlap among the three categories. Hydrocarbon-based small molecule surfactants refer to synthetic surfactants like the "soaps" which consist of a hydrophilic polar group and a lipophilic hydrocarbon chain containing about 8–20 carbon atoms in the molecule. Most commercial silicone-in-water (Si/W) emulsions are made using this group of surfactants. High molecular weight copolymers can be effective emulsion stabilizers. Some high molecular weight copolymers act in a similar manner to low molecular weight surfactants and display the same type of aggregation and adsorption behavior. Others act simply as thickeners to the continuous phase. Still others form an insoluble third phase residing at the surface of the emulsion droplet whereby providing a barrier to droplet coalescence. Silicone emulsifiers are typically block copolymers that contain a polydimethylsiloxane backbone to which one or more lipophilic or hydrophilic groups are attached through Si–C or Si–O–C bonds. The dimethylsiloxane portion is sometimes referred to as the siliphile. Silicone copolymers used as emulsifiers range in size from trisiloxane copolymers that contain only three siloxane units in the siloxane backbone, to copolymers that contain hundreds of siloxane units. Apart from these three main categories of emulsifiers, natural lipids, proteins, and solid particles can also be used to stabilize silicone emulsions, but they are not commonly used in commercial silicone emulsion products.

2.3.1 Hydrocarbon-Based Small-Molecule Surfactants for Silicone Emulsions

Surfactants are commonly divided into the anionic, cationic, nonionic, and amphoteric (or zwitterionic) type depending on the charge of the hydrophilic moiety. We summarize here which situation each of these types of surfactant is used for silicone emulsions.

2.3.1.1 Anionic

Silicone emulsions that are used in cleansing products like shampoos and detergents are usually prepared using anionic surfactants. This is because cleansing formulations tend to be composed of anionic surfactants. Anionic surfactants that can be used to prepare silicone emulsions include primary or secondary alkane sulfonates, alkylaryl sulfonates, alcohol sulfates, ethoxylated alcohol sulfates, and fatty acid salts (soaps). The sodium or ammonium salts of alkylbenzenesulfonic acids in particular are frequently used. Alkylbenzenesulfonic acids are used as catalysts for condensation-based silicone emulsion polymerization. The acid itself is not an effective surfactant, but once neutralized, the salt becomes a good stabilizer for Si/W emulsions.

2.3.1.2 Cationic

Silicone emulsions that are used for treating textiles are often prepared using cationic surfactants. The rationale is that many textile surfaces are negatively charged, so the attraction between the positively charged silicone droplets in the emulsion and the negatively charged substrate increases deposition of the silicone. Many hair conditioners and skin care products also incorporate silicone emulsions prepared with cationic surfactants for the same reason. Alkyltrimethylammonium halides are by far the most common cationic surfactants used to prepare silicone emulsions. Alkyltrimethylammonium hydroxide, formed when alkyltrimethylammonium halide is combined with a base in the aqueous emulsion, is an effective surfactant catalyst for condensation-based silicone emulsion polymerization.

2.3.1.3 Nonionic

Nonionic surfactants are the most frequently used surfactants in preparing silicone emulsions. Since the resulting droplets have a neutral surface charge, the product can be incorporated into a wide range of base formulations. Commonly used nonionic surfactants for silicone emulsions include ethoxylated fatty alcohols, alkylphenols, glycerol, sorbitol, and fatty amines.

2.3.1.4 Amphoteric

Amphoteric, or zwitterionic, surfactants such as lecithin and betaines are not as commonly employed in silicone emulsions, primarily because they are relatively ineffective in stabilizing Si/W emulsions, especially when used alone.

2.3.1.5 Combination

Frequently, more than one surfactant is used to stabilize an emulsion. Emulsions prepared using a combination of different surfactants often display better stability and can tolerate a wider range of conditions than if only one surfactant is used. This is also the case with silicone emulsions. Commercial emulsion products need to remain stable over a wide range of storage and transportation conditions. It is not uncommon for emulsion manufacturers to place a stringent requirement of emulsion stability over a temperature range of $-15°C$ to $+50°C$ in order to accommodate both winter and summer conditions. In certain applications, emulsion stability with variations in pH and electrolyte concentration is also desired. To meet this level of stability, an ionic surfactant is often used in combination with a nonionic surfactant; or, two nonionic surfactants of different hydrophile–lipophile balance (HLB) values are used. The reasons that mixed surfactant systems impart greater stability depend on the specific system. Aggregation and adsorption at interfaces from mixtures of surfactants in many cases deviate from what is predicted based on ideal mixing. The adsorption at interfaces from the mixed systems can be significantly greater than the adsorption of either surfactant by itself [86]. Thus, mixed surfactants can be more efficient in stabilizing an emulsion. However, it is important to point out that behavior of mixed surfactants cannot always be predicted. When the individual surfactants are similar in chemical nature and molecular structure, the mixture is expected to exhibit intermediate behavior. When the individual surfactants are very different, the behavior of the mixture is difficult to predict. The generally accepted method for calculating HLB value of two mixed surfactants is to take the weighted average of the two individual HLB values [87,88]. Unfortunately, such an "average" concept does not always apply to emulsion behavior. This topic will be expanded upon in Section 2.3.4.

2.3.2 HYDROCARBON-BASED POLYMERIC STABILIZERS FOR SILICONE EMULSIONS

High molecular weight copolymers can be effective emulsion stabilizers, some of which act as surfactants while others act as thickeners. Commonly used thickeners for Si/W emulsions include the various types of cellulose derivatives and gums such as Methocel® cellulose ethers and Keltrol® Xanthan Gum. Cross-linked acrylate copolymers, for example, Carbopol® and Pemulen®, are also used to thicken Si/W emulsions.

Polyvinyl alcohol (PVA) is a very effective aqueous emulsion stabilizer. PVA is synthesized via the alcoholysis of a polyvinyl acetate precursor. With a high degree of alcoholysis, PVA is water soluble and thickens water. In the aqueous solution, PVA is not surface active, as it does not self-aggregate like a surfactant, though it forms microgel particles in the bulk solution that consists of paracrystalline and amorphous domains [89]. It is not obvious whether PVA partitions at the O–W interface with an excess concentration and thus stabilizes the emulsion by providing a barrier to coalescence, or if it merely thickens the aqueous phase and slows down the collision rate of the emulsion droplets. It has been demonstrated that water-soluble thickeners often concentrate at the interface to form an interfacial film [90]. PVA emulsified silicone emulsions are used in several applications. In paper release coatings, emulsions of curable silicones are made using PVA as the sole emulsifier [91]. PVA is believed to hydrogen bond to cellulose [92] and therefore help anchor the silicone to the paper substrate. In architectural paints and exterior building renders, PVA emulsified silicone resins are used as binders or cobinders [93]. Here PVA simultaneously binds with pigments and silicone resins. Silicone elastomer emulsions made using PVA dry upon water evaporation into coherent films that have the characteristic feel of PVA, which is semi-hard, slippery, and completely tack-free [94]. An interesting and telling phenomenon arises when a drop of water is placed on top of such a film and is allowed to stand. The drop becomes cloudy, and if the drop is removed with a paper towel, the underlying film feels like silicone rubber to the finger touch. This indicates that the PVA segregates on the top of the film, which is somewhat surprising, considering that silicone has a lower surface tension than PVA. In skin care applications, the tack-free finger feel of the PVA film is desirable; therefore, PVA emulsified silicone emulsions can be advantageous [81]. In construction applications, PVA has been used to encapsulate hydrolytically reactive silanes and silicones in an emulsion to prevent the reactive ingredients from leaking into the aqueous phase prior to application [93]. In one method, reactive silane is emulsified into an aqueous PVA solution followed by mixing with a mineral to form a granulated powder. The powder is then mixed with cementitious materials to render the latter waterproof [95].

The class of triblock copolymers of polyethyleneoxide–polypropyleneoxide–polyethyleneoxide, denoted as $(EO)_x(PO)_y(EO)_x$, are unique surfactants. They form micelles and lyotropic liquid crystals [96] in water in a similar fashion as common surfactants of low molecular weight. EO–PO block copolymers self-aggregate in water due to the hydrophobicity of the polypropyleneoixide (PPO) segment. While oligomers of propyleneoxide are water soluble, polypropyleneoxides become immiscible with water above a certain molecular weight. Aggregates formed from these copolymers with water have been shown to selectively solubilize various polar and polarizable species [97–101], and this feature has been projected to have utility in pharmaceutical applications. These block copolymers, known as Poloxamers and sold under the trade names Pluronic® and Synperonic®, are available with molecular size ranging from oligomers containing a few repeating units of EO and PO to polymers containing more than a hundred repeating units. When x is significantly greater than y, $(EO)_x(PO)_y(EO)_x$ can be effective emulsifiers for Si/W emulsions. Polypropyleneoxide is immiscible with alkanes or siloxanes when the alkane or siloxane is above a certain molecular weight. To give an idea, polypropyleneoxide of molecular weight roughly 400 g/mol is completely immiscible with alkanes greater than dodecane or dimethylsiloxane containing more than five siloxane units. It was discovered that silicone gums can be emulsified using high molecular weight, water-soluble Pluronics but not with common surfactants of low molecular weight [102]. It is not clear how these block copolymers configure at the silicone–water interface. Given the immiscibility of PPO with PDMS, the PPO portion is not expected to partition with any appreciable degree into the silicone oil phase.

The inverse counterpart, $(PO)_x(EO)_y(PO)_x$ triblock copolymers, do not stabilize water-in-silicone emulsions. This is presumably due to the limited partitioning of the emulsifier into the silicone continuous phase (see discussion in Section 2.3.4).

2.3.3 SILICONE EMULSIFIERS

A brief overview of silicone surfactants is given in Chapter 1, Section 1.5 of this volume. For a more comprehensive description of the structure and properties of silicone surfactants, see Reference 103. As pointed out earlier, silicone surfactants are not often used as emulsifiers for Si/W emulsions. The vast majority of commercial Si/W emulsions are produced using hydrocarbon-based emulsifiers. On the contrary, W/Si emulsions are commonly prepared using silicone emulsifiers. Occasionally, silicone surfactants are used in O/W emulsions, but in these cases, the silicone surfactants are used primarily for the purpose of imparting sensory benefits. Often in these cases, additional hydrocarbon-based co-emulsifiers or thickeners are needed in order to sustain the emulsion stability. There is a large volume of patents and articles on silicone surfactants, a good portion of which mention the use of silicone surfactants as emulsifiers, but specific examples tend to be for the W/Si kind [104–108]. Reports demonstrating examples of Si/W emulsions prepared using silicone surfactants as the sole emulsifier are scarce [109–111]. It has been well established that silicone surfactants can be both efficient and effective in lowering the surface and interfacial tensions [112–118]. (Surfactant efficiency refers to the amount of surfactant it takes to lower the surface tension of water by a certain fixed level. Surfactant effectiveness refers to the extent the surfactant lowers the surface tension of water [112].) But are silicone surfactants effective *emulsifiers*? To answer that, we take a look at the role of the emulsifier.

The role of the surfactant as an emulsifier is often described as to lower the interfacial tension and thereby reduce the stress necessary to overcome Laplace pressure in order to deform and break up a droplet. However, lowering the interfacial tension is not the essential role of the emulsifier. The primary function of the emulsifier is to stabilize the newly created interface and prevent the droplets from recombining (more discussion is provided in Section 2.5.1.2). The *effectiveness of an emulsifier* is therefore manifested in two aspects:

1. The droplet size achievable from a given emulsification process
2. The stability of the resultant emulsion

Note that these two aspects are necessarily related. The achievable droplet size is itself a reflection of the emulsion stability/instability during the emulsification process. To the best of the authors' knowledge, there has not been a study that directly compares silicone surfactants to hydrocarbon-based surfactants in their effectiveness *as emulsifiers* for Si/W emulsions. It had been speculated that because of the fluidity of the siloxane portion, a solid or liquid crystalline barrier would not form in the silicone surfactant boundary layers at the oil–water interface, and therefore the emulsion droplets are susceptible to coalescence [119]. Emulsion stability, or rather, instability, is a complicated phenomenon involving many contributing factors. Assuming emulsion ripening does not occur, which is generally the case for aqueous emulsions of polymeric silicones, the ultimate irreversible instability in an emulsion occurs with coalescence (fusing together) of the droplets. The process leading to coalescence can be described to consist of two stages: film thinning and film rupture. Film thinning depends on the hydrodynamics of the film flow and the interaction forces between the droplets. Film rupture, occurring when the film separating the two opposing droplets is sufficiently thin, depends on the rheological properties of the thin film and the mechanical properties of the boundary surfactant layer. The process of coalescence has been extensively studied and the interested reader is referred to reviews by Binks [120] and Kalbalnov [121]. Suffice it to say that we suspect that if an inherent difference exists between silicone surfactants and hydrocarbon-based surfactants in terms of the effectiveness in stabilizing O/W emulsions, likely it is in the mechanical properties of the adsorbed surfactant layer at the interface.

In the following, we first report the result of a recent study comparing silicone surfactants with common hydrocarbon-based surfactants in the effectiveness to emulsify and stabilize Si/W emulsions (Section 2.3.3.1). We then describe a unique advantage of using silicone surfactant to stabilize certain O/W emulsions (Section 2.3.3.2). We end the section with a look at why silicone surfactants are particularly effective in stabilizing W/Si emulsions (Section 2.3.3.3).

2.3.3.1 A Direct Comparison between Silicone Surfactants and Hydrocarbon-Based Surfactants in the Effectiveness as Emulsifiers for Silicone-in-Water Emulsions

The study compares five commercial surfactants: two fatty alcohol ethoxylates and three silicone polyethers. All five surfactants are frequently used in commercial products. The nominal structures are listed in Table 2.1. Also provided in the table are the surfactant–aqueous binary phase types at room temperature along with the composition ranges corresponding to the phases. The abbreviations in Table 2.1 are as follows. For surfactant structure: C stands for alkyl chain, E stands for ethyleneoxide unit, $-CH_2CH_2O-$, and the MDTQ silicone nomenclature is explained in Chapter 1. All ethoxy chains are terminated with OH. For composition: S = surfactant, W = water. For phase: W = aqueous solution (may contain an exceedingly small amount of surfactant), L = micellar solution, L_α = lamellar liquid crystal phase, H_1 = hexagonal liquid crystal phase.

Four sets of emulsions, each set consisting of five samples, were prepared. Each sample is labeled with a number and a letter. The number, 1–5, corresponds to the surfactant number in Table 2.1. The letter, A–D, encodes the method of emulsification and the silicone oil content. The silicone oils used were PDMS of 75 and 60,000 cSt viscosity; both have no miscibility with water or any of the surfactants. Samples in sets A and B were prepared by blending together the low viscosity silicone oil, surfactant and water, and then subjecting the blend to high intensity sonication using an ultrasonic disperser (Misonix S-4000) following the procedure described elsewhere [122]. Samples in sets C and D were prepared by homogenizing a mixture of the high viscosity silicone oil, surfactant and water using a high speed mixer (SpeedMixer™ DAC 150 FVZ), which inverts the mixture from an oil-continuous emulsion to an oil-dispersed emulsion, and then diluting the inverted emulsion with additional water to the final composition [123].

TABLE 2.1

Surfactant Structure and Room Temperature Binary Phases

Surfactant	Structure	Surfactant Fraction in the Binary Mixture, S/(S + W)	Phase
1	$i\text{-}C_{10}E_7$	0.01–0.54	$W + L_\alpha$
		0.54–0.68	L_α
		0.74–1	L
2	$n\text{-}C_{13}E_{12}$	0.01–0.38	L
		0.42–0.73	H_1
		0.79–1	L
3	$MD'(E_7)M$	0.01–0.60	$W + L_\alpha$ (vesicle)
		0.60–0.76	L_α
		0.84–1	L
4	$MD_9D'_4(E_{12})M$	0–1	L
5	$E_{12}D_{15}E_{12}$	0.01–0.50	$W + L_\alpha$ (vesicle)
		0.50–0.90	L_α
		0.92–1	L

The samples were kept under ambient conditions, undisturbed, other than when measurements were made. Measurements were made immediately after the emulsions were prepared (fresh) and then repeated at 2 and 4 months. Measurements entailed visual inspection with the naked eye and measurement of emulsion droplet size. For the latter, a drop of the emulsion from the top of the sample was removed by a pipette and diluted in water. The diluted emulsion was immediately measured using a Malvern Mastersizer 2000 instrument. Table 2.2 summarizes the results. To facilitate the foregoing analysis, the surfactant fraction in the polar phase of the final emulsion (for all samples) and the intermediate emulsion during inversion (for sets C and D) is also listed in the table. All quantities are weight based. Note that in all the samples, the surfactant-to-oil ratio remains at 0.05. In the table, emulsion stability is noted as no change, creamed but no coalescence, or bulk phase separation. No change means that no observable difference was detected when compared with the fresh emulsion, either with the naked eye or based on emulsion droplet size measured. Cream was detected by a loss of the white intensity toward the bottom of the sample and was confirmed by measuring the droplet size distribution of the emulsion at the top. Cream is a gravitational effect; it may or may not be followed by droplet coalescence. If the emulsion is creamed but no coalescence of the droplets has occurred, the distribution of the droplet size at the top layer will be found to shift toward the larger side but not exceed the largest size in the fresh sample. In addition, if the creamed sample is remixed by toggling the container back and forth, and if coalescence is absent, the droplet size distribution will become the same as that of the fresh sample. Bulk phase separation means that the sample has separated into oil and water layers visible to the naked eye. Below is a summary of the observations on each set; some elaboration is given to sets B and C to demonstrate emulsion appearance and droplet size distribution.

Set A: All samples in this set did not display any change at 2 months. After 4 months, minor creaming was observed but no coalescence or phase separation was detected.

Set B: Creaming but no coalescence was detected in all samples at 2 months. After 4 months, the same held true for four of the samples, but 3B showed bulk phase separation. To illustrate some details, a diagram for how the samples in this set were handled is provided in Figure 2.1. Figures 2.2 and 2.3 show photographs and droplet size distributions, respectively, of the samples taken at the various stages corresponding to what is illustrated in Figure 2.1.

Set C: No significant change was observed in any of the samples at 2 or 4 months (Figures 2.4 and 2.5). The oil phase fraction in these emulsions was so high that the emulsions were gel emulsions, and as such droplet mobility was restricted. This set thus provides an excellent test for stability against coalescence.

Set D: Samples in this set showed similar behavior as in set B, except that creaming was significantly more noticeable. 3D exhibited bulk phase separation similar to 3B.

The five emulsifiers evaluated in this study form a narrow and yet meaningful series suitable for the intent of the study. The five emulsifiers are all relatively small

TABLE 2.2
Emulsion Composition and Stability

| Sample | Composition of Final Emulsion | | | Surfactant Fraction in Polar Phase, $S/(S+W)$ | | Emulsion Stability | |
	Si	S	W	Final Emulsion	During Inversion	2-Month	4-Month
1A	0.60	0.03	0.37	0.075	—	No change	Creamed but no coalescence
2A	0.60	0.03	0.37	0.075	—	No change	Creamed but no coalescence
3A	0.60	0.03	0.37	0.075	—	No change	Creamed but no coalescence
4A	0.60	0.03	0.37	0.075	—	No change	Creamed but no coalescence
5A	0.60	0.03	0.37	0.075	—	No change	Creamed but no coalescence
1B	0.30	0.015	0.685	0.02	—	Creamed but no coalescence	Creamed but no coalescence
2B	0.30	0.015	0.685	0.02	—	Creamed but no coalescence	Creamed but no coalescence
3B	0.30	0.015	0.685	0.02	—	Creamed but no coalescence	Bulk phase separation
4B	0.30	0.015	0.685	0.02	—	Creamed but no coalescence	Creamed but no coalescence
5B	0.30	0.015	0.685	0.02	—	Creamed but no coalescence	Creamed but no coalescence
1C	0.90	0.045	0.055	0.45	0.56	No change	No change
2C	0.90	0.045	0.055	0.45	0.56	No change	No change
3C	0.90	0.045	0.055	0.45	0.56	No change	No change
4C	0.90	0.045	0.055	0.45	0.56	No change	No change
5C	0.859	0.043	0.098	0.30	0.30	No change	No change
1D	0.20	0.01	0.79	0.0125	0.56	Creamed but no coalescence	Creamed but no coalescence
2D	0.20	0.01	0.79	0.0125	0.56	Creamed but no coalescence	Creamed but no coalescence
3D	0.20	0.01	0.79	0.0125	0.56	Creamed but no coalescence	Bulk phase separation
4C	0.20	0.01	0.79	0.0125	0.56	Creamed but no coalescence	Creamed but no coalescence
5D	0.20	0.01	0.79	0.0125	0.30	Creamed but no coalescence	Creamed but no coalescence

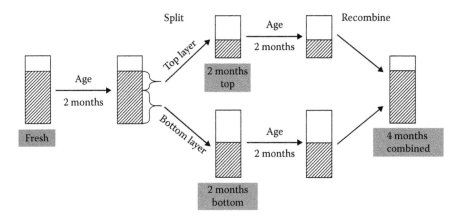

FIGURE 2.1   Diagram illustrating sample handling for 1B–5B.

molecule surfactants. This avoids the complicated viscoelastic effects associated with polymeric amphiphiles at the interface upon film thinning. The polar parts of the five surfactants are either EO_7 (surfactant 1 and 3) or EO_{12} (surfactants 2, 4, and 5). This allows for a comparison of the same hydrophile at two different hydrophilic lengths. Charge stabilization is absent. Therefore, the difference observed with regard to stability against coalescence should reflect mainly the mechanical properties of the adsorbed surfactant layer at the interface, and in particular help elucidate whether silicone surfactants demonstrate an inherent weakness due to the fluidity of the siloxane chains. The apolar parts of the hydrocarbon-based surfactants are branched C_{10} (surfactant 1) and linear C_{13} (surfactant 2) alkyl. The apolar parts of the silicone surfactant are methylsiloxanes. However, the three siliphiles differ in their overall molecular size and structure. The first silicone surfactant has the smallest siliphile, a trisiloxane. The second and third both have 15 siloxane units in the backbone, but one has a rake structure attached with four pendent polar chains of EO_{12}, and the other has an ABA structure with two terminal chains of EO_{12}. This particular set of 5 surfactants was chosen because their surfactant–water binary phase behaviors represent what is commonly encountered in commercial emulsion products. Therefore, the result is of practical interest. The surfactant–water binary phases in the four sets of emulsions involve the following four types: aqueous solution containing an exceedingly small concentration of surfactant (virtually pure water) (W), aqueous micellar solution (L_1), hexagonal (H_1), and lamellar (L_α) liquid crystals. The silicone oil used in this study is immiscible with any of the above phases, therefore the equilibrium phases of the emulsion system consist of the silicone oil phase coexisting with the surfactant–water binary phase, and hence the nature of the polar phase of the emulsion is determined solely by the surfactant-to-water ratio.

Results (Table 2.2) indicate that macroscopic phase separation occurred only in emulsions made with surfactant 3. However, this occurred only with emulsions at low oil content (samples 3B and 3D) but not with emulsions at high oil content (3A and 3C). Emulsions made with the other four surfactants were substantially stable against coalescence in the 4-month period, regardless of oil content. Among the

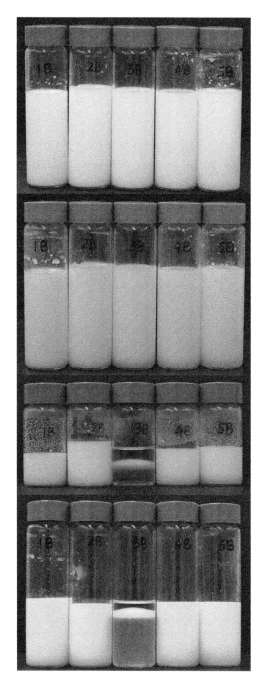

FIGURE 2.2 Photographs of 1B–5B immediately after the samples were prepared (*top row*), after 2 months (*second row*), top portion after additional 2 months (*third row*), and bottom portion after additional 2 months (*bottom row*).

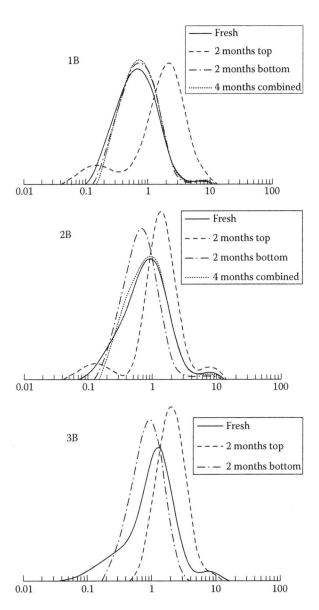

FIGURE 2.3 Droplet size distribution of 1B–5B immediately after the samples were prepared (solid line), after 2 months top portion (dashed line), after 2 months bottom portion (dash-and-dotted line), and after 4 months top-and-bottom combined (dotted line). The abscissa is particle size in microns; the ordinate is volume percent, all scaled to the same total area under the curve.   (*Continued*)

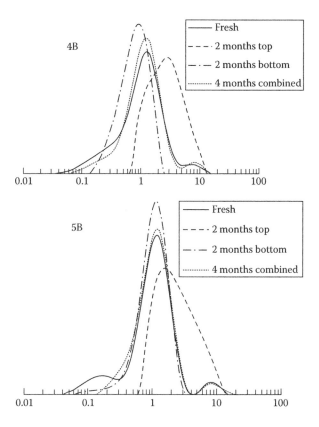

FIGURE 2.3 (*Continued*) Droplet size distribution of 1B–5B immediately after the samples were prepared (solid line), after 2 months top portion (dashed line), after 2 months bottom portion (dash-and-dotted line), and after 4 months top-and-bottom combined (dotted line). The abscissa is particle size in microns; the ordinate is volume percent, all scaled to the same total area under the curve.

four surfactants, there does not seem to be an inherent difference between the two silicone surfactants and the two hydrocarbon surfactants in their effectiveness to emulsify and stabilize Si/W emulsions. Since the effectiveness of an emulsifier is manifested in the achievable droplet size and the stability of the resultant emulsion, we analyze these two aspects—achievable droplet size and stability—separately using the above results.

The results from set B provide a good comparison regarding the achievable droplet size, because the emulsification procedure is fixed *and* the rheology of the liquid phases involved during the emulsification process is roughly the same across all five emulsions, which means that the droplets formed are expected to be of the same size unless the effectiveness of the emulsifiers are different. (The polar phase in each sample of set B is composed of 0.015/0.685 surfactant/water, therefore the phase acts

FIGURE 2.4 Photographs of 1C–5C immediately after the samples were prepared (*top row*) and after 4 months (*bottom row*).

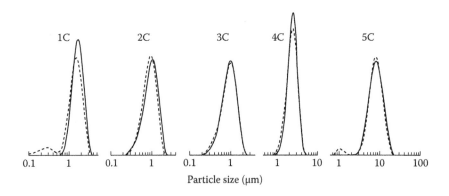

FIGURE 2.5 Droplet size distributions of 1C–5C immediately after the samples were prepared (dashed line) and after 4 months (solid line). The ordinate is volume percent. All curves scaled to the same total area under the curve.

as a low-viscosity Newtonian fluid.) Indeed, all five fresh samples in set B showed comparable droplet size distribution and mean droplet size (Figure 2.3).

The five gel emulsions from set C, on the other hand, displayed a more significant difference in the mean droplet size (Figure 2.5). The difference in the droplet size from set C is mainly due to the difference in the rheology of the polar phase during inversion. Rheology of the liquids during emulsification significantly impacts the attainable droplet size (see discussion in Section 2.5.1.2). Using the phase information provided in Table 2.1 and the polar phase compositions provided in Table 2.2, it is seen that the polar phases during inversion for the five samples in set C are

1. 1C–lamellar liquid crystal
2. 2C–hexagonal liquid crystal
3. 3C–mostly lamellar phase (93% lamellar + 7% aqueous solution)
4. 4C–micellar solution
5. 5C–about 60% lamellar and 40% aqueous solution

The droplet size achieved is therefore understandably larger for 4C and 5C, as both samples have a significantly lower polar phase viscosity compared to the other three samples, which involve liquid crystal phases. Thus, if the rheology effects are set aside, based on the results from set B, there does not seem to be a marked difference among the five surfactants in their impact on achievable droplet size.

To discuss emulsion stability, it is necessary to separate the event of creaming from the event of coalescence as the factors involved are different. Creaming is a gravitational effect; its rate depends on the difference in density between the dispersed phase and the continuous phase, the droplet size, the viscosity of the continuous phase, and the internal phase volume fraction [124]. Assuming no flocculation is at play, creaming does not relate to the property of the adsorbed surfactant at the interface. The observed severity of creaming among the samples in this study followed the trend D > B > A, and there was no creaming at all in the gel samples of set C. These results are consistent with the dependence of creaming rate on the viscosity of the continuous phase and internal phase volume fraction. Therefore, creaming profiles cannot be used to differentiate the five surfactants.

The only remaining measure that may differentiate the five surfactants in terms of their stabilization effect is the observation of stability against coalescence leading to ultimate phase separation. For this, only 3B and 3D displayed phase separation. Since phase separation occurred only in emulsions stabilized with the trisiloxane surfactant *and* with a low dispersed phase content (20%–30%), we suspect that the instability is due to the degradation of the surfactant over time. Silicone surfactants are known to hydrolyze in aqueous solutions and subsequently lose their surface activity [114,119,125]. The hydrolysis occurs at the siloxane backbone. The rate of hydrolysis increases under acidic or basic conditions [114,126]. It has been shown that hydrolysis occurs more readily with the surfactant in an unassociated state than in the aggregates [119,125]. This is presumably because in the aggregates, the siloxane chains are shielded from exposure to water. This may explain why only 3B and 3D phase separated while 3A and 3C did not. Using the phase information provided

TABLE 2.3

Phases Present and Their Respective Weight Fractions in the Continuous "Phase" of 3A–3D

Sample	W	L_α
3A	0.88	0.12
3B	0.96	0.04
3C	0.250	0.75
3D	0.98	0.02

in Table 2.1 and the component compositions provided in Table 2.2, the specific phases present in the continuous "phase" of the emulsion and their respective weight fractions can be calculated for the four emulsions stabilized with the trisiloxane surfactant (Table 2.3). The continuous "phase" in each of the four samples is actually composed of two coexisting phases: an aqueous phase (W) and a lamellar phase. The aqueous phase contains an extremely low concentration of the surfactant; the surfactant is mostly present as unassociated molecules. Since 3D and 3B contain the largest amount of the aqueous phase, it is reasonable to expect that hydrolysis of the trisiloxane surfactant occurs to the greatest extent in these two samples. The pH of 3D, 4D, and 5D were measured to be 5.5, 4.3, and 5.0, respectively, which are slightly acidic (this is common in nonionic surfactant stabilized emulsions prepared in open air). The two higher molecular weight silicone surfactants are apparently more stable against hydrolysis, presumably because higher molecular weight silicone surfactants have lower CMC, and the aggregation state shields the siloxane backbone from water. Larger silicone surfactants have been found to be stable in neutral aqueous solutions for years [113,119].

To summarize, the results from this study suggest that unhydrolyzed silicone surfactants are just as effective in emulsifying and stabilizing Si/W emulsions as hydrocarbon-based surfactants. The main drawback in using silicone surfactants as emulsifiers is the hydrolytic instability associated with low molecular weight silicone surfactants in the aqueous solution. Hydrolysis of the siloxane backbone leads to a loss of surface activity, and the emulsion subsequently phase separates. This points to the direction of using high molecular weight silicone surfactants as emulsifiers. However, in order to have a siliphile–hydrophile balance suitable for stabilizing Si/W emulsions, a great amount of polar-functional groups have to be grafted onto the large siloxane chain, which is difficult in synthesis due to steric hindrance. For this reason, high molecular weight silicone surfactants usually have a relatively low hydrophile-to-siliphile ratio and therefore tend to stabilize water-in-silicone emulsions.

2.3.3.2 Using Silicone Surfactant in Oil-in-Water Emulsions Containing Water-Miscible Cosolvent

Although it is not commonplace to use silicone surfactants as emulsifiers for O/W emulsions, there are applications involving aqueous-based formulations where silicone surfactants provide a superior stability to their hydrocarbon counterparts.

An example is in water-based paints and coatings. Water-based paints and coatings often contain coalescing agents that help the paint or coating dry to a coherent film by plasticizing the polymer binder (e.g., acrylic, polyurethane) during water evaporation. Typical coalescing agents are glycols, glycol ethers, esters, and alcohols. However, the presence of such solvents can destabilize O/W emulsions that are used as additives in water-based paints and coatings. The presence of coalescing agents significantly increases the CMC of the surfactants. It was discovered that O/W emulsions stabilized by silicone surfactants are unaffected by the presence of coalescing agents [127]. Indeed, O/W emulsions that are designed for coating systems containing coalescing agents are best prepared using silicone surfactants in order to keep the emulsion stable in the coating formulation [128].

The phenomenon that O/W emulsions stabilized by silicone surfactants remain stable in the presence of water-miscible cosolvent has also been observed in other applications, such as in alcohol-containing hair sprays. Gee and Vincent discovered that aqueous silicone emulsions prepared using a combination of hydrocarbon surfactant and silicone surfactant could remain stable when as much as 50% ethanol was added to the emulsion [129]. Vincent et al. showed that the stability provided by silicone surfactant in the presence of alcohol was not limited only to emulsions of silicone oils, but also applied to aqueous emulsions of lipophilic oils [130]. The phenomenon was investigated by Wang et al. using the atomic force microscopy (AFM) [131,132]. The authors compared a series of silicone surfactants of varying structures with fatty alcohol ethoxylates and Pluronic surfactants. In their experiments, the surfactant was allowed to adsorb onto an n-octadecyltrichlorosilane (OTS) modified silicon wafer surface, and surface force was measured between the surfactant-adsorbed OTS surface and the AFM tip in water. Ethanol was then added to the AFM liquid cell and the force profile was recorded. Both the maximum repulsive force and the thickness of the adsorbed surfactant layer were obtained from the force profile. The study showed that the steric repulsion provided by silicone surfactant persisted in the presence of up to 80% ethanol in the polar medium. On the other hand, the repulsion provided by alcohol ethoxylate surfactant or the Pluronic surfactant diminished at about 30%–40% ethanol. The force profiles were also compared with emulsion stability observations as well as with interfacial tension data. All results pointed to the conclusion that in the presence of alcohol cosolvent, silicone surfactants did not desorb from the interface, whereas hydrocarbon-based surfactants did. The authors attributed the persistent adsorption of the silicone surfactant in the presence of alcohol to silicone being hydrophobic, but not lipophilic [131,132].

2.3.3.3 Silicone Surfactants for Water-in-Silicone Emulsions

Silicone surfactants are commonly used to stabilize W/Si emulsions. In fact, they are the preferred emulsifiers. These surfactants typically have a large siloxane backbone containing from 100 to 400 siloxane repeating units, and only a few mole percent of the siloxane units are grafted with polar groups. Siliphile-rich silicone surfactants are more effective in stabilizing W/Si emulsions than lipophile-rich surfactants. Silicone surfactants also provide an unrivaled sensory benefit.

W/Si emulsions prepared using silicone surfactants first entered the market during the 1970s. Gee and Keil disclosed the preparation of W/Si emulsions using a mixture of hydrocarbon-based surfactants and silicone surfactants [104]. W/Si emulsions prepared with silicone surfactants quickly gained popularity in the antiperspirant [105,106,108,133] and cosmetic [107,134,135] markets. Chapter 9 provides a more detailed account of the development of this line of products. Multiple emulsions of the W/O/W [133] and O/W/O [136] kind can also be made using silicone surfactants. Silicone ter-block copolymers where both lipophilic alkyl chains and hydrophilic alkyleneoxide groups are attached to the same siloxane backbone can stabilize W/O emulsions containing a mixture of lipophilic oil and silicone oil [133,137]. Silicone ter-block copolymers are quite popular in personal care and cosmetic formulations.

High molecular weight silicone surfactants stabilize W/Si emulsions by adsorbing at the silicone oil–water interface forming a viscoelastic film. In W/O emulsions, electrostatic stabilization is absent. Stabilization against droplet coalescence is achieved by the adsorption of polymeric surfactants, insoluble macromolecules or other network forming structures at the interface. The presence of such species lends strong viscoelasticity to the interfacial film and thus impedes film thinning and film rupture. The use of proteins in the food industry is a good example [138]. Desorption of macromolecules from the interface is energetically more costly than desorption of small molecule surfactants. In many cases, the adsorption of polymeric surfactants or insoluble macromolecules is irreversible, hence film rupture is prohibited. Dahms and Zombeck showed that high molecular weight silicone surfactants had high adsorption efficiency at the oil–water interface [133]. Dimitrova et al. demonstrated that the adsorption of the silicone surfactants at the interface of W/Si emulsions is multilayered instead of monolayered [139].

Perhaps the most revealing study of how silicone surfactants stabilize W/Si emulsions is that by Anseth and Hill [140]. The authors studied the interfacial rheology of the silicone oil–water interface in the presence of polymeric silicone surfactants. A series of polymeric silicone surfactants were investigated. The silicone surfactants each contained about 400 dimethylsiloxane units and 4–10 methylsiloxane units grafted with polyalkyleneoxide pendent groups. The polyoxyalkylene chains differed in the number and combination of ethyleneoxide (EO) and propyleneoxide (PO) units. These silicone surfactants represent common silicone emulsifiers used in W/Si emulsions for personal care applications. They are typically delivered in low molecular weight silicone carrier fluids such as decamethylcyclopentasilxoane (D_5), although they are only partially miscible with D_5. Miscibility between the polymeric silicone copolymer and D_5 occurs when the mixture is rich in the copolymer. With increasing amount of D_5, excess D_5 separates out. The authors found that when these copolymers were swelled with D_5 and contacted with water, a third phase spontaneously formed. This third phase was rich in the copolymer. It contained moderate amount of D_5, but only a small amount of water. The copolymer-rich phase co-existed with excess silicone oil (D_5) and adsorbed at the silicone–oil–water interface in the emulsion. Contact angle measurement between the copolymer-rich phase, water, and silicone oil showed hysteresis. This suggests that upon contact with water and silicone oil, the copolymer-rich phase became locally nonuniform. Presumably, the copolymer molecules in the copolymer-rich phase rearranged to have the hydrophilic part (EO and PO) exposed to water and the siliphilic part exposed to the silicone oil. Since these copolymers are immiscible with water and partially miscible with D_5, it stands to reason that the copolymer-rich phase wets (partitions) more into the D_5 continuous phase than into water, therefore the emulsion type is W/Si. The formation of these silicone copolymer-rich third phases was apparently *in situ* and gradual, as evidenced by the change of the interfacial rheological responses with time. Both elastic and viscous moduli were shown to increase with time, and after a period of time the interface reached a highly viscoelastic state [140]. The study revealed that stabilization of the W/Si emulsion droplets is not due to a monolayer of the polymeric surfactant adsorbed at the droplet surface but rather the formation and irreversible adsorption of a third phase at the interface. Once the third phase is formed, the emulsion is in essence like a Pickering emulsion, as pointed out by the authors [140].

In a similar study, Sakai et al. reported stabilization of W/Si emulsions using what the authors referred to as silicone hybrid polymers. The silicone hybrid polymers contained a silicone backbone modified with hydrocarbon chains and hydrolyzed silk peptides. The hybrid polymers were described to be soluble in neither the silicone oil phase nor water but formed a third phase in the presence of water and silicone oil. The third phase resided between water and the silicone oil phase, stabilizing the W/Si emulsion. The authors named these hybrid polymers "active interfacial modifiers" [141].

High molecular weight silicone surfactants can also stabilize water-in-lipophilic oil emulsions by the same mechanism. Gašperlin et al. demonstrated that stable emulsions of water-in-petrolatum could be prepared by using merely a small amount (1%) of silicone surfactant. The authors recognized that the siloxane portion of the surfactant was both hydrophobic and lipophobic and therefore the surfactant formed a third phase [142].

2.3.4 Methods to Select Emulsifiers for Silicone Emulsions

The previous sections have introduced the broad categories of emulsifiers used for silicone emulsions. In this section, we provide some perspective on how to select the correct emulsifier for a targeted silicone emulsion, as a general principle. There are indeed no special rules for formulating silicone emulsions; principles that apply to emulsions in general also apply to silicone emulsions. The only unique aspects silicone presents to emulsion formulation are the following. First, in most silicones, the PDMS backbone is hydrophobic but not lipophilic. Second, silicone oil has a wide range of molecular weight. Miscibility of the silicone oil with a surfactant phase depends on the molecular weight of the silicone in addition to the nature of the functional groups attached to the silicone. These aspects will become more obvious in the forthcoming discussions. In general, when selecting emulsifiers for a targeted emulsion, the first consideration is given to the miscibility profile of the candidate emulsifier with respect to the two phases to be emulsified, because the relative state of miscibility determines the type of emulsion that will result, and inherently, the emulsion stability. Second consideration is given to the desired droplet size, which also affects emulsion stability. For commercial products, additional considerations may include suitability of the emulsifier for the specific application, regulatory status of the emulsifier, cost, etc. Here we elaborate upon the first two considerations: the effect of emulsifier on emulsion type and on emulsion droplet size.

2.3.4.1 Bancroft Rule, HLB, and Other Related Concepts for Predicting Emulsion Type

Bancroft in 1913 described a correlation between the emulsion type and the preferential miscibility of the emulsifier with the oil and water phases [143]. This correlation has since become the best known principle in emulsion science. The generally accepted version of the Bancroft rule—which is not necessarily Bancroft's original words—states that the phase in which an emulsifier is more soluble constitutes the continuous phase [144]. It turns out the interpretation of "more soluble" is not simple when it comes to surfactants. Surfactants form molecular solutions at concentrations below their critical micelle concentrations. Above the critical micelle concentration, the surfactant forms aggregates of a defined structure: micelle in a W–S binary system and microemulsion in a W–S–O ternary system. Depending on what is meant by "more soluble," the commonly stated Bancroft rule may or may not hold in all cases. Binks demonstrated that for the system water–$C_{12}E_5$–heptane, the critical aggregate concentration (CμC) in the oil phase is always higher than that in the aqueous phase, regardless of whether the temperature is below or above the phase inversion temperature (PIT) [145]. Here the aggregate is microemulsion droplet. At temperatures below PIT and a surfactant concentration below the CμC in the oil phase but above the CμC in the water phase, the emulsion type is O/W even though the total amount of surfactant present in the oil phase is greater than the total amount of surfactant present in the aqueous phase [145]. Therefore, the Bancroft rule appears to be violated. Friberg had also demonstrated similar facts in different systems [146,147]. Therefore, the continuous phase of an emulsion is not necessarily the phase that contains the higher concentration of the surfactant but the phase that the surfactant forms self-associated structures. As Binks pointed out, Bancroft's rule applies to surfactant aggregates rather than the total amount of surfactant present in the system [148].

In Bancroft's original description, the adsorbed surfactant layer at the emulsion droplet surface was said to act as a "flexible, vertical diaphragm, which separates two liquids A and B, and which is wetted by each." Bancroft argued that "the surface tensions of the two sides of the wetted diaphragm will not be the same as a rule. Owing to this difference in the surface tensions, the diaphragm will bend so that the side with the higher surface tension becomes concave" [143]. This depiction of the adsorbed surfactant layer as a diaphragm having two interfacial tensions, one with the oil and the other with water, suitably describes a situation where the surfactant is present as a third phase physically located between the oil and the water phases. Bancroft's rule was developed at a time when surfactant aggregation behavior had not been understood. We now know that emulsions can form not only with two-phase (O and W) systems where the surfactant is solubilized in the continuous phase, but also with three-phase (O, W, and SP) systems where the surfactant is predominantly present as a third phase (SP—surfactant phase), immiscible with the oil and water. The W/Si emulsions stabilized by high molecular weight silicone surfactants described earlier are three-phase emulsions. In short, the continuous phase of an emulsion is one that preferentially wets the surfactant *aggregates* (two-phase system) or the surfactant *phase* (three-phase system).

The concept described earlier constitutes a first guidance to selecting an appropriate emulsifier for a targeted emulsion. It is, however, not a quantitative one. Often times, bench formulators look for quantitative parameters that can help one pick the best suitable emulsifier. Several attempts to define such a parameter have been made since the Bancroft rule. The first is the HLB method introduced by Griffin in 1949 [87]. Griffin recognized that there was a general correlation between the emulsion type on the one hand and the relative size and strength of the hydrophilic and lipophilic moieties of the surfactant on the other. This observation led Griffin to establish a scale called the hydrophile–lipophile balance (HLB). Griffin first arbitrarily assigned an HLB value of 1 for oleic acid and 20 for potassium oleate. He then prepared emulsions using a standard formulation with fixed oil and composition ratios while varying only the emulsifier. Visual observations of emulsion stability were made in comparison to the two reference emulsions made with oleic acid and potassium oleate. Based on the observed stabilities of the emulsions, a series of emulsifiers were scaled and each emulsifier was assigned an HLB value. In a similar manner, Griffin used the already scaled emulsifiers to scale oils; each oil was then assigned an optimum HLB value. He found that if HLB of the surfactant is in the range 4–6, the emulsion type is W/O; if HLB of the surfactant is in the range 8–18, the emulsion type is O/W [87].

The empirical work Griffin carried out was obviously tedious, as Griffin wrote, "The magnitude of this type of study is understood when you realize that each of these values (the HLB value) was derived from approximately 75 emulsions" [87]. To simplify the work, Griffin later proposed that HLB values could be estimated based on molecular formula [88]. For example, the HLB value for a fatty alcohol ethoxylate can be estimated as the weight percent of ethylene oxide in the molecule divided by five. Similar estimates were also established for other types of surfactant structure. HLB values for an extensive range of emulsifiers are now available in handbooks and from emulsifier manufacturers. Most of them are obtained from calculations based on nominal molecular formula. Even though the estimation method based on molecular formula has found reasonable success in predicting the emulsion type in many situations, discrepancies exist, and the reason has been explained [149]; we come back to this later.

Several subsequent treatments have been developed in attempts to better estimate the HLB values and to define a concept from a more fundamental approach. Davies developed a method to calculate HLB number by summing over hydrophilic and lipophilic "group numbers" associated with structural groups of the surfactant [150]. Winsor developed an "R" ratio, which is a ratio of the interaction energy between the surfactant molecule and the oil molecule to the interaction energy between the surfactant molecule and the water molecule [151]. In an attempt to provide a quantitative basis to the Winsor "R," Beerbower and Hill proposed a "cohesive energy ratio" (CER) concept [152]. The CER was defined as the ratio between the adhesion energy of the lipophilic part of the surfactant to the oil phase and the adhesion energy of the hydrophilic part of the surfactant to the polar phase. The adhesion energies were in turn calculated in terms of the solubility parameters. Salager developed a correlation between the emulsion type and the "surfactant affinity difference" (SAD) [153]. SAD is essentially the difference in the standard chemical potentials of the surfactant in water and in the oil. A zero in SAD corresponds to an "optimum formulation" where ultralow interfacial tension is attained, and the scenario matches with a Winsor R equal to 1. The emulsion is W/O when SAD is positive and O/W when SAD is negative. These treatments and various others have been extensively covered in the literature [150–154]. Despite the large amount of effort in pursuing an ultimate "parameter" of the emulsifier to predict emulsion type, none of the methods has so far gained common use in industrial practice except for Griffin's HLB; and in the latter case, it is the method based on molecular formula rather than the empirical approach that is being practiced.

For silicone emulsions, the HLB method works only in predicting Si/W emulsions using hydrocarbon-based surfactants. That is to say, that surfactants with HLB values greater than 10 tend to produce Si/W type of emulsions as opposed to W/Si. On the other hand, hydrocarbon surfactants with low HLB values (4–6), that are predicted to stabilize water-in-lipophilic oil emulsions, generally do not produce stable water-in-silicone oil emulsions. Complications also arise when silicone surfactants are involved. Especially complicated is when the silicone surfactant contains both hydrophlic and lipophilic groups attached to the siloxane backbone; we now have three distinct moieties: hydrophile, lipophile, and siliphile.

Recognizing this limitation, O'Lenick and Parkinson added another dimension to the conventional HLB scale, which they called the "3D-HLB" [155]. As surfactants that contain two types of moieties (hydrophilic and lipophilic) were scaled in a one-dimensional line in the Griffin HLB method, surfactants that contain three types of moieties (hydrophilic, lipophilic, and siliphilic) were scaled in a two-dimensional triangle in the O'Lenick 3D-HLB method. A right triangle was thus constructed with each side of the triangle scaled 0–20. The two legs of the right triangle represent each a balance between silicone oil and water, and a balance between silicone oil and lipophilic oil. The hypotenuse represents a balance between lipophilic oil and water. Any composition of a silicone surfactant can be put on the triangle with the coordinate values being the weight present of the respective group in the surfactant divided by five. To correlate emulsion type with 3D-HLB value, a series of emulsions were prepared using a standard formulation varying only in the structure of the silicone emulsifier, and stability of the emulsions were observed. Six regions, W/Si, Si/W, W/O, O/W, Si/O, and O/Si ("O" stands for lipophilic oil), were identified in the triangle thus correlating emulsion type with 3D-HLB value. The model has the beauty of simplicity. However, to the best of the authors' knowledge, it has not been widely adapted in the silicone industry.

It is important to recognize that when HLB (or 3D-HLB) values are established based on observations of emulsion stability, the value reflects the status of *the emulsion*. The HLB number is not a material property of the emulsifier itself. It is meaningful only with reference to a certain emulsion formulation. Calculating HLB values based on surfactant molecular formula neglects the impact from the nature of the oil, isomeric structure of the surfactant, temperature, salinity, etc. A meaningful HLB that correlates with emulsion type is necessarily an HLB concept associated with the emulsion system, that is, a system HLB, and not an HLB of an isolated surfactant. It is also important to recognize the size effect of the emulsifier. Experimental observations have clearly established that stable emulsions are derived from emulsifiers that form aggregates or an insoluble third phase. Aggregation or precipitation (to form a third phase) is very much size driven.

It is clear from this discussion that what predicts emulsion type is the aggregation property of the emulsifier in the emulsion system. To this end, a most profound observation was made by Binks [145]: surfactants that display an equilibrium association structure with a positive spontaneous curvature (defined as hydrophilic portion pointing outward) favor the formation of O/W emulsion, and surfactants that display an equilibrium association structure with a negative spontaneous curvature (defined as hydrophilic portion pointing inward) favor the formation of W/O emulsion. This correlation has been explained by Kabalnov and Wennerstrom in terms of a thermally activated film rupture model relating emulsion stability against coalescence to the bending elasticity of the surfactant aggregate [156]. In general, surfactants that form aqueous micellar solutions (or aqueous microemulsions), liquid crystal phases of the normal structure, and the lamellar phase favor the formation of O/W emulsions. Surfactants that form inverse association structures favor the formation of W/O emulsions. We have found that this correlation applies to the silicone system as well. This will be demonstrated in Section 2.5.2 when we describe emulsion phase inversion.

2.3.4.2   Phase Behavior and Attainable Emulsion Droplet Size

The above discussions pertain to emulsion type. Predictions based on the Bancroft rule, the HLB value, or the nature of the equilibrium phases, however, do not say anything about the attainable emulsion droplet size. In practice, once a general type of the emulsifier is chosen for a targeted emulsion, selection among similar emulsifiers of that type is aimed to minimize the emulsion droplet size. In Section 2.5.1, we describe the dependence of attainable droplet size on formulation and process parameters. We will show that for a given process, the attainable droplet size is most dramatically influenced by the rheology of the fluids involved during homogenization. For an aqueous emulsion of a given oil, the attainable droplet size can be, to the largest extent, controlled by the phase behavior of the polar phase containing the surfactant and water during homogenization. Therefore, surfactant phase behavior not only determines the ultimate emulsion type but also governs the attainable emulsion droplet size. When selecting the emulsifier for a targeted emulsion, it is necessary to consider how the targeted use level of emulsifier will affect the system's phase behavior *at the pertinent stages* of the emulsification process. We have already used the phase behavior to analyze experimental results in Section 2.3.3.1. We will further demonstrate the impact of phase behavior on emulsification results in Section 2.5 when we describe specific methods in specific silicone systems. Before proceeding to Section 2.5, it is necessary to provide a description of the general phase behavior of the W–Si–S system.

2.4   GENERAL PHASE BEHAVIOR OF THE W–Si–S SYSTEM

Most of the published work regarding emulsion phase behavior has been focused on systems where the oil has a similar chemical nature and molecular size as the lipophilic portion of the surfactant. For such systems, there exist composition ranges where the oil, water, and surfactant are mutually solubilized. Silicone oils, on the other hand, are hydrophobic but not lipophilic. The phase behavior of water–silicone oil–surfactant is therefore different from that of water–lipophilic oil–surfactant.

Published phase diagrams with silicone as the oil phase are mostly limited to low molecular weight silicones. Many of the silicone emulsions sold on the market, however, involve polymeric silicones. Luckily, for the silicone emulsion chemists, the phase behavior of aqueous emulsions of polymeric silicones is rather simple. This is because polymeric silicones that contain no polar functionality are generally insoluble in the water–surfactant phase. Rouviere et al. showed that when PDMS of 10 and 50 cP were equilibrated with liquid crystal phases composed of water and an alkylphenolethoxylate surfactant, there was no swelling of the liquid crystal phases by the silicones [157]. Messier et al. reported that while a lamellar phase formed from water and an alcohol ethoxylate surfactant could be swollen by an oligomeric dimethylsiloxane bearing terminal hydroxyl groups, the same lamellar phase was not swollen at all when a much larger silicone was used [158]. Binks and Dong examined a system comprising water–$C_{12}E_3$–silicone (PDMS, 50 cSt) in the PIT (phase inversion by temperature) range. Typically for a system comprising water–C_xE_y–lipophilic oil in the PIT range, a middle phase is found, which contains most of the surfactant and an appreciable amount of both the oil and water. In the system, water–$C_{12}E_3$–silicone, however, Binks and Dong found that the middle phase contained only water and surfactant, and the volume of the silicone phase remained constant above and below the inversion temperature [159]. This indicates that there is no mutual solubilization between the silicone and the surfactant phase.

Our own observations are that PDMS greater in molecular size than D_5 has no solubility in the aqueous solution or liquid crystal phases of surfactant; this is true for both hydrocarbon-based and silicone-based surfactants. Even with hydrophobic surfactant (low HLB value), PDMS larger than D_5 has no or very limited miscibility with the surfactant phase. For instance, at room temperature D_5 is completely miscible with oleic acid, but PDMS of 50 cSt viscosity is completely immiscible with oleic acid. Similarly, D_4 is partially miscible with a silicone surfactant of the structure $MD'(EO_7OH)M$, but PDMS of 50 cSt is completely immiscible with the same silicone surfactant. Even when the silicone contains some polar-functional groups grafted to the siloxane chain, as long as the amount is limited to only a few mole percent and the silicone is polymeric, it is insoluble in water and in the surfactant phases. For instance, an aminosilicone of the structure $MD_{98}D'_2\left[C_3H_6NHC_2H_4NH_2\right]M$ is immiscible with water and most water-dispersable surfactants. If the amine is protonated, a small amount of surfactant and water can be solubilized in the silicone phase, but the silicone is not solubilized in the surfactant liquid crystal phases. However, the protonated aminosilicone can be solubilized in an aqueous surfactant micellar solution; this will be described in more detail in Section 2.6.

The limited solubility of polymeric silicone in the water–surfactant phase simply means that the equilibrium phase behavior for the W–Si–S system is determined solely by the W–S binary phase behavior. The silicone phase coexists as an additional phase to the aqueous surfactant binary phase. This is demonstrated in Figure 2.6. Figure 2.6a presents a generic W–S binary phase diagram for a nonionic surfactant. The phase diagram shows a general liquid crystal region and the clouding region at

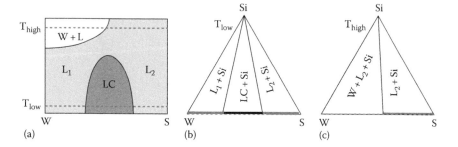

FIGURE 2.6 Generic phase diagrams for (a) W–S binary system showing regions of micellar (L_1), inverse micellar (L_2), and liquid crystal (LC) phases and the clouding region at elevated temperature; (b) W–Si–S ternary system at T_{low}; and (c) W–Si–S ternary system at T_{high}.

elevated temperature. Figure 2.6b and c present corresponding phase diagrams for the W–Si–S system at low and high temperatures, respectively.

At temperatures below clouding, the emulsion system consists of the silicone oil (Si) coexisting with a polar phase formed from water and the surfactant, Figure 2.6b. When the surfactant concentration is low, the polar phase is an aqueous micellar solution (L_1). When the surfactant concentration is medium, the polar phase is a liquid crystal phase (LC). When the surfactant concentration is high, the polar phase is an inverse micellar solution (L_2). At temperatures above clouding, Figure 2.6c, the surfactant dehydrates from water. Therefore, at low surfactant concentration, the emulsion system consists of the silicone oil coexisting with pure water (W) and a surfactant solution containing a moderate amount of water (L_2). At high surfactant concentration, the pure water phase goes away.

The temperature dependence of the phase behavior depicted in Figure 2.6 can be further illustrated by the following experiment reported by Liu [160]. In the experiment, an emulsion was first prepared at room temperature with brine (10 mM NaCl), tetraethyleneglycol dodecyl ether ($C_{12}E_4$), and PDMS (viscosity 50 cSt). The emulsion was then heated under a constant stir, and the conductivity was measured as a function of temperature. The result is shown in Figure 2.7. The emulsion initially formed at room temperature was Si/W, and it remained stable up to 40°C. Between 40°C and 70°C, the conductivity of the emulsion fluctuated dramatically. At above 70°C, the emulsion phase separated into visible bulk silicone and water. The steep increase in conductivity confirms the presence of bulk water phase.

The emulsion behavior in this example can be understood by examining the water–surfactant binary phase behavior of the system. The temperature dependence of the binary system water–$C_{12}E_4$ was reported by Mitchell et al. [161]. At the W:S ratio of 74:6, the binary phases for water–$C_{12}E_4$ are

20°C–50°C	$W + L_\alpha$
50°C–54°C	$L_3 + L_\alpha$
54°C–56°C	L_3
56°C–70°C	$W + L_3$
>70°C	$W + L_2$

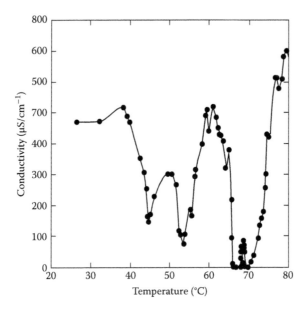

FIGURE 2.7 Conductivity versus temperature of an emulsion consisting of brine (10 mM NaCl)–PDMS–$C_{12}E_4$ at a ratio of 74:20:6.

At room temperature, the ternary system of the emulsion at hand must be, there-fore, brine coexisting with L_α and silicone. The lamellar phase is present at a very small fraction in the emulsion and is presumably adsorbed at the Si–W interface. Not surprisingly, the emulsion type is Si/W. At above 70°C, the ternary system must be brine coexisting with L_2 and silicone. The inverse micellar solution, L_2, wets neither brine nor silicone. Therefore, no stable emulsion is formed. Based on the binary phase behavior, one would expect the Si/W emulsion be stable up to 50°C where a sponge phase, L_3, appears. The conductivity result indicates that the emul-sion became unstable above 40°C. This could be caused by the presence of NaCl, which is known to depress the cloud point and decrease PIT. In the intermediate tem-perature range 50°C–70°C, the binary phase of water–$C_{12}E_4$ is lamellar coexisting with the sponge phase, the sponge phase alone, or water coexisting with the sponge phase, depending on the specific temperature. The ternary system is likewise with an additional coexisting silicone phase. The formation of the sponge phase may cause the fluctuation in the conductivity of the emulsion; a similar phenomenon had been previously reported by others [162].

In short, the phase behavior of a binary water–surfactant system can be used to predict the type of emulsion that will form upon adding silicone. Polymeric, non-functional dimethylsiloxanes have little to no solubility in surfactant aggregates, which means they do not interfere with the binary water–surfactant phase behavior. The silicone coexists as an additional phase to the binary water-surfactant phase, and the specific type of that binary water-surfactant phase determines the emulsion type. At low temperature, the hydrophilic portion of the surfactant is hydrated with water and thus the spontaneous curvature of the surfactant is positive. Hence, the emulsion formed is Si/W. With increasing temperature or salinity, the surfactant becomes increasingly dehydrated. However, since polymeric silicone molecules do not penetrate into the surfactant lipophilic chains, a negative curvature is not formed. Hence, a stable W/Si emulsion does not form.

The immiscibility of the silicone phase with the aqueous surfactant phase has a profound practical implication for silicone emulsion process. First, the phase behavior of the polar phase, and thus the emulsion type, is controlled solely by the amount of water in the system. Second, since the precise nature of the polar phase determines the rheology of the polar phase as well as the interfacial tension between the polar phase and the silicone phase, which in turn governs the emulsion droplet size formed, the emulsion droplet size is also controlled solely by the amount of water in the system. This links the complete outcome of the emulsion to an easily controllable parameter—the amount of water—in the emulsification process. We will use this concept to guide the discussion on emulsification of silicones in Section 2.5 and demonstrate the convenience of using water content to manipulate the emulsification result.

2.5   EMULSIFICATION OF SILICONES

Methods for emulsification used in industry follow three general categories:

1. Direct emulsification, also referred to as high energy dispersion method
2. Phase inversion
3. Membrane, microchannel, and microfluidics

The first two routes are routinely used in the production of silicone emulsions. Methods in the third category are based on forcing the dispersed phase into the continuous phase through uniform pores, channeled holes, or capillaries [163]. They are, in principle, applicable to producing silicone emulsions as for any other emulsions, but due to their limited production rate are not employed in silicone emulsion productions.

The discussions in this section focus on the preparation of silicone-in-water emulsions. The emulsification of linear silicones, which are liquid at room temperature regardless of molecular weight, is straightforward. Typically low viscosity silicone fluids are emulsified using the direct emulsification method, and high viscosity silicone fluids are emulsified by phase inversion. These two methods are described first in Sections 2.5.1 and 2.5.2. Challenges arise when the silicone has very high molecular weight or the siloxane chains are significantly cross-linked. Emulsification of high molecular weight silicone gums, silicone elastomers, and silicone resins are described separately in Sections 2.5.3 through 2.5.5. Low molecular weight silicones, reactive silicones, and silanes present a different set of challenges for achieving good emulsion stability. These emulsions are described in Section 2.5.6.

2.5.1 DIRECT EMULSIFICATION

2.5.1.1 Process

As the name implies, direct emulsification entails adding the intended dispersed phase to the continuous phase under mixing to arrive directly at the targeted emulsion type. Thus, if a Si/W emulsion is to be made, silicone oil is added to the aqueous phase containing the emulsifier. The process typically includes two main steps. First, a crude emulsion is prepared in a mixing vessel by adding the oil phase to the aqueous phase. The crude emulsion, also called the "premix," contains coarse oil drops ranging from tens to hundreds of microns and is stable only for a short period of time. In the second step, the premix is passed through an energy-intensive, high-shear device to reduce the droplet size. The process is schematically illustrated in Figure 2.8.

There are many types of high-shear devices used for emulsion production. Examples include the Ultra-Turrax® disperser, colloid mill, the Greerco® Pipeline Mixer, and the Homomic Line Mill (HLM). These devices are based on a rotor stator mechanism [164] and are used for a wide range of fluids. The Rannie and Gaulin, the Sonolator™, and the Microfluidizer® are high pressure homogenizers that generate highly inertial flows created by valves and nozzles [165]. High pressure homogenizers are suitable for homogenizing low viscosity fluids. In the manufacturing setting, sometimes two different high-shear devices are arranged in sequence. Alternatively, the emulsion can be circled through the same high-shear device continuously until a desired final droplet size is achieved. The output of the highly sheared, fine emulsion can then go into a second vessel where dilution water or other additives are added. Polymerization can also be carried out in the second vessel.

In certain batch processes, the two steps, premix and high shear, can be carried out in one vessel without having to pass the emulsion through an additional device.

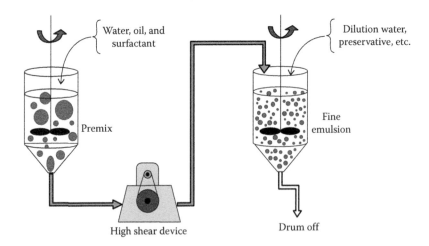

FIGURE 2.8   Schematic illustration of the direct emulsification process.

In this case, powerful impeller or dispersing blades are inserted in the vessel that can directly produce fine emulsions. Planetary mixers manufactured by Charles Ross & Son, and MYERS, which are often referred to as "change can" mixers, are an example of high shear vessels. Direct emulsification can also be carried out in a continuous fashion where all ingredients are metered in and the mixture is passed through a high-shear device without the premix step. While the continuous process may be advantageous in terms of process time, it may not be suitable for complicated formulations such as multiple emulsions.

2.5.1.2 Attainable Droplet Size

The emulsion droplet size and distribution that results from a homogenization process depends on many factors including, on the process side, the type of the homogenizer and the operating conditions, and on the formulation side, the viscosities of the dispersed and the continuous phase, emulsifier type and concentration, and dispersed phase volume fraction. Although the emulsion produced from a given formulation and process is generally reproducible, a quantitative prediction of droplet size distribution is difficult because the state of droplet deformation is not experimentally accessible. Established models are mostly built based on idealized situations. A comprehensive review on droplet deformation and breakup under various conditions can be found elsewhere [166,167]. Here we comment on two aspects that can be easily considered during formulation design and which greatly influence the outcome of the emulsification process: the impact of viscosity and the role of the emulsifier. The discussion here is not specific to the nature of the oil; therefore, we make no specific reference to silicone, but it certainly applies to silicone emulsions.

2.5.1.2.1 Impact of Viscosity

The dispersed and continuous phase viscosities have a profound and complicated impact on the attainable droplet size. In general, the higher the dispersed phase viscosity, the more resistant the droplet is to deformation. Meanwhile, the higher the continuous phase viscosity, the greater the stress transmitted on the droplet to deform and disrupt it. Therefore, the attainable droplet size generally decreases with decreasing dispersed phase viscosity and increasing continuous phase viscosity. The ratio between the two viscosities, η_D/η_C, is important too. A useful parameter is the Weber number (We, if the stress is due to inertial force), or the Capillary number (Ca, if the stress is due to viscous force). These numbers are defined as the ratio between the stress exerted on a droplet and the Laplace restoring pressure. Droplet breakup occurs when the Weber number or Capillary number exceeds a critical value, We_{cr} or Ca_{cr}, respectively. For a given shear device, We_{cr} or Ca_{cr} is shown to be a function of the viscosity ratio, η_D/η_C [166,167]. A general trend is shown in Figure 2.9. Understandably, the attainable droplet size follows a similar functional dependence on the viscosity ratio η_D/η_C. The minimum attainable droplet size therefore corresponds to, roughly, η_D/η_C between 0.1 and 5.

Different high-shear devices have different effectiveness for a given set of fluids to emulsify. For aqueous emulsions of silicone fluids having a viscosity below ca.

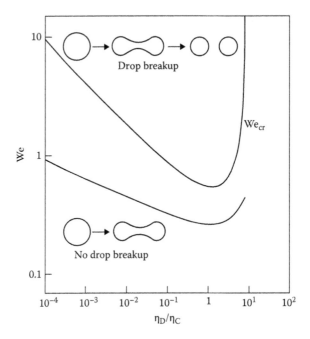

FIGURE 2.9 Schematic illustration of dependence of critical Weber or Capillary number for drop breakup on viscosity ratio. Different types of flow have different critical curves.

5000 cSt, an average droplet size of 0.3 μm can be routinely produced using most high-shear devices in manufacturing scale. Since the flow pattern is highly inhomogeneous within any high-shear chamber, the droplet size is typically polydispersed. If the oil phase is homogeneous, the droplet size distribution tends to be monomodal, and the width of the distribution usually decreases with decreasing average droplet size.

There is an upper viscosity limit beyond which high pressure homogenizers become ineffective. Typically, the viscosities of both the oil phase and the aqueous phase should be kept below 5000 cP if a high pressure homogenizer is used. Rotor-stator type of homogenizers and propeller mixers can handle much higher viscosities. As Figure 2.9 suggests, the viscosities of the oil phase and the aqueous phase should be kept at similar orders of magnitude. The viscosity of the continuous phase can be manipulated by the concentration of the emulsifier in the continuous phase, or alternatively by the dispersed phase volume fraction. A common practice in making O/W emulsions is to first prepare a thick emulsion where a portion of the water is deliberately withheld. The thick emulsion is sheared to reach a targeted droplet size, and then the remaining water is added to dilute the thick emulsion to a final composition. Withholding water increases the dispersed phase volume fraction as well as the surfactant concentration in the continuous phase; both factors raise the effective viscosity of the emulsion, which in turn increases the effective stress exerted on the droplet to deform and disrupt into smaller droplets. The following

TABLE 2.4

**Prototype Pre-Mix Formulations for a Si/W
Emulsion; Quantities in Unit of Part by Weight**

Formulation	1	2
Water	32	15
i-C$_{10}$E$_7$	6	6
Sodium C$_{13-17}$-sec-alkanesulfonate	0.4	0.4
PDMS (350 cSt)	100	100

example demonstrates the effect of water content during homogenization on the final emulsion droplet size.

Two emulsions were prepared with the premix formulation shown in Table 2.4. The two premixes differ only in the water content. The premix was prepared by first mixing water and the surfactants in a vessel using a propeller stirrer and then adding to the aqueous surfactant mixture the PDMS fluid under a constant mixing speed. The premix was passed through a Sonolator to produce a fine emulsion. In the first case, the premix was passed through the Sonolator twice, once at 1750 psi and a second time at 2000 psi. In the second case, the premix was passed through the Sonolator only once, at 1400 psi. The droplet sizes obtained in the two emulsions are shown in Figure 2.10. As can be seen, water level has a significant impact on the resultant droplet size. Even though the first emulsion (dashed curve) was sheared more intensively than the second, the second emulsion (solid curve) has a smaller

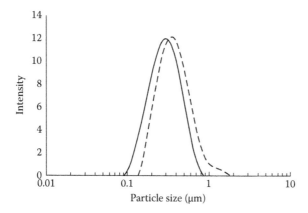

FIGURE 2.10 Droplet size distribution of two silicone emulsions prepared using a high pressure homogenizer. Dashed curve corresponds to the emulsion produced with the greater amount of water and the solid curve corresponds to the emulsion produced with the smaller amount of water. The volume/surface average (or Sauter mean) are: solid curve, $d_{3,2}$ = 0.245 µm; dashed curve, $d_{3,2}$ = 0.318 µm.

average droplet size along with a narrower distribution. Both emulsions in this example can be diluted with additional water without changing the droplet size.

There is a lower limit to the water content during homogenization. Emulsion viscosity increases with increasing dispersed phase volume fraction. Beyond a certain dispersed phase volume fraction, the viscosity rises sharply. Highly viscous emulsions are difficult to process and experience a depressed flow turbulence in the homogenizer, which then results in large droplet sizes [166]. In addition, when the amount of water is reduced to a level at which liquid crystals are formed, the emulsion becomes too thick to process through a high pressure homogenizer. In these cases, a rotor-stator mixer can be used instead. In Section 2.5.2.2, we will see that the presence of liquid crystals is ideal for emulsifying a highly viscous oil phase using catastrophic phase inversion.

2.5.1.2.2 Interfacial Tension and the Role of the Emulsifier

The attainable droplet size also depends on the interfacial tension. This dependence is due to the Laplace pressure, which opposes deformation. The Laplace pressure arising from a curved interface is

$$\Delta P = \gamma \left(\frac{1}{R_1} + \frac{1}{R_2} \right) \tag{2.1}$$

where
R_1 and R_2 are the radii of curvature
γ is the interfacial tension

The attainable droplet size, d, varies with interfacial tension according to

$$d \propto \gamma^p \tag{2.2}$$

where p is 1 for viscous forces and 3/5 for inertial forces [167]. Therefore, the presence of a surfactant can facilitate the formation of the emulsion by lowering the interfacial tension to reduce the stress necessary to overcome Laplace pressure. However, lowering the interfacial tension is not the essential role of the emulsifier. The primary function of the emulsifier is to stabilize the newly created interface and prevent the droplets from recombining. Walstra and Smulders gave a convincing example by comparing the emulsification of triglyceride in water without surfactant to the emulsification of paraffin oil in water in the presence of surfactant. The equilibrium interfacial tensions in the two cases are equal. The authors pointed out that only the second system forms an emulsion [167, p. 74]. In Section 2.5.3, we show that silicone gum can be emulsified using polymeric surfactants but not with low molecular weight surfactants, though polymeric surfactants typically give higher interfacial tension than low molecular weight surfactants. Clearly interfacial tension is not the primary factor in the emulsion formation. It has been shown that surfactant adsorption time, including diffusion time, generally exceeds droplet disruption time [168]. During the disruption of a droplet, almost no surfactant molecules adsorb at the newly created interface. Surfactant adsorbs between two breakup events. This is particularly true for polymeric surfactants. Brösel and Schubert showed that droplet disruption with and without surfactant is not significantly different [168]. If stabilization is poor, droplets will recoalesce, and the disruption result is negated. Lobo and Svereika demonstrated that the obtained droplet size depends mostly on the ability of surfactants to stabilize the droplets against coalescence rather than their ability to reduce the interfacial tension [169]. Thus, the primary concern in selecting an appropriate emulsifier is emulsion stability.

Aside from providing emulsion stability and lowering the interfacial tension, the surfactant exerts another influence on the outcome of the emulsion by affecting the rheological properties of the polar phase. The rheology can drastically impact the emulsion droplet size. The resultant droplet size varies with the viscosity of the continuous phase according to

$$ d \propto \frac{1}{\eta_C^q} \tag{2.3} $$

where q is 1 for laminar flow and 1/2 for turbulent flow [167]. The viscosity of the continuous phase changes with the concentration of the surfactant. Even with a constant surfactant-to-oil ratio, the continuous phase can be tuned from an aqueous micellar solution to a viscoelastic liquid crystal gel solely by adjusting the water content. In contrast, variation in the interfacial tension brought by using different surfactants has a limited effect on droplet size. As an example, the interfacial tension between PDMS and pure water is 45 mJ/m^2. The lowest interfacial tension in a PDMS-in-water emulsion stabilized with the most effective surfactant is about 2 mJ/m^2. This scale of change (from 45 to 2) is relatively small compared to the possible scale of change (many orders of magnitude) in the viscosity of the polar phase. While reducing interfacial tension is important, manipulating the phase behavior of the continuous phase is far more effective in reducing the emulsion droplet size.

Finally, it is noteworthy to point out that emulsification is an energy intensive but inefficient process. The amount of energy that is converted into the interfacial energy is typically less than 0.1%, and often less than 0.01%, of the total energy consumed by the homogenization process. The vast majority of the energy is dissipated as heat.

2.5.2 EMULSIFICATION BY PHASE INVERSION

Emulsion phase inversion refers to the phenomenon that the dispersed and the continuous phases swap upon a certain change in the conditions. For example, a water-in-oil emulsion becomes an oil-in-water emulsion. Phase inversion can be deliberately triggered as a method to prepare emulsions [170]. Emulsification by phase inversion generally results in a small and narrowly distributed droplet size. To produce an emulsion by the phase inversion method, an initial emulsion opposite to the intended type is first prepared by dispersing the intended continuous phase in the intended dispersed phase. A trigger is then applied to induce a phase inversion. Depending on

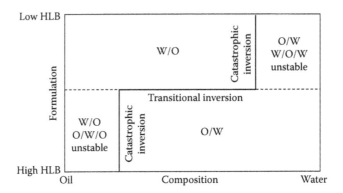

FIGURE 2.11   Diagram illustrating emulsion phase inversion.

the trigger used, phase inversion is further categorized into two types: transitional inversion and catastrophic inversion. Transitional inversion is induced by changing formulation variables such as temperature, salinity, and surfactant type, whereas catastrophic inversion is induced by changing the water-to-oil ratio. The two types of inversion display different characteristics during the inversion process and can be suitably employed in different situations.

Emulsion phase inversion has been extensively studied, and a large volume of literature is available [170–187]. The two types of inversion have been often discussed with the aid of a diagram, like the one illustrated in Figure 2.11. The abscissa represents the oil-to-water ratio, and the ordinate represents a generic formulation variable [185]. A review of the fundamental mechanisms of the two types of phase inversion and the applicability of each in industrial practice can be found elsewhere [160]. Here we give a brief description of the two inversion mechanisms in order to provide sufficient background before discussing the specific details of the emulsification of silicones via phase inversion.

2.5.2.1   Transitional Inversion

2.5.2.1.1   Mechanism

Transitional inversion is regarded as being caused by a change in the surfactant affinity toward the oil and the water phases as a result of changing a formulation variable. PIT (phase inversion by temperature) is a prominent example of transitional inversion. PIT was first reported by Shinoda who compared the clouding phenomenon associated with aqueous solutions of nonionic surfactant to emulsion inversion as a result of temperature change [188]. Aqueous solutions of water and alcohol ethoxylate surfactant exhibit a closed miscibility gap. The lower consolute, known as the cloud point, corresponds to the temperature above which the aqueous surfactant solution separates into a water phase containing an extremely small amount of surfactant and a surfactant phase containing some water. Solutions of hydrocarbon oil and alcohol ethoxylate surfactant also exhibit clouding at a temperature referred to as the haze point. The haze point is the upper consolute of the miscibility gap below which the oil surfactant solution phase separates into an oil phase and a surfactant phase. Shinoda observed that for a system consisting of water, hydrocarbon oil, and alcohol ethoxylate surfactant there exists a transitional temperature range; above this range the system forms a W/O emulsion when stirred, and below this range the system forms an O/W emulsion [171]. The close association of the PIT phenomenon with the aqueous solution cloud point and the oil solution haze point led to the general misconception that the surfactant's preferential solubility in water and oil reverses at PIT. This misconception was later corrected by Binks [145]. It turns out that the surfactant that does not diffuse from one phase to the other at PIT; rather, a third phase forms in a temperature range associated with the cloud point and the haze point. This third phase, referred to as the surfactant phase or "the middle phase" (denoted in Shinoda's original publication as D), contains a large portion of the total surfactant in the system and solubilizes a large amount of both water and oil. The ternary phase behavior in the temperature range around the cloud point, PIT, and haze point was described in detail by Shinoda and Friberg [189]. In the ensuing studies, it was revealed that the PIT process is a phase separation process rather than surfactant migration from one phase to the other [147,190]. The process is illustrated in Figure 2.12 and explained below based on the result reported by Friberg et al. on a system water–hexadecane–$C_{12}E_4$ [147].

At high temperature, Figure 2.12a, the system consists of an oil phase, O_m, in equilibrium with a water phase, W. The oil phase contains a moderate amount of water and surfactant. The surfactant is present in the form of unassociated molecules as well as inverse micelles. Water is solubilized in the surfactant micelles forming

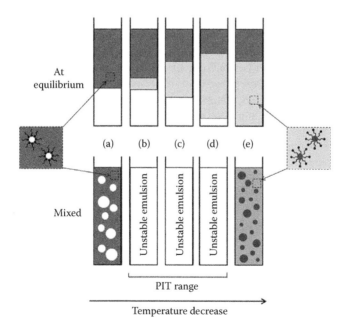

FIGURE 2.12 Schematic illustration of the equilibrium phase behavior (*top row*) and the corresponding emulsion morphology (*bottom row*) in a PIT process for a generic system containing equal amount of oil and water. Dark shade represents oil phase; white shade, water phase; and gray shade, surfactant (middle) phase.

water-in-oil microemulsion droplets, as illustrated in the insert on the left. The water phase contains virtually no surfactant and oil. If the system is stirred, a water-in-oil (W/O_m) emulsion forms. As the system cools and enters the PIT range, Figure 2.12b, a middle phase separates out of the oil phase. The middle phase consists mostly of oil, and a moderate amount of surfactant and water. The retained oil phase, lessened in volume, still contains a moderate amount of surfactant but lost most of the water in the original oil phase. The volume of the water phase is slightly reduced compared to the original water phase, but otherwise consists, still, of virtually pure water. As temperature continues to decrease, Figure 2.12c and d, the middle phase expands greatly in volume solubilizing more oil and much more water. The oil phase composition remains roughly the same. The water phase, significantly reduced in the volume, contains yet still virtually pure water. In the entire PIT range (Figure 2.12b through d), the only surfactant aggregate is present in the middle phase. The interfacial tension between the middle phase and the oil phase at the beginning stage of the PIT range, Figure 2.12b, and the interfacial tension between the middle phase and the water phase at the ending stage of the PIT range, Figure 2.12d, are both exceedingly small (one should see that this is necessarily so). If the system at this range is stirred, an emulsion forms but is extremely unstable. Finally, as temperature decreases below the PIT range, Figure 2.12e, the oil phase increases its volume but otherwise keeps roughly the same composition, while the original water phase disappears. The middle phase becomes the new aqueous phase containing less oil and most of the water in the system. The surfactant is present in the new aqueous phase in the form of unassociated molecules and normal micelles. Oil is solubilized in the surfactant micelles forming oil-in-water microemulsion droplets, as illustrated in the insert on the right. If the system is stirred, an oil-in-water (O/W_m) emulsion forms.

It is important to point out that during the entire process, the surfactant concentration is higher in the oil phase than in the water phase or in the final aqueous phase. The only difference is that no micelles/microemulsion droplets are present in the oil phase at low temperature (Figure 2.12b through e), and no micelles/microemulsion droplets are present in the water phase at high temperature (Figure 2.12a through d). This is consistent with the discussion in Section 2.3.4 regarding the interpretation of the Bancroft rule. It is not the preferential solubility per se that determines which phase becomes the continuous phase of the emulsion, but the presence of the surfactant aggregate. The inversion point corresponds to a balanced condition where the aggregation structure (here in the middle phase) has an average curvature of zero.

Transitional inversion can also be induced by change in solution salinity, pH, and by addition of another surfactant. A common characteristic of all transition inversions is that the state of the surfactant in the emulsion is never too far from that of the equilibrium; thus, the inversion process is reversible. The emulsion type along the inversion path is consistent with the equilibrium microemulsion type. When the system at equilibrium consists of water (W) and water-in-oil microemulsion (O_m), the emulsion formed upon mixing is W/O_m. When the system at equilibrium consists of oil (O) and oil-in-water microemulsion (W_m), the emulsion formed is O/W_m. When the system at equilibrium consists of water (W), oil (O), and a surfactant phase in the form of a bicontinuous microemulsion (D), the emulsion formed upon mixing is exceedingly unstable.

It can be easily seen that in order for transitional phase inversion to occur, the surfactant needs to be soluble and form aggregates in water under one condition, and under another condition, be soluble and form aggregates in the oil. If the oil is a silicone, this happens if and only if the silicone oil is of low molecular weight [159]. Polymeric silicone oils, in the absence of polar-functional groups, are generally immiscible with common surfactants, whether the surfactant has a high or low HLB value and whether the hydrophobic portion is lipophilic or siloxane. For polymeric silicone oils, a transitional interval is absent because the system instantly phase separates once the condition for forming surfactant aggregates in the water phase is passed (Figure 2.6). W/Si emulsions can be made with polymeric silicone surfactants, but the surfactant is present not in the silicone continuous phase but as a third phase, as described in Section 2.3.3.3, and the emulsification cannot be carried out through the phase inversion route. Note that there is nothing peculiar about the chemical nature of the silicone oil; the immiscibility with the surfactant is solely due to the polymeric size of the silicone oil molecule. Studies on transitional inversion typically deal with low molecular weight hydrocarbon oils because paraffins of C_{18} or greater are solid at room temperature. Silicones are only "deceiving" because they can be very high in molecular weight and still remain liquid at room temperature, a feature fundamentally rooted in the low intermolecular forces between siloxane chains (see Chapter 1).

2.5.2.1.2 Nanoemulsion Made via Microemulsification Path

An impressive phenomenon arises in the PIT process when the system contains a sufficient amount of surfactant. That is, with enough surfactant the middle phase can solubilize all the oil and water, and the entire system becomes a single phase microemulsion. This phenomenon has attracted a great deal of attention in both the academia and industry [191], especially for its potential application in oil recovery [192]. However, a regrettable drawback for microemulsion in many applications is that it is confined to a limited composition and temperature range, and if the application condition falls out of this range, the microemulsion phase separates thereby becoming a cloudy emulsion. For instance, in consumer products, most formulations contain many ingredients. If the microemulsion is to be used as an additive, it may phase separate because the final composition is no longer the same. For this reason, kinetically stabilized emulsions are preferred. If optical clarity is desired, the emulsion needs to have small enough droplet size. Emulsions having small droplet size (<100 nm) are fashionably called "nanoemulsions." One technique for producing a nanoemulsion is to pass the system through a single-phase microemulsion condition before arriving at the final two-phase emulsion. This has been demonstrated using PIT [193] as well as other modes of transitional inversion [194,195]. In fact, it has been shown that the system does not necessarily have to go through a complete phase inversion. The system can start anywhere in the transitional inversion interval so long as it passes the microemulsion phase solubilizing all the intended dispersed phase of the targeted nanoemulsion [193,196]. Specifically, an O/W nanoemulsion can be prepared by starting the composition at a condition where it forms a microemulsion phase that solubilizes all the oil, and then moving the system into a two-phase condition corresponding to an oil-in-water microemulsion phase (W_m) coexisting with excess oil phase (O). During this process, the oil phase separates out from the initial microemulsion phase as very small droplets. A "trick" is involved, however, in the preparation of nanoemulsions via the microemulsion path. When the initial microemulsion is a bicontinuous type, the nanoemulsion formed near the microemulsion condition is highly unstable [171], even though the interfacial tension involved is exceedingly small (10^{-2}–10^{-3} mJ/m^2) [197]. Therefore a rapid quench is necessary to move the system quickly and far away from the microemulsion condition. Several methods have been described in the literature [198–200].

Figure 2.13 illustrates a case for making silicone-in-water nanoemulsion via the microemulsion path using temperature as the controlling condition. In this case, a composition was formed by combining 24.75 parts of D$_4$, 12.75 parts of a silicone

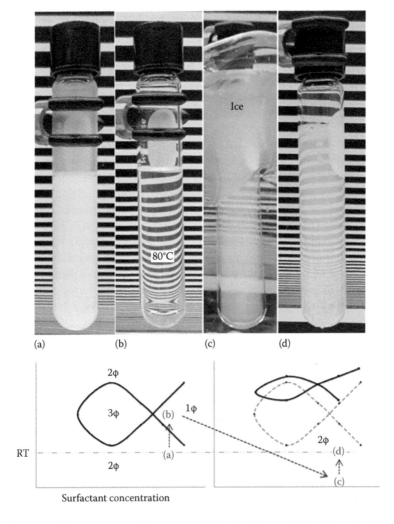

FIGURE 2.13 Silicone nanoemulsion prepared via the microemulsification path using temperature as the controlling condition.

surfactant with a nominal structure MD$'$(EO$_7$)M, and 37.5 of parts water in a glass tube. The mixture formed a cloudy white emulsion when shaken (Figure 2.13a). When heated to about 80°C, the sample formed a microemulsion and became clear (Figure 2.13b). If the sample was allowed to stand and cool to room temperature, it would immediately turn cloudy white. It would turn clear again if heated. When the microemulsion formed at 80°C was rapidly placed into an iced bath, the system became a nanoemulsion that was translucent with a bluish haze. This nanoemulsion was stable only in the iced bath but would turn cloudy white if brought to room temperature. The process was so far reversible; that is, the sample would turn clear if heated and become cloudy if cooled to room temperature (which further proves that the microemulsion is an equilibrium system). To arrive at a nanoemulsion at room temperature, 25 parts of an aqueous solution containing 3% C$_{12}$E$_{23}$ were added to the nanoemulsion in the iced bath and the content was quickly mixed (Figure 2.13c). This led to a nanoemulsion that remained stable when brought to room temperature (Figure 2.13d). The bottom diagram in Figure 2.13 illustrates the process path on a conceptual (not measured) phase diagram. The addition of C$_{12}$E$_{23}$ at the end of the process shifted the phase diagram toward higher temperature, thus effectively moving the final composition further away from the microemulsion condition.

In the method of preparting nanoemulsions via the microemulsion path, the initial microemulsion does not necessarily have to be the bicontinuous type. Solans et al. have demonstrated two-phase nanoemulsions obtained from diluting lamellar or cubic single phases [194,195]. In one system, a nanoemulsion was prepared by diluting a cubic phase microemulsion formed from an aqueous solution of potassium hydroxide, hexadecane, oleic acid, and decylethylene glycol dodecyl ether [194]. We followed the same procedure using D$_4$ in place of hexadecane and arrived at a clear microemulsion that had a hard gel consistency (Figure 2.14a). However, when this

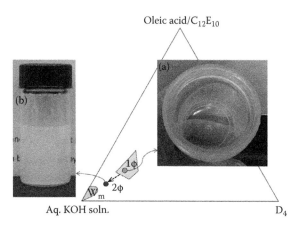

Oleic acid/C$_{12}$E$_{10}$

Aq. KOH soln.

D$_4$

FIGURE 2.14 Silicone emulsion prepared via the microemulsification path using pH as the controlling condition. The gel microemulsion (a) contains 15.0 parts D$_4$, 30 parts of a 3:7 mixture of oleic acid and C$_{12}$E$_{10}$, and 55.0 parts of an aqueous solution containing 3.5% KOH. The two-phase emulsion (b) was obtained by mixing 6 parts of (a) with 2 parts of the same KOH solution.

gel microemulsion, was diluted, the system turned cloudy (Figure 2.14b). The cloudiness is indicative of the presence of droplets that are significantly larger than the nanometer scale associated with a cubic phase. The process path is illustrated on the phase diagram (Figure 2.14).

In both cases described above, low molecular weight silicone is necessary to obtain a microemulsion (Figures 2.13b and 2.14a). Higher molecular weight PDMS cannot be solubilized in the surfactant phase.

2.5.2.2 Catastrophic Inversion

In the catastrophic inversion process to make an O/W emulsion, water is added to the oil phase. The surfactant can be included in the aqueous phase or alternatively added to the oil phase prior to water addition. Often in large scale production, the preferred method is to mix the surfactant with the oil first. This is because mixing surfactant with water produces liquid crystal, and the presence of liquid crystal gel prevents proper mixing to reach a homogeneous phase. The amount of water added is often only a small fraction of the oil phase, and the reason for this will become clear in a moment. The emulsion initially formed is oil-continuous. As shear is applied, the emulsion inverts to an oil-dispersed one. The emulsion formed after inversion is typically very thick; it is a high internal phase fraction emulsion (HIPE) [123]. The thick emulsion is then diluted to the desired final composition. In a continuous process, oil, surfactant, and water are successively flown in, and the mixture is passed through an in-line shear device outputting an inverted, oil-dispersed HIPE, which is then diluted in a vessel. The amount of water to affect inversion to occur is referred to as the inversion water.

Catastrophic inversion is particularly effective when emulsifying viscous oils. Very often it takes less water to trigger an inversion if the oil phase has a higher viscosity than if it has a lower viscosity [174,178]. Generally, a minimum threshold amount of inversion water is needed to trigger an inversion; this will also become clear in a moment. Catastrophic inversion can be carried out in mixers using impeller blades, rotor stators, or in extruders, but not with high pressure homogenizer. Extrusion in particular allows extremely viscous or even viscoelastic oil phases to be emulsified [201].

When processed carefully, emulsions produced by catastrophic inversion can be quite narrowly dispersed in droplet size. Figure 2.15 shows an aqueous emulsion of a polymeric silicone prepared by the catastrophic inversion to give an idea of how uniform the droplets can be.

2.5.2.2.1 Mechanism

Catastrophic inversion is generally regarded to be caused by an increase in the volume ratio of the dispersed to the continuous phase such that the dispersed phase becomes sufficiently voluminous and the inversion takes place in an abrupt manner [173]. Since the volume fraction corresponding to random close packing of monodispersed spheres is 0.64 and that of ordered close packing is 0.74, one would expect inversion due to volume constraint occurs at a dispersed phase volume fraction in that range, or even greater, considering that emulsion droplets are polydispersed and deformable. However, in most practical situations, inversion takes place with a much smaller volume fraction of the dispersed phase, for example, less than 10%. Liu et al.

FIGURE 2.15 Optical micrograph of an aqueous emulsion of PDMS prepared by catastrophic inversion.

have shown that only 2% aqueous phase is needed to induce phase inversion to produce a silicone emulsion that contains 98% silicone oil [202]. The reason that inversion can be triggered with so little dispersed phase is because the *effective* dispersed phase volume is much larger than the actual amount of the dispersed phase added. This is illustrated in Figure 2.16. As shown in the figure, the increase in the *effective* volume of the dispersed phase, here, marked as "W," is not only caused by the addition of water, but more significantly by the formation of multiple emulsions enclosing part of the oil. The enclosure of the oil from the continuous phase into the inner drops of the multiple emulsion "beefs up" the volume of the dispersed phase and leads to inversion occurring with only a small amount of water added. It is imperative at this point to clarify that the initial dispersed phase, labeled in the figure as "W," is actually not water. Although water is being added to the system, the initial dispersed phase is a mixture of water and surfactant. The surfactant, which was initially mixed with the oil phase, migrates to mix with water. The precise nature of this mixture, "W", depends on the content of the water relative to the surfactant. It is a surfactant solution if water content is low, a liquid crystal phase if water and surfactant contents are comparable, and an aqueous micellar solution if water content is high. The nature of "W" critically determines the outcome of the inverted emulsion. Liu et al. demonstrated that for O/"W"/O multiple emulsion to form, which is a necessary step during inversion from "W"/O to O/"W," the polar phase must contain surfactant aggregates with a normal structure [202]. Often, this polar phase is a liquid crystal phase [123]. Most commercial emulsions contain only a few percent of surfactant; the amount of water it takes to trigger inversion is also only a few percent. This almost guarantees that the polar phase during the inversion process is a liquid crystal phase. As such the inverted product is often a gel emulsion due to both a high internal phase volume fraction and the presence of liquid crystal in the continuous phase [123]. When the

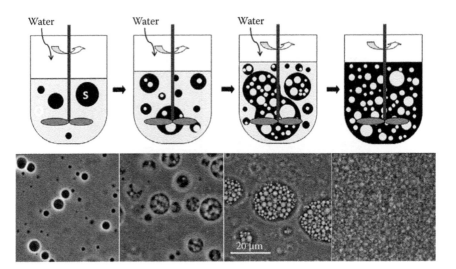

FIGURE 2.16 Illustration of a typical catastrophic inversion from W/O to O/W. Bottom row illustrations are micrographs taken from an inversion process where water was added to a silicone oil containing 5% of a hydrophilic nonionic surfactant. Scale bar represents 20 μm and all micrographs are scaled to the same magnification. (Reprinted from Liu, Y., in: *Encyclopedia of Surface and Colloid Science*, 2nd edn., Somasundaran, P, Ed., Taylor & Francis, New York, 2013. With permission.)

inverted emulsion is further diluted with water, the result is a flowable O/W emulsion, where W becomes an aqueous micellar solution. The diluted emulsion has the same droplet size distribution as in the gel emulsion upon inversion. The oil droplets in the final diluted emulsion are the same oil droplets present in the primary emulsion of the multiple emulsion during inversion (Figure 2.16).

It is no coincidence that this type of inversion works more effectively with high viscosity oil than low viscosity oil. This can be understood by realizing that the final emulsion droplet size is set at the stage where the multiple emulsions are formed during the inversion process. After inversion is complete, addition of more water to the continuous phase merely dilutes the composition. The pertinent event that sets the final oil droplet size is the deformation and breakup of the oil that is engulfed by the enclosure process that leads to the formation of the multiple emulsion. The idea was presented by Sajjadi et al. [180] and is illustrated in Figure 2.17. In other words, the same process of droplet deformation and break-up occurring in a direct emulsification also occurs in a catastrophic inversion, except that it takes place at the internal phase of the multiple emulsion. The continuous phase that exerts shear to deform and disrupt the oil droplet is the intermediate polar phase—liquid crystal in most cases—formed by the surfactant and water *during* inversion. Therefore, the process parameters that impact the final oil droplet size are energy intensity *during inversion*, interfacial tension between the oil phase and the *intermediate* polar phase, the viscosities of the oil phase and the *intermediate* polar phase, and the ratio of the viscosities. Since attainable droplet size scales inversely with the viscosity of the continuous phase, and since droplet breakup is more efficient when the

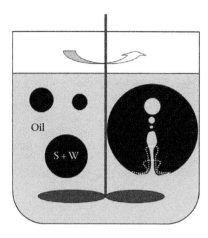

FIGURE 2.17 Schematic illustration of droplet formation during emulsion catastrophic phase inversion from an oil-continuous emulsion to an oil-dispersed emulsion. Droplets are not drawn to scale.

viscosities of the dispersed phase and the continuous phase are similar, catastrophic inversion works more effectively when the oil phase is viscous and the intermediate polar phase during inversion is a liquid crystal. Liu et al. showed that inversion with an aqueous micellar solution as the intermediate polar phase results in a much larger oil droplet size than inversion with liquid crystal phases. The authors also showed that different liquid crystal phases result in different droplet sizes, and the difference is attributed to the impact of rheology on droplet breakup during inversion [202].

Although catastrophic inversion is triggered by change in the water-to-oil volume ratio, a seemingly extensive variable, while transitional phase inversion is triggered by change in intensive variables such as temperature, salinity, surfactant, etc., the underlying driving force is fundamentally the same. We showed earlier that in transitional inversions, the surfactant aggregation structure changes sign of the spontaneous curvature. Liu et al. showed that in catastrophic inversions, too, whether and when inversion occurs is ultimately governed by the aggregation structure of the surfactant. The latter influences the stability and hence the prevailing emulsion type. Change in the water-to-oil volume ratio at a constant amount of surfactant in the system changes the surfactant concentration, and that in turn changes the nature of the surfactant aggregates. Liu demonstrated that inversion from a silicone oil-continuous emulsion to a silicone oil-dispersed emulsion does not occur when the polar phase (W + S) is an inversed micellar solution, even if a great amount of the inversed micellar solution is included. When inversion is attempted by adding water to the oil phase containing the surfactant, inversion occurs at the phase boundary between the inverse micellar solution and the liquid crystal phase [203]. The idea is illustrated in Figure 2.18. This illustrates the decisive role surfactant phase behavior plays in inversion.

Catastrophic inversion differs from transitional inversion in that it occurs with the surfactant physically located in the "wrong" phase and therefore the system is far from the equilibrium state. This can be seen by the following. Hydrophilic surfactant

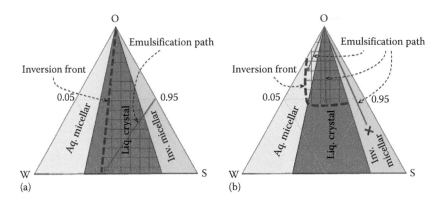

FIGURE 2.18 Diagram illustrating catastrophic inversion features. (a) Inversion is brought by adding water to the oil phase containing a fixed amount of surfactant. (b) Inversion is brought by adding to the oil phase a premixed water and surfactant mixture at a constant W:S ratio. Thick dashed line indicates "inversion front," which is the minimum amount of polar phase volume fraction needed to complete inversion under a fixed mixing condition. Gridded regions are where multiple emulsions are formed. The three colored phase domains correspond to the surfactant aggregation structure being aqueous micellar solution (pink), liquid crystal (red), and inversed micellar solution (yellow). No inversion occurs in the inversed micellar domain. Note that both triangles depict only the top portion of the ternary phase diagram with oil phase volume fraction greater than 0.9.

self-aggregates in the water phase and hydrophobic surfactant self-aggregates in the oil phase. Thus, a system containing hydrophilic surfactant favors the formation of O/W emulsion, whereas a system containing hydrophobic surfactant favors the formation of W/O emulsion. However, as the diagram in Figure 2.11 shows, W/O or O/W/O is found to form in the lower left region that corresponds to the system containing hydrophilic surfactant, and O/W or W/O/W is also found to form in the upper right region that corresponds to the system containing hydrophobic surfactant. This happens because of the volume constraint. When there is not enough water present, it simply cannot form the continuous phase, even if the surfactant is hydrophilic. For this reason, the W/O or O/W/O emulsion at the lower left region and the O/W or W/O/W at the upper right region of Figure 2.11 are referred to as "abnormal" emulsions [181], and they are not stable. In catastrophic inversions, since the system is far from the equilibrium, many process parameters can influence temporary emulsion structure and introduce inversion hysteresis [175,184]. In Section 2.5.4, we demonstrate one case of severe inversion hysteresis where silicone resin is emulsified by the catastrophic inversion.

2.5.2.2.2   Gel Emulsion Made via Catastrophic Inversion

The high internal phase fraction emulsions (HIPEs) that immediately result from the catastrophic inversion process have their own beauty. Because of the high viscosity, HIPEs do not suffer from sedimentation and creaming. HIPEs are also more cost-effective to transport without the excess water. Handling of such highly concentrated emulsions is not entirely impermissible as these emulsions

FIGURE 2.19 A silicone gel emulsion (on a spatula) containing 90% PDMS obtained from catastrophic inversion. (Reprinted from Liu, Y., in: *Encyclopedia of Surface and Colloid Science*, 2nd edn., Somasundaran, P., Ed., Taylor & Francis, New York, 2013. With permission.)

are typically shear-thinning [202]. HIPEs are also useful to template synthesis and create highly porous low density materials [204,205]. A novel application of silicone HIPEs was disclosed by Gibas et al. for an application in wound dressing where the semi-solid nature of the gel emulsion allows pattern coating of the silicone adhesives onto an absorbent substrate [206]. It was demonstrated that only a very small amount of surfactant (<1% per gel emulsion) was needed to produce such a gel emulsion [206].

One neat property with silicone gel emulsions is that the gel emulsion can be crystal clear. Figure 2.19 shows a representative clear silicone gel emulsion. The optical clarity is often mistakenly interpreted as the emulsion having a very small droplet size. In reality, the transparency is solely due to a match in the refractive index. Silicone has a refractive index 1.39–1.46, which is between that of water (1.33) and that of the surfactant (1.48–1.50). Therefore the silicone phase and the polar phase in the gel emulsion have a high chance to match their refractive indices. The particular gel emulsion shown in Figure 2.19 has a droplet size close to 1 μm. The clear gel emulsion becomes a white emulsion when diluted with additional water.

2.5.3 EMULSIFICATION OF SILICONE GUMS

Silicone gums are linear silicones that have viscosities in the range of tens to hundreds of millions centipoises at room temperature. The fluid is so thick that it can be picked up by the fingers and rolled into a ball. Silicone gums are useful in many applications. In manufacture of tires, they are used as band ply lubricants. In coatings and leather treatment, they are used as top coat additives to provide slip and anti-mar. To emulsify silicone gum, or any gum for that matter, into an aqueous emulsion is challenging. The reason can be easily seen from the following calculation. The emulsion shown in Figure 2.15 was prepared from a PDMS fluid of 10,000 cSt. It took 30 seconds of high shear (3500 RPM using a DAC 150 Speed Mixer™ in a lab scale

(100 g sample size). According to Walstra and Smulders, the droplet deformation time before breakup scales with the dispersed phase viscosity [167]. If everything else being equal, the time it takes to break up a gum of 10 million cSt using the same equipment would be about 1000 times longer. This amounts to 8 h of nonstop high shear to arrive at the same result. This, of course, cannot be realized in practice; process cost aside, the heat generated will simply fry the equipment.

It has been found, surprisingly, that certain polymeric surfactants can do the trick. Fenton and Keil described silicone gum emulsions prepared using a "unique siloxane copolymer dispersing agent" [127]. Evans and Liles described the emulsification of silicone gums using high molecular weight PEO-PPO-PEO triblock copolymers [102]. High molecular weight silicone polyethers are also effective in emulsifying silicone gums [128]. It is unclear how these polymeric surfactants facilitate the emulsification of silicone gums, while low molecular weight surfactants cannot.

2.5.4 EMULSIFICATION OF SILICONE ELASTOMERS

Silicone elastomers are also difficult to be emulsified into aqueous emulsions. Unlike silicone gums, silicone elastomers are cross-linked silicones and have essentially infinite molecular weight. Silicone elastomers are gels and are not soluble in any solvents or in low molecular weight silicones. But they can be swelled in solvents or low molecular weight silicone fluids. The extent of swelling decreases as the cross-link density increases. If the cross-link density is low, the elastomer–carrier fluid mixture can be emulsified into an aqueous emulsion. The obtained emulsion droplet size is fairly large, ranging from a few microns to tens and hundreds of microns. Figure 2.20 shows an aqueous emulsion of a silicone elastomer. The irregular shaped dispersed particles are

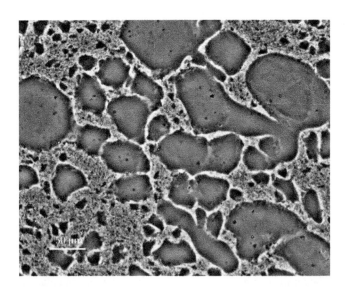

FIGURE 2.20 Optical micrograph of an aqueous emulsion of a silicone elastomer gel. Scale bar represents 50 μm.

the elastomer gel swelled with the carrier fluid. Typically in these emulsions, small (submicron) spherical droplets formed from the excess carrier fluid can be discerned (not easily seen Figure 2.20; when diluted, the small droplets become more discernible). Aqueous emulsions of silicone elastomer are often used in cosmetic and skin care products. The elastomer gel provides a unique sensation to the skin (see Chapter 9).

2.5.5 EMULSIFICATION OF SILICONE RESINS

Silicone resins are highly cross-linked network-structured polysiloxanes (see Figure 1.1). There are two general types of silicone resins: one based on mainly the T building blocks, and the other based on M and Q units (see Figure 1.2 for the structural notation MDTQ). T-based resins are also called silsesquioxanes. Both types of resins can be liquid or solid at room temperature, depending on the degree of condensation. T-based resins are primarily used in industrial coatings to enhance the thermal stability, weatherability, and hydrophobicity of organic coatings. They are also used in alkyd, epoxy, and acrylic paints as performance modifiers and additives. T resins are commonly delivered either neat or as solutions in aromatic solvents. MQ resins are primarily used to modify the rheology of PDMS fluids and are used in release liners for organic adhesives as well as in silicone pressure-sensitive adhesives. MQ resins are also used in making silicone elastomers, silicone surfactants, and silicone antifoams. MQ resins are delivered as solutions in aromatic solvents or as MQ/PDMS blend. Solid MQ resins are also available in the flake form. Traditionally, there had not been much demand in the market for silicone resins to be delivered in aqueous emulsions. Most industrial coatings are either solvent-based or neat. With increasing environmental awareness, there is an increased need for waterborne silicone resins.

Solid resin cannot be emulsified unless it is first dissolved in a solvent, or melted. For liquid resins, in most cases, the emulsification process is the same as in emulsifying a linear silicone fluid. Difficulty arises when the silicone resin molecule contains an appreciable amount of silanol (SiOH) groups. The presence of silanol groups in silicone resin favors the formation of the water-in-silicone emulsions. Two examples are given below; these examples also serve to illustrate hysteresis in catastrophic inversion as mentioned in Section 2.5.2.2.1.

In the first example, a silicone oil phase was formed that comprised a 1:1 (by weight) mixture of a linear PDMS of 450 cP and a MQ resin. The MQ resin has a average molecular weight of 4700, an M:Q ratio of 0.9, and an average of about six SiOH groups per molecule. To the silicone oil phase was added 8%, per oil phase, of a surfactant mixture consisting 1:1 (by weight) $C_{12}E_4$ and $C_{12}E_{23}$. The content was mixed using a Speed Mixer DAC 150 FVZ operated at 3500 RPM for 22 seconds, arriving at a surfactant-in-silicone oil emulsion. Water was then added to the emulsion in increments of 5% per oil phase, each addition followed by mixing at 3500 RPM for 22 s. The operation was repeated until 100% water had been added (W:Si = 1:1). At all stages, the emulsion was oil-continuous. If the same procedure was followed using PDMS without the MQ resin, the emulsion inverted to oil-dispersed with less than 10% of the water.

In the second example, the same oil phase and surfactant mixture were mixed using a Lightnin™ mixer equipped with a Cowles blade. Water was added in increments of 5% per oil phase while mixing was maintained under a constant speed

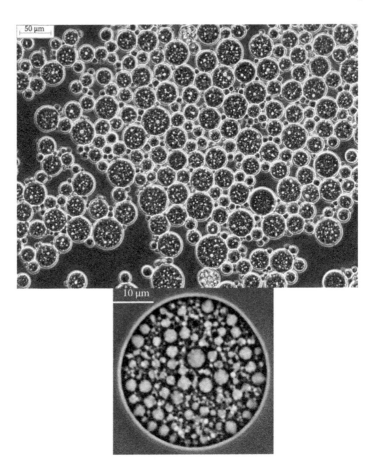

FIGURE 2.21 Optical micrographs of a W/Si/W multiple emulsion where the silicone phase is a solution mixture of PDMS and MQ resin. Scale bars at upper left corner represent 50 μm (*top image*) and 10 μm (*bottom image*). The bottom image is a zoomed view of one of the multiple droplets.

of 500–1000 RPM. The emulsion remained oil-continuous until at 50% water (W:Si = 0.5), when the emulsion abruptly changed in the texture and inverted into a water-continuous emulsion. When examined under the microscope, the emulsion was found to be a W/Si/W multiple emulsion (Figure 2.21).

When the water addition in the above experiments was carried out in a rapid fashion instead of incrementally, inversion from oil-continuous to oil-dispersed occurred more readily at smaller water-to-oil ratio. The inversion was not always repeatable, though. The slower the water addition, the more resistance to inversion. Even after inversion took place and an aqueous-continuous emulsion was formed, prolonged stirring would cause the system to invert back to oil-continuous. This signals a significant level of inversion hysteresis.

The above phenomenon has also been encountered when emulsifying certain T-based resins that contain an appreciable amount of silanol groups. We suspect that

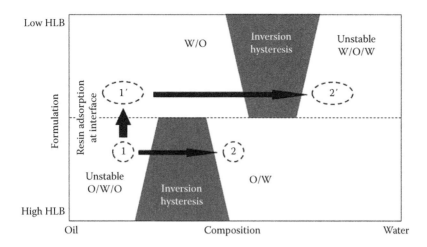

FIGURE 2.22 Inversion resistance and hysteresis associated with emulsification of silicone resin.

resistance to inversion as well as inversion hysteresis are both due to partitioning of the silicone resin into the surfactant film at the silicone–water interface. MQ resin is miscible with PDMS; however, in the presence of a silicone–water interface, the silanol bearing MQ molecules likely migrate from the silicone bulk phase to the silicone–water interface. The situation is similar to pentanol partitioning at lipophilic oil–water interface, even though pentanol shows no particular surface activity in the aqueous solution without the presence of oil. Since MQ resin is fairly hydrophobic, the combination of the surfactant and the partitioning MQ resin at the interface may very well behave like an overall hydrophobic surfactant favoring the formation of water-in-silicone emulsion. The observed inversion hysteresis suggests that the adsorption of the MQ resin molecules from the silicone bulk phase to the interface takes time. Figure 2.22 illustrates what might have happened during the emulsification process of the MQ/PDMS system. Normally for a PDMS oil phase, emulsification takes place along the path 1 → 2. The adsorption and partition of the silanol bearing resin molecules change the composition of the interfacial film toward more hydrophobic (1′). Now emulsification takes place along the path 1′ → 2′, landing in the W/O/W multiple emulsion zone.

Normally, W/O/W or O/W/O type of multiple emulsions are unstable when prepared from one type of surfactant. This is because a W/O/W (or O/W/O) double emulsion drop has two interfaces, W/O and O/W. If the surfactant is suitable for stabilizing one interface, it is unsuitable for stabilizing the other interface. Surprisingly, the multiple emulsions prepared with silanol bearing MQ resin were found to be exceedingly stable. Barnes et al. disclosed that W/Si/W double emulsions made using a silicone MQ resin were stable for several years under ambient conditions [207]. One possible reason for the unusual level of stability observed is that there may be a different level of partitioning of the MQ molecules at the two interfaces. The inner interface W/Si may contain a relatively higher amount of MQ molecules than the outer interface Si/W.

Should a simple Si/W emulsion be achieved from a silicone phase that contains MQ resin, inversion resistance must be overcome; and there is one method disclosed to help with that [208]. By adding a polar-functional organopolysiloxane to the system, inversion to simple Si/W emulsions can proceed readily. We speculate that, like the MQ resin, polar-functional organopolysiloxane may preferentially adsorb at the interface, such that the overall film present at the interface now becomes favorable for the formation of silicone-in-water emulsion.

2.5.6   EMULSIFICATION OF LOW MOLECULAR WEIGHT SILICONES, REACTIVE SILICONES, AND SILANES

While silicone gums and elastomers are challenging to emulsify because of their very high viscosity (gum) and viscoelasticity (elastomer), low molecular weight silicones, reactive silicones, and silanes pose another type of challenge to emulsification. The challenge is due to emulsion stablity. Ripening in these emulsions is rapid; sometimes creaming also develops relatively fast. First, we look at the situation for creaming and then ripening.

2.5.6.1   Creaming

Creaming or sedimentation is a gravity effect. When the density of the dispersed phase is less (greater) than that of the continuous phase, the emulsion droplets float (settle) to the top (bottom). Emulsions of PDMS and silane tend to cream and emulsions of silicone resins tend to sediment. The rate of creaming or sedimentation, as expressed by the net velocity of the emulsion droplet, can be estimated using a model of monodispersed hard spheres suspended in a fluid by

$$v = \frac{\left|\rho_C - \rho_D\right| \times g \times d^2}{18\eta_C} f(\phi) \tag{2.4}$$

where

v is volume-averaged velocity of the particles
ρ's are densities of the dispersed and continuous phases
g is gravitational constant
d is diameter of the particle
η_C is viscosity of the continuous phase
$f(\phi)$ is a function of the dispersed phased volume faction, ϕ [124]

$f(\phi)$ decreases with increasing ϕ. It has a value close to 1 at small ϕ; the value drops to below 0.01 at ϕ greater than 0.6.

The specific gravity of PDMS increases with molecular weight, ranging from 0.76 for the smallest silicone, hexamethyldisiloxane, to a plateau value of 0.97 for PDMS of viscosity greater than 100 cSt. Organo-functional silicones based on PDMS backbone such as silicone polyethers and aminosilicones often have specific gravities very close to 1. Silicone resins are typically heavier. Their specific gravities range from 0.98 for low molecular weight liquid resins to 1.25 for high molecular weight solid resins. Specific gravity for silanes ranges from, on the light end, 0.88 for octyltriethoxysilane to, on the heavy end, 1.06 for phenyltrimethoxysilane. The rates of creaming for PDMS and for liquid resins are therefore slow; the rates of creaming for low molecular weight silicones and many silanes are appreciable. To give a feel for the time scale, an aqueous emulsion of 40% octyltriethoxysilane having an average droplet size of 0.3 μm in the absence of a thickener will show no visible creaming within a year or two, if no ripening occurs. But the same emulsion having an average droplet size of 1 μm will display obvious creaming within a month. An aqueous emulsion of 70% PDMS having an average droplet size of 2 μm and in the absence of flocculation will not show creaming for years.

2.5.6.2 Ripening

Ripening, also called Ostwald ripening, or coarsening, refers to the diffusive migration of the dispersed phase from the smaller droplets to the larger ones. This process is driven by the difference in the chemical potential of the different sized droplets. A quantitative expression for the rate of Ostwald ripening is provided by the Lifshitz–Slezov–Wagner (LSW) theory [209,210],

$$\omega = \frac{d}{dt}\left(r_c^3\right) = \frac{8\gamma D c_\infty V_m}{9RT} \qquad (2.5)$$

where

r_c is the critical radius of a droplet, which is neither growing nor decreasing in size
γ is the interfacial tension between the dispersed phase and the continuous phase
D is the diffusion coefficient for the dispersed phase in the continuous phase
c_∞ is the bulk solubility of the dispersed phase in the continuous phase
V_m is the molar volume of the dispersed phase

Aside from the parameters in Equation 2.5, increasing the dispersed volume fraction and the presence of surfactant micelles are also known to increase the rate of ripening. Micelles are believed to function as carriers for the diffusion of the dispersed molecules through the continuous phase. However, the experimentally measured increase in the rate of ripening was found to be much smaller than what is predicted based on the solubilized concentration in the micelles [211]. In any case, micellar-added ripening is not pertinent to aqueous emulsions of nonpolar silicones that have a molecular size greater than D_5, because large, nonpolar silicones cannot be solubilized in the surfactant micelles (see Section 2.6).

There is no reported data, to the authors' awareness, on the rate of Ostwald ripening in aqueous emulsions of silicones. In practice, measurements of ripening are difficult since it is a challenge to separate destabilization solely due to ripening from destabilization due to coalescence. However, one can get a sense of the rate of Ostwald ripening by comparing aqueous emulsions of silicones to aqueous emulsions of alkanes; the latter have been well studied. First, we notice that the rate of ripening is solubility limited and size controlled (Equation 2.5). The diffusion coefficient depends on size of the solute; specifically, $D \sim V_m^{-0.6}$ [212]. Plugging this into Equation 2.5 gives

$$\omega \propto \gamma c_\infty V_m^{0.4} \qquad (2.6)$$

at a fixed temperature. The characteristic time for Ostwald ripening can then be expressed as

$$\tau = \frac{r^3}{\omega} \propto \frac{r^3}{\gamma c_\infty V_m^{0.4}} \tag{2.7}$$

Equation 2.7 indicates that the larger the droplet size, the slower the ripening. In reality, only submicron size emulsion droplets undergo significant Oswald ripening. The interfacial tension in an aqueous emulsion of silicone is similar to the interfacial tension in an aqueous emulsion of alkane, both on the order of a few mN/m^2. This leaves aqueous solubility and molar volume of the oil as the only variables to compare the rate of ripening between the two cases. To this end, data in the literature shows that the rate of ripening in aqueous emulsions of alkanes decreases exponentially with the number of carbon atoms in the alkane [213]. This is not surprising since the solubility of alkanes in water obeys the same exponential dependence (see Figure 1.5). For alkanes of C_{16} or larger, Ostwald ripening is too slow to be of practical concern, even for droplet sizes as small as 0.1 μm [211]. Figure 1.6 shows that the solubility of n-$C_{15}H_{32}$ in water is roughly the same as the solubility of MD_3M in water. The molar volume of MD_3M is about 1.5 times that of n-$C_{15}H_{32}$. This suggests that ripening in an aqueous emulsion of PDMS is of practical concern when the PDMS molecule is composed of only a few siloxane units.

2.5.6.2.1 Ostwald Ripening in Aqueous Emulsions of Low Molecular Weight Dimethylsiloxanes

Aqueous emulsions of low molecular weight dimethyl siloxanes, such as D_4, D_5, MM, MDM, MD_2M, and MD_3M, are exceedingly unstable. The instability is presumably due to ripening. No emulsions of such low molecular PDMS are sold commercially. It has been discovered, however, that aqueous emulsions of low molecular PDMS could be stabilized by using a combination of silicone polyether surfactants and anionic surfactants [214]. It was demonstrated that emulsions of D_5 prepared using a silicone polyether surfactant in combination with a minor amount of a sodium lauryl ether sulfate were stable upon storage and water dilution while emulsions of the same D_5 prepared with either surfactant alone separated after a short period of time. The particular silicone surfactant reported is known to form lamellar liquid crystal [215], but how this prevents Ostwald ripening when used in combination with a minor amount of an anionic surfactant is unclear.

2.5.6.2.2 Ostwald Ripening in Aqueous Emulsions of Hydrolysable Silicones and Silanes

Alkoxy-functional siloxanes and alkoxysilanes are often used to waterproof building materials and other various substrates. These siloxanes and silanes are frequently delivered in the form of aqueous emulsion for the ease of application. The Si–OR group hydrolyzes in water in the presence of acid or base. When this happens, the resultant hydrolazate has a significantly increased solubility in water and therefore can readily diffuse through the aqueous phase leading to Ostwald ripening. Typically, as hydrolysis proceeds, droplet size increases with time and the emulsion displays creaming. Eventually, a bulk oil layer appears. Even though destabilization due to Ostwald ripening cannot be differentiated from destabilization due to coalescence by visual inspection or by monitoring droplet size change, occurrence of Ostwald ripening can be corroborated by a change in the hydrolytic state of the reactive silane or siloxane; the latter can be measured by NMR. To minimize ripening, one method is to use a pH buffer in the aqueous phase to slow down the hydrolysis [216]. Another method is to incorporate a species soluble in the oil phase but insoluble in the aqueous phase. This creates a reverse osmotic pressure and hence retards Ostwald ripening [217]. For the latter, one can use a hydrophobic surfactant that is soluble in the oil phase [218]. Alternatively, a silicone having a molecular weight high enough to have virtually no solubility in the aqueous phase (PDMS of 10 cSt suffice) and yet soluble in the alkoxysilane or siloxane can be used. In fact, low molecular weight condensate of the alkoxy functional silane or siloxane itself can be an effective retarder to Ostwald ripening. Still, another approach is to keep the water level low in the emulsion and dilute the emulsion, if needed, just prior to application. The overall hydrolysis/condensation reaction consumes water, and by limiting the water available for hydrolysis the emulsion can be kept stable [219].

2.5.6.2.3   Composition Ripening

Composition ripening also occurs in emulsions containing silanes and siloxanes that have significant solubility in the aqueous phase. Composition ripening occurs when two emulsions of different dispersed phases are mixed, where the different dispersed phases are miscible in each other, and at least one of them has an appreciable solubility in the continuous phase. The following example demonstrates a case of composition ripening in an emulsion of silicone and silane.

A mixture was prepared by combining in a bottle 6 parts of n-octyltriethoxysilane (n-OTES), 4 parts of a linear PDMS of viscosity 10 cSt, 0.5 parts of a surfactant of the nominal structure i-$C_{10}E_7$, and 6.2 parts of water. The mixture was subjected to sonication using an ultrasonic disperser (Misonix S-4000) forming an initial emulsion. This initial emulsion showed a monomodal droplet distribution centered around a Sauter mean, $d_{3,2} = 0.350$ µm. To this initial emulsion was added 10 additional parts of n-OTES. The content was immediately shaken by hand and then the sample was allowed to gently mix on a rotating wheel. Droplet size was measured periodically. Results are shown in Figure 2.23. As can be seen, immediately after n-OTES was added to the initial emulsion, the system consisted of the initial small droplets plus very large drops (5–100 µm); the latter were presumably formed by the added n-OTES (black dashed line). After only 1 hour, the large drops had substantially shrunk and the smaller ones had grown. This is indicated by a significant decrease in the right-hand-side peak and a concomitant shift in the left peak toward right (blue dashed line). After 3 days the right-hand-side peak completely disappeared and the left peak had further grown (green dashed line). After 1 week the emulsion stabilized at a $d_{3,2} = 0.527$ µm.

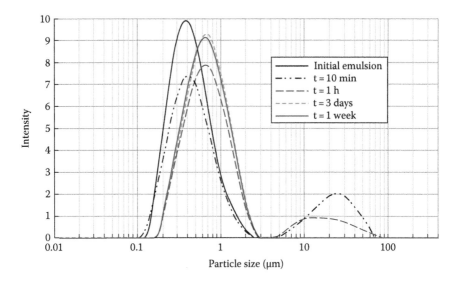

FIGURE 2.23 Droplet size distribution of a silane emulsion under composition ripening. A silane is added at time zero to an initial emulsion containing in the dispersed phase a solution mixture of silane and PDMS.

In this example, the free energy gain for *n*-OTES to diffuse from the larger droplets containing pure *n*-OTES to the smaller droplets that contain a solution mixture of *n*-OTES and PDMS evidently offsets the energy loss due increased interfacial energy. The process in fact goes against the direction of Ostwald ripening.

2.6 SILICONE MICROEMULSIONS

Microemulsions are single phase solutions comprising oil, water, and amphiphile [47]. As such, microemulsions are thermodynamically stable. Optical clarity, long term stability, and droplet size are often used to judge whether a system is an emulsion or a microemulsion. However, this can lead to the wrong conclusion, as certain emulsions can also display all of the above characteristics but are not microemulsions. The defining element of a microemulsion is that the system contains only one phase. A substantial number of patent disclosures and commercial products claiming silicone microemulsions are actually silicone emulsions with small droplet size.

The distinction between emulsions and microemulsions is not just a matter of semantics. There are practical reasons for understanding the difference. The process requirements to produce these two types of products are very different. Microemulsions form spontaneously, independent of emulsification path. Therefore, the emulsification processes critical to emulsions, such as those discussed so far in the chapter, are irrelevant. However, the process needs to be designed to achieve good mixing in order to allow the system reach equilibrium within in a tolerable time frame.

Solubilization is obviously essential to the formation of a microemulsion. A microemulsion containing water solubilized in the oil surfactant solution is referred to as water-in-oil droplet microemulsion and denoted by O_m. Conversely, a microemulsion containing oil solubilized in the aqueous surfactant solution is referred to as oil-in-water droplet microemulsion and denoted by W_m. If both oil and water are solubilized in a surfactant phase, the system may be a bicontinuous microemulsion or, at higher surfactant concentrations, a swelled liquid crystal phase.

There is not a large amount of published literature on silicone microemulsions. In the following, we describe two types of silicone microemulsions based on the nature of the silicone.

2.6.1 MICROEMULSIONS OF LOW MOLECULAR WEIGHT DIMETHYLSILOXANES

Siloxanes such as hexamethyldisiloxane (MM), octamethyltrisiloxane (MDM), decamethyltetrasiloxane (MD_2M), octamethylcyclotetrasiloxane (D_4), and decamethylcyclopentasiloxane (D_5) are customarily called low molecular weight PDMS, even though the prefix "poly" as in the abbreviation letter "P" is not very meaningful here. These siloxanes are also referred to as volatile silicones due to the relatively low boiling points ($<200°C$). Low molecular weight dimethylsiloxane provides unique application benefits due to its volatility and tendency to spread. For example, in skin and hair care applications volatile silicones have a light, pleasant feel during application and then evaporate after application, leaving no residue on the skin or hair. Low molecular weight silicones can form microemulsions with water and surfactant.

Hill et al. established the ternary phase behavior of a series of water–dimethylsiloxane–silicone surfactant systems where the dimethylsiloxane is D_4, D_5, MM, and MD_2M, and the silicone surfactant has the structure $MD'(EO_x)M$, where

$$D' = -O_{1/2}Si(CH_3)(R)O_{1/2}-$$
$$R = -CH_2CH_2CH_2-O(EO)_xH$$
$$EO = -CH_2CH_2O-$$
$$x = 6{-}18$$

These ternary systems form various single phases. The phase behavior in these systems resembles that in the water–C_iE_j–alkane systems, though the exact conditions where microemulsions form may not be identical. In both types of systems, the microemulsion domain shifts to higher temperature and higher surfactant concentration with increasing polyether chain length, decreasing salt (NaCl) concentration, and increasing molecular weight of the oil [55].

Wolf et al. found that droplets as well as bicontinuous microemulsions can be formed from MM, water, and a mixture of nonionic and anionic hydrocarbon-based surfactants [220,221].

Silas et al. studied the effect of a cationic surfactant, didodecyldimethylammonium bromide, on the miscibility gap of the binary systems water–C_iE_j and silicone–C_iE_j, and on the critical temperature and surfactant efficiency of the ternary system water–silicone–C_iE_j [56]. The critical temperature is the temperature at the "fish" tail where the three-phase domain joins the microemulsion phase (see bottom diagram of Figure 2.13). Surfactant efficiency is defined as the minimum surfactant concentration to solubilize an equal amount of oil and water. The silicones they investigated are D_4, MM, and MD_2M. The authors showed that adding the cationic surfactant expanded the lower miscibility gap of the silicone–C_iE_j binary system and correspondingly raised the critical temperature of the water–silicone–C_iE_j ternary system, though surfactant efficiency was increased. They also showed that critical temperature increased dramatically with increasing molecular weight of the silicone oil; interestingly, the surfactant efficiency increased as well [56].

It is helpful to keep two important molecular aspects in mind when comparing microemulsion phase behavior between the silicone system and the lipophilic oil counterpart. First, in terms of the oil phase, the molar volume of a silicone is significantly larger than the molar volume of an alkane having the same number of repeating units. For instance, the molar volume of MD_2M is similar to that of $C_{20}H_{42}$. Castellino et al. determined the equivalent alkane carbon number (EACN) for a silicone of 0.65 cSt (equivalent to MM) and a silicone of 3 cSt (equivalent to an average of ca. MD_5M). Their results showed that that the silicone of 0.65 cSt behaved like n-dodecane and the silicone of 3 cSt behaved like n-pentadecane. The authors had expected a higher EACN number for the 3 cSt silicone. They attributed the low value of EACN for the 3 cSt silicone to the possibility that the 3 cSt PDMS was composed of a mixture of different molecular weight fractions and that only the smaller molecular weight fraction participated in the microemulsion formation [222]. Figure 1.6 showed that the aqueous solubility of MM is close to that of n-octane, and the aqueous solubility of MD_3M is close to that of n-pentadecane.

In terms of the surfactant phase, Hill [223] pointed out that the hydrophobicity of the trisiloxane group in a trisiloxane polyether surfactant as reflected by CMC values [113] is comparable to that of the linear $C_{12}H_{25}$ group in an alcohol polyether. However, the trisiloxane group is larger in volume than $C_{12}H_{25}$. As such, even though the general patterns of aqueous phase behavior of the trisiloxane surfactants are similar to those of the C_iE_j surfactants, the phase behavior is shifted toward longer EO groups [223]. Silicone surfactants containing greater than three siloxane units have still larger apolar volumes and therefore require correspondingly larger hydrophiles to balance.

Microemulsions formed by water, silicone polyether, and low molecular weight dimethylsiloxanes have been extensively patented by Hill et al. [48–52]. Expanded microemulsion compositions composed of nonaqueous polar solvent, lipophilic oil, silicone surfactant selected from a range of silicone polyethers, and additional personal care active ingredients are also patented [224]. O'Rourke and Stevenson patented perfumed oil-in-volatile silicone microemulsions [225]. The use of silicone microemulsions as an additive to the injection fluid to improve the yield from oil and gas wells has been disclosed [226–229].

The preparation of microemulsions of low molecular weight silicone is exceedingly easy. All that is required is to combine the ingredients and mix. Due to the low viscosity involved, the system reaches equilibrium relatively quickly.

Many applications, however, benefit from high molecular weight silicones, and at the same time require the system to have small droplet size. High molecular weight PDMS do not form microemulsions with water and common surfactants. Halloran et al. disclosed a method where reactive, low molecular weight silicones are polymerized in a microemulsion to arrive at small droplet size latexes of high molecular weight [230]. Brook et al. reported the making of transparent elastomeric hydrogels encoded with a bicontinuous structure by photo-polymerizing acrylic monomers in the aqueous compartment of a bicontinuous microemulsion and subsequently cross-linking a reactive silicone and silane in the oil compartment. The bicontinuous microemulsion in their system was stabilized with an acrylate-functional silicone polyether surfactant [231]. Alternatively, microemulsions can be formed from polymeric silicones if the silicone contains certain amount of polar-functional groups attached to the siloxane backbone. This is described next.

2.6.2 MICROEMULSIONS OF POLAR-FUNCTIONAL POLYMERIC SILICONES

The solubilization of silicone in an aqueous surfactant phase increases dramatically when the silicone molecule contains polar-functional groups. Low molecular weight, polar-functional silicones can be water soluble and surface active. Many useful high molecular weight polar-functional silicones have a degree of polymerization greater than 100, but only a few mole percent of the siloxane units on the molecule are grafted with polar-functional groups. These silicone polymers are insoluble in water. However, the presence of the polar groups makes it possible for the silicone polymer to form microemulsion with water and surfactant. Among these polar-functional silicone polymers, those that are grafted with amino functional groups (see Table 1.1), commonly referred to as aminosilicones, have been used in textile applications since the 1960s [232]. Fibers, finished fabrics, and tissue papers that have been treated with aminosilicones are more pliable and feel soft to touch [67]. They also display better antistatic character, water and soil resistance, crease resistance, and compression recovery. High molecular weight silicones grafted with other polar-functional groups such as carboxyl and polyoxyalkylene have been found to deliver similar benefits [233,234]. In skin and hair care, high molecular weight polar-functional silicones show enhanced conditioning benefit [235]. It was noticed from the very beginning that polar-functional silicones were readily emulsified. In fact, in an example of US3303048, it was shown that an emulsion of the aminosilicone was formed without using any surfactant when sufficient acid was present in the aqueous phase. Microemulsions of an aminosilicone were first disclosed in US4620878. Later, Chrobaczek and Tschida in US5057572 pointed out that aminosilicones form microemulsions only in the presence of acid and that the amount of acid in the systems is particularly crucial. Other polar-functional silicones can also form microemulsions [58,236–238]. In WO2010110047, the inventors pointed out the large solubilization capacity in microemulsions of carboxyl grafted silicones.

Literature on the ternary phase behavior of systems containing water, surfactant, and water-immiscible, polar-functional polymeric silicones is sparse. Gee reported the phase diagram of a system consisting of water, a nonionic surfactant of the C_iE_j structure, and an aminosilicone [239]. The aminosilicone has a nominal structure of $MD_xD_y'(NH_2RNHR')M$, where R and R' are alkyl spacers, the weight-average molecular weight is approximately 7500 and the amino content is 0.8%. The author reported five single phases: an aqueous micellar phase containing less than 5% aminosilicone oil, a hexagonal liquid crystal phase, two isotropic phases rich in the aminosilicone oil, and a lamellar liquid crystal phase. The hexagonal phase extended far into the phase diagram and contained as much as 50% aminosilicone oil.

The author described that a portion of the hexagonal phase disappeared after 1 day. The lamellar phase was surprisingly located near the oil corner, disconnected with the water–surfactant axis. This phase was described to become vanishingly small after 1 month [239]. We have later examined the same system and found no single phase in the entire composition range.

Gräbner et al. reported a similar system and showed that aminosilicone formed a microemulsion when enough acetic acid was added [240]. The aminosilicone in their system corresponds to a structure $MD_{200}D'_{2-3}(NH_2C_2H_4NHC_3H_6)M$. The surfactant used was $i\text{-}C_{13}E_5$. The authors did not provide a phase diagram of the system but showed that an optically clear and isotropic mixture formed that contained 19% aminosilicone and 10% surfactant. This mixture was highly viscous. Their cryo-TEM images did not clearly resolve the microstructure, but a diluted sample showed spherical and cylindrical droplets of approximately 25 nm, which were presumably microemulsion droplets. To explain the large solubilization, the authors proposed a model depicting a core-and-palisade structure where the siloxane backbone of the aminosilicone constituted the core, and the surfactant molecules made up the palisade layer with the amino side groups from the silicone penetrating into the palisade. At higher surfactant-to-water ratio, cryo-TEM images revealed a coexistence of microemulsion droplets with unswollen lamellar/vesicle bilayers [240]. This suggests that the aminosilicone can be solubilized into the surfactant micelles but not into the surfactant bilayers—a significant point that we will come back to later.

Sharma et al. reported a study of the system consisting of water, a silicone of the structure $MD_{400}D'_{10}(EO_{19}/PO_{19})M$, an ethoxylated (containing 20 EO groups) glycerin isosterate, and a random copolymer of PEO/PPO pentaerythritoltetramethyl ether. The authors found a large microemulsion domain solubilizing as much as 40% of the silicone per microemulsion. When the silicone was mixed with 5% oleic acid, the microemulsion extended even further into the oil corner reaching as high as 80% silicone content. The microemulsion was observed to be flow birefringent. The authors concluded that the microemulsion structure must be bicontinuous at the high silicone content and droplet at the low silicone content [241].

In order to better understand the phase behavior and microemulsion formation in systems containing water-immiscible, polar-functional polymeric silicones, we recently conducted a study in which phase diagrams of two representative systems were constructed. The two polymeric silicones investigated (provided by Dow Corning Corporation) have the nominal structures as illustrated in Figure 2.24. The nonionic surfactant used (Bio Soft® EC690 from Stepan Company) has the nominal structure $n\text{-}C_{12}E_7$. Phase equilibration was obtained by initially heating the compositions in sealed test tubes to 80°C to allow adequate mixing of the melted liquids, then cooling the samples to room temperature at a rate of 0.2°C/h, followed by equilibration at room temperature for a year. Phase structure was identified by using a combination of polarized optical microscopy, cryo-transmission electron microscopy (TEM), and small angle x-ray scattering (SAXS).

Figure 2.25 shows phase diagram of the system $MD_{95}D'_2(EO_7)M\text{–water–}C_{12}E_7$ at 25°C. A large aqueous microemulsion domain was identified that extended to a composition containing 40% silicone oil. In comparison, no observable amount of silicone oil was solubilized in the hexagonal and lamellar liquid crystal phases

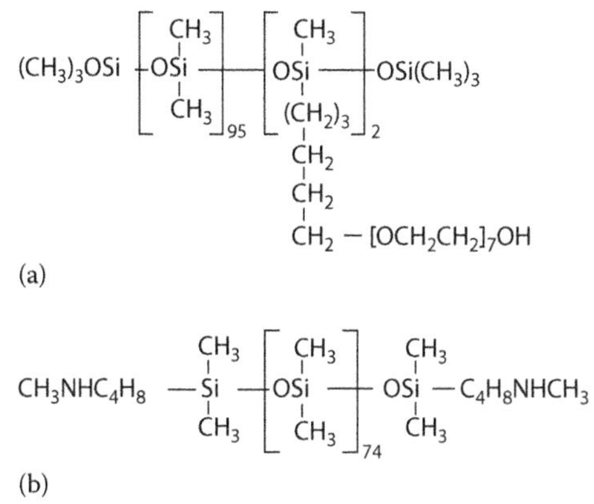

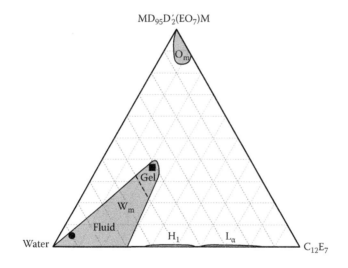

(a)

(b)

FIGURE 2.24 Nominal structure of the silicone oil used in the study: (a) polyether functional silicone, (b) aminosilicone.

FIGURE 2.25 Phase diagram of $MD_{95}D'_2(EO_7)M$-water-$C_{12}E_7$ at 25°C. Notations are as follows: W_m, aqueous microemulsion; H_1, hexagonal; L_α, lamellar; O_m, inverse microemulsion. Dashed line indicates microemulsion transition from a flowable fluid to a gel. The solid circle and square represent compositions analyzed by cryo-TEM and SAXS, respectively, see text.

(heights of the H_1 and L_α phases in the phase diagram are exaggerated to give visibility). The inverse micellar solution (microemulsion) had limited amount of water and surfactant solubilized.

The aqueous microemulsion was a thin clear liquid when the composition contained high water content. The viscosity of the microemulsion increased as the oil and surfactant content increased. The solution became a nonflowable gel when the combined oil and surfactant content reached above 50%. The composition where the solid square symbol lies was in fact a hard clear gel that felt elastic when gently poked. There was not an abrupt onset of gelation; rather, the viscosity increased in a continuous manner. Nonetheless the transition from a flowable state to gel state occurred within a very narrow composition range. The dashed line in Figure 2.25 is not to indicate a phase boundary but merely to indicate the approximate location where the solution changed from a flowable fluid to a gel. No phase separation occurred within the shaded W_m domain. To investigate the colloidal structure in this domain, two compositions were further characterized. A dilute microemulsion was examined by cryo-TEM and a concentrated microemulsion was analyzed by SAXS. Results are shown in Figures 2.26 and 2.27, respectively. It can be seen that the dilute microemulsion had a droplet structure with a size of approximately 10 nm. The concentrated gel microemulsion displayed a strong but broad diffraction peak corresponding to a Bragg spacing of 19.7 nm. No secondary peak was detected beyond uncertainty.

Figures 2.28 and 2.29 show phase diagrams of the ternary system $M'(NHCH_3)$ $D_{74}M'(NHCH_3)$–water–$C_{12}E_7$. Since the behavior of aminosilicone in the presence of water is sensitive to acid, phase diagrams were established for two different acid levels present in the system. In the first system, 1.09 mol of lactic acid were added per mole of the aminosilicone $M'(NHCH_3)D_{74}M'(NHCH_3)$. This gave a pH of approximately 7 in the dilute state (water > 50%). In the second case, 9.77 mol of lactic acid were added per mole of the aminosilicone. This gave a pH of approximately 3 in the dilute state. We expect in the second system the amine was fully protonated.

Figure 2.28 shows that no microemulsion was formed at pH of 7. On the other hand, a large microemulsion domain appeared when a sufficient amount of acid was incorporated (Figure 2.29). The microemulsion solubilized as much as 58% aminosilicone oil per microemulsion. Again, no discernible amount of silicone oil was solubilized in the hexagonal and lamellar liquid crystal phases (heights of the H_1

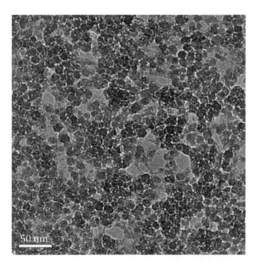

FIGURE 2.26   Cryo-TEM images of the composition corresponding to the solid circle in the phase diagram in Figure 2.25: $MD_{95}D_2'(EO_7)M$: $C_{12}E_7$: water = 0.0517: 0.0497: 0.8991.

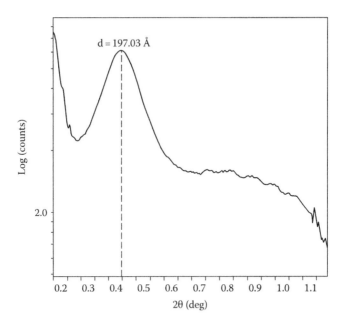

FIGURE 2.27 SAXS spectrum for the composition corresponding to the solid square in the phase diagram in Figure 2.25: $MD_{95}D'_2(EO_7)M$: $C_{12}E_7$: water = 0.3724:0.2209: 0.4067.

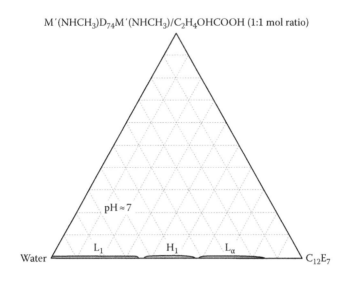

FIGURE 2.28 Phase diagram of $M'(NHCH_3)D_{74}M'(NHCH_3)/C_2H_4OHCOOH$ (1:1 molar ratio)–water–$C_{12}E_7$ at 25°C. Notations are as follows: L_1, aqueous micellar solution; H_1, hexagonal; L_α, lamella.

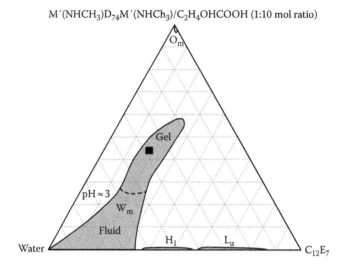

FIGURE 2.29 Phase diagram of M′(NHCH₃)D₇₄M′(NHCH₃)/C₂H₄OHCOOH (1:10 molar ratio)–water–C₁₂E₇ at 25°C. Notations are as follows: W_m, aqueous microemulsion; H_1, hexagonal; $L_α$, lamella; O_m inverse microemulsion. Dashed line indicates microemulsion transition from a flowable fluid to a gel. The solid square represents composition analyzed by SAXS, see text.

and $L_α$ phases in the phase diagram are exaggerated to give visibility). The inverse micellar solution (microemulsion) had a very limited amount of water and surfactant solubilized.

As in the case of the aqueous microemulsion formed with polyether grafted silicone, the aqueous aminosilicone microemulsion was a thin clear liquid at high water content. Viscosity gradually and continuously increased with increasing content in the oil and the surfactant. When the combined oil and surfactant content reached beyond approximately 40%, the microemulsion became a nonflowable gel. The composition where the solid square symbol lies was a clear hard gel that felt elastic when gently poked. This composition was examined by SAXS (Figure 2.30). Results show a broad diffraction peak corresponding to a Bragg spacing of 21.7 nm. A very weak hump at a higher angle was detected.

These results indicate that polar-functional polymeric silicones that are otherwise immiscible with water form aqueous microemulsions in the presence of a surfactant when the polar-functional group is sufficiently polar. The aqueous microemulsion can solubilize a great amount of the silicone. The liquid crystal phases on the other hand are not swelled by the silicone but coexist with it.

It has been known that aqueous surfactant solutions can solubilize apolar or weakly polar species that otherwise have extremely limited solubility in water. However, the solubility of oil in an aqueous micellar solution decreases exponentially with increasing oil molecular size [242]. Aqueous microemulsions of organic oils are generally limited to low molecular weight (≤C₁₆) paraffin and esters [191]. Nonfunctional dimethylsiloxane forms microemulsion in aqueous surfactant solution only when the

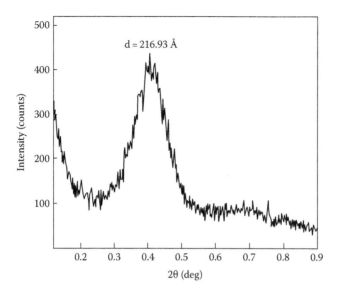

FIGURE 2.30 SAXS data for the composition corresponding to the solid square in the phase diagram in Figure 2.30: M'(NHCH$_3$)D$_{74}$M'(NHCH$_3$)/C$_2$H$_4$OHCOOH (1:10 molar ratio): C$_{12}$E$_7$: water = 0.1870: 0.4357: 0.3773.

molecule is small (e.g., less than a pentamer). The polymeric silicone oils in the present systems are not only solubilized in the aqueous micellar solutions, but the total amount of silicone oil solubilized reaches as high as 58%. For paraffin oils, this level of solubilization is realized only in the bicontinuous microemulsion. In aqueous droplet microemulsions, solubilization capacity is typically less than 50% [146,147]. The amount of surfactant needed for solubilizing the polymeric silicones in the present systems seems to be rather moderate. The surfactant lean side of the microemulsion domain in Figure 2.25 has a nearly constant surfactant-to-oil ratio of 1:2 by weight. The aminosilicone microemulsion shows an even lower surfactant-to-oil ratio, less than 1:3 at the concentrated end of the microemulsion domain (Figure 2.29).

To understand this unusual level of solubilization, insight into the structure of the microemulsion is helpful. The microemulsion at low oil concentration in the system containing the polyether grafted silicone is evidently a droplet microemulsion (Figure 2.26). The aminosilicone microemulsion was not examined by cryo-TEM, but we expect the structure be the same. For the concentrated gel microemulsions, the broad but strong diffraction peak from SAXS in both cases (Figures 2.27 and 2.30) indicates an ordered microemulsion structure. We believe the microemulsion droplets are in a closely packed configuration. It is known that colloidal particles that are suspended in a fluid and interact through a steep repulsive potential can display phase behavior progressing from "colloidal fluid" at low concentration to "colloidal crystal" at high concentration [243–245]. "Crystallization" has been shown to occur at a colloidal volume fraction close to 0.53 [244,245]. The transition from fluid to gel in the present systems occurred at roughly a combined oil-and-surfactant volume fraction of about 0.5 in the microemulsion of the polyether grafted silicone and about

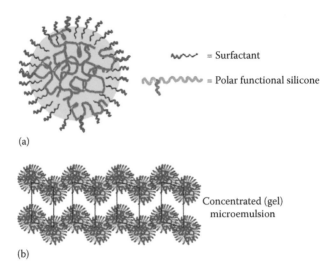

(a)

ᗰᗯᗯ = Surfactant

ᗰᗰᗯᗯ = Polar functional silicone

Concentrated (gel)
microemulsion

(b)

FIGURE 2.31 Proposed core–shell aggregation structure of aqueous microemulsion of polar-functional polymeric silicone. (a) depicts a single aggregate; (b) depicts aggregates in a close packed arrangement.

0.45–0.5 in the microemulsion of the aminosilicone. These values are close enough to the reported value for colloidal crystallization. Therefore, it is plausible that the gel microemulsion is a closely packed droplet microemulsion.

Gräbner et al. [240] proposed a molecular model for the aggregation structure of an aminosilicone microemulsion. The model depicts a micellar structure swollen with the aminosilicone. The core of the micelle contains the dimethylsiloxane portion of the aminosilicone. The surfactant (a C_iE_j type) forms a palisade layer around the micelle. Embedded in the palisade layer are protonated amine groups attached to the silicone backbone. Figure 2.31a provides a schematic illustration of the model. This structure allows the polar groups from the silicone to be mixed with hydrophilic portion of the surfactant layer and be exposed to water. Figure 2.31b illustrates the situation when the silicone oil swollen micelles closely pack to form a gel microemulsion.

The hypothesis of the gel microemulsion being a closely packed droplet microemulsion can be corroborated by a simple back-of-the-envelope calculation. Take the composition indicated by the solid square in the phase diagram of Figure 2.25. Using the core–shell model as depicted in Figure 2.31a and assuming

1. The close packing in the gel emulsion follows a body centered cubic structure (an arbitrary assumption)
2. The thickness of the shell is 1.5 nm (a reasonable assumption)
3. The densities of all components are 1 g/cm³

we obtain a droplet diameter of 27 nm with a core diameter of 24 nm, and a droplet center-to-center distance of 28 nm. This means that the distance between outer surfaces of two closest neighbor droplets is 1 nm, a situation commensurate with a strong repulsive interaction between droplets that can lead to a long range spatial order. The calculated droplet size and distance are also in rough agreement with the measured spacing from SAXS.

The calculation gives additional interesting numbers: Each droplet is composed of approximately 2900 surfactant molecules and 520 silicone polymers. This extent of micelle swelling can be appreciated by realizing that the (empty) surfactant micelle likely has an aggregation number in the hundreds. ($C_{12}E_6$ aqueous micelle has an aggregation number roughly 1000 [246] and a $C_{12}E_7$ micelle is expectantly smaller.) Note that the droplet size may depend on the composition of the microemulsion. This aggregation model necessarily implies a dependence of size on composition. The observed droplet size of a dilute microemulsion (Figure 2.26) is evidently smaller than the measured (by SAXS) or the calculated (by core–shell model) droplet size of a concentrated gel microemulsion.

With this model, we are now in a position to understand the unusually high solubilization capacity of the surfactant micellar solution for polar-functional polymeric silicones. Polar-functional silicone molecules copartition with surfactant molecules to form micellar/microemulsion aggregates. The partition is energetically favorable because the polar groups from the siloxane molecule become mixed with the hydrophilic portion of the surfactant layer and exposed to water. The partitioning process is also made possible due to the extreme flexibility of the siloxane backbone. Such partitioning, along with the large siloxane backbone, significantly flattens the curvature and expands the core volume of the micelles. As a result, the solubility of polymeric silicone in the aqueous micelles increases from zero (with no polar group partitioning) to 2–3 times the mass of the surfactant. If partitioning is not present, such solubilization is not possible. In the typical situation where there is a disparity in the size involved, large solubilizates cannot be solubilized in compartments made by small solubilizers, even if they are chemically compatible, whereas small solubilizates can be solubilized in compartments made by large solubilizers [247,248]. Despite the extreme size disparity in the present system, solubilization of the large molecules (silicone) in the aggregates of the small molecules (surfactant) is made possible due to the partitioning of the polar-functional groups from the silicone molecule on the micelle surface. This also explains why aminosilicones do not form microemulsion without the amine being protonated. A certain level of ionic strength is necessary for the partitioning to occur.

While the polar-functional silicone is solubilized in the aqueous micellar solution to a great extent, virtually none is solubilized in the hexagonal or lamellar liquid crystal phases. This can be explained by the fact that in order to accommodate the large siloxane backbone, the core of the aggregate needs to be significantly expanded. This is geometrically possible only with a spherical aggregate. Swelling of the oil compartment of a hexagonal or lamellar aggregate by large oils will necessarily destroy the geometric constraint. Therefore, solubilization of a silicone oil with a large siloxane backbone in the hexagonal or lamellar phase is not likely.

In conclusion, water-insoluble, polar-functional polymeric silicones can form aqueous microemulsions with surfactant and water. The microemulsion droplet is a mixed micelle. The core of the micelle contains the surfactant lipophilic groups swelled with the nonpolar siloxane backbone. The core is surrounded with a shell consisting of the polar-functional groups from the silicone polymer and the surfactant hydrophilic groups. The mixed aggregation structure allows for a large quantity of silicone oil to be solubilized in the micelles.

2.7 SUMMARY

Silicone emulsions and microemulsions are used in a wide range of applications. This chapter provided a brief overview of the technology, development, and applications of silicone emulsions and microemulsions over the past 70 years. The discussion on emulsifiers and emulsification methods centered around the profound impact of phase behavior on emulsion type, stability, and droplet size. The aqueous emulsion behavior of polydimethyl-siloxane (PDMS)—the most common silicone oil—can be easily understood and controlled by realizing that PDMS is immiscible to water and most conventional surfactants, and therefore the emulsion phase behavior is solely dictated by the phase behavior of the water–surfactant binary system. On the other hand, polar-functional polymeric silicones can be solubilized in an aqueous micellar solution to form silicone microemulsions. Several special cases of challenges in the emulsification of silicones were described, demonstrating the use of general principles of emulsion science.

ACKNOWLEDGMENTS

The authors thank Dr. Qian J. Feng for providing the conductivity data in Figure 2.7, David Evans for constructing the phase diagram in Figure 2.14, Dr. Xiaohua Liu and Dr. Ginam Kim for obtaining the cryo-TEM image in Figure 2.26, and Linda Sauer for obtaining the SAXS data in Figures 2.27 and 2.30. The SAXS characterization was carried out in the Institute of Technology Characterization Facility, University of Minnesota, under partial support from NSF through the NNIN program. The authors also thank Professor Stig E. Friberg for valuable discussions. The authors are grateful to the management, especially to Dr. Heidi M. Van Dort, at the Dow Corning Corporation, for support while preparing this manuscript.

REFERENCES

1. Shah, R. K., Shum, H. C., Rowat, A. C., Lee, D., Aüesti, J. J., Utada, A. S., Chu, L. Y. et al. *Mater. Today* 2008, *11*(4), 18–27.
2. Fryd, M. M., Mason, T. G. *Langmuir* 2013, *29*, 15787–15793.
3. Warrick, E. L. *Forty Years of Firsts*. McGraw-Hill, New York, 1990.
4. Hommel, M. C., Currie, C. C. US2666685, 1954.
5. Westinghouse Electric Int. Co. GB596833, 1948.
6. Gagarine, D. M., Repokis, H., Clemson, S. C. US2750305, 1956.
7. Volkmann. R. J., Baecker, H. J. US2755194, 1956.
8. Hyde, J. F., Wehrly, J. R. US2891920, 1959.
9. Findlay, D. E., Weyenberg, D. R. US3294725, 1966.
10. Weyenberg, D. R., Findlay, D. E., Cekada, J., Bey, A. E. *J. Polym. Sci. Polym. Symposia* 1969, *27*, 27–34.
11. Ona, I., Ozaki, M., Tanaka, O. US4784665, 1988.
12. Wolfgruber, M., Deubzer, B., Frey, V. US4857582, 1989.

13. Graiver, D., Tanaka, O. US4999398, 1991.
14. Graiver, D., Tanaka, O. US5817714, 1998.
15. Craig, D. H. US 5726270, 1998.
16. Gee, R. P. US5852110, 1998.
17. Craig, D. H., Morgan, W. F. US6214927, 2001.
18. Gee, R. P. US6316541, 2001.
19. Cekada, J. Jr., Weyenberg, D. R. US3433780, 1969.
20. Ona, I., Ozaki, M. US4935464, 1990.
21. Gee, R. P., Liu, Y. US7875673, 2011.
22. Axon, G. L. US3360491, 1967.
23. Huebner, D. J. Saam, J. C. US4567231, 1986.
24. Johnson, R. D., Saam, J. C., Schmidt, C. M. GB2056473, 1983.
25. Liles, D. T. US5026769, 1991.
26. Willing, D. N. US4248751, 1981.
27. Dalle, F., Marteaux, L. US6013682, 2000.
28. Cauvin, S., Gordon, G. V., Liles, D. T. US8492477, 2013.
29. Saam, J. C., Wegener, R. L. US4273634, 1981.
30. Huebner, D. J., Saam, J. US4568718, 1986.
31. Berg, D. T., Joffre, E. J. US5674937, 1997.
32. Berg, D. T., Hides, L. G. A., Joffre, E. J., O'neil, V. K., Tselepis, A. J., Wolf, A. T. F. US5777026, 1998.
33. Liles, D. T. *Polym. Mater. Sci. Eng.* 1992, *66*, 172–173.
34. Liles, D. T., Joffre, E. J. *Polym. Prepr.* 1998, *39*, 526–527.
35. Hanada, T., Morita, Y. US4594134, 1986.
36. Liles, D. T., Morita, Y., Kobayashi, K. *Polym. Prepr.* 2001, *42*, 240–241.
37. Hilliard, J. R. US3898300, 1975.
38. Sasaki, I., Yamamoto, N., Yanagase, A. US4994523, 1991.
39. Cavivenc, E., Richard, J. US5618879, 1997.
40. Mautner, K., Deubzer, B. US5223586, 1993.
41. Hellstern, A. M., Mitchell, L. L., Halley, R. J. US5106900, 1992.
42. Kim, G., Sousa, A., Meyers, D., Shope, M., Libera, M. *J. Am. Chem. Soc.* 2006, *128*, 6570–6571.
43. Kim, G., Sousa, A., Meyers, D., Libera, M. *Microsc. Microanal.* 2008, *14*, 459–468.
44. Lind, D., Meyers, D., Shope, M. US7767747, 2010.
45. Badour, L. R., Meyers, D., Kim, G. WO2011087767, 2011.
46. Hasinovic, H., Friberg, S. E., Rong, G. *J. Colloid Interface Sci.* 2011, *354*, 424–426.
47. Lindman, B. In: *Microemulsions*, Monzer, F. (Ed.), CRC Press, Boca Raton, FL, 2009, pp. xv–xx.
48. Hill, R. M. US5623017, 1997.
49. Hill, R. M. US5705562, 1998.
50. Hill, R. M. US5707613, 1998.
51. Hill, R. M., Lin, Z. US6616934, 2003.
52. Feng, Q. J., Lin, Z., Hill, R. M. US6998424, 2006.
53. Li, X., Washenberger, R. M., Scriven, L. E., Davis, H. T., Hill, R. M. *Langmuir* 1999, *15*, 2267–2277.
54. Li, X., Washenberger, R. M., Scriven, L. E., Davis, H. T., Hill, R. M. *Langmuir* 1999, *15*, 2278–2289.
55. Hill, R. M., Li, X., Davis, H. T. In: *Silicone Surfactants*, Hill, R. M. (Ed.), Marcel Dekker, New York, 1999, pp. 313–348.
56. Silas, J. A., Kaler, E. W., Hill, R. M. *Langmuir* 2001, *17*, 4534–4539.
57. Gee, R. P. US4620878, 1986.
58. Ona, I., Ozaki, M. US4857212, 1989.

59. Chrobaczek, H., Tschida, G. US5057572, 1991.
60. Joyner, M. *Text. Chem. Color* 1986, *18*(3), 34–37.
61. Andriot, M., Chao, S. H., Colas, A. R., Cray, S. E., deBuyl, F., DeGroot, J. V., Dupont, A. et al. In: *Silicon-Based Inorganic Polymers*, Nova Science Publishers, New York, 2009, p. 84.
62. Clarke, D. E., Creutz, S., F., Henault, B., Small, S. US 6251850, 2001.
63. Winder, L., Refalo, S., Raleigh, M., Corona, A., Burns, A., Dinniwell, A. EP1204793, 2007.
64. Inoue, Y., Momii, K. US5254621, 1993.
65. Takanashi, M., Matsumoto, M. US5916687, 1999.
66. Parvinzadeh, M., Hajiraissi, R. *Tenside Surf. Deterg.* 2008, *45*, 254–257.
67. Skinner, M. W., Qian, C., Grigoras, S., Halloran, D. J., Zimmerman, B. L. *Textile Res. J.* 1999, *69*, 935–943.
68. Narula, D. *J. Am. Leather Chem. Assoc.* 1995, *90*(3), 93–98.
69. Butler, D. W. *Surf. Coatings Int. Part B Coatings Trans.* 1997, *80*(5), 230–234.
70. Mohamed, M., Liles, D. T. *Pitture e Vernici, European Coatings*, 2011, *87*(4), 17–20.
71. Liles, D. T. *Proceedings of 39th International Waterborne High-Solids Powder Coatings Symposium*, 2012, School of Polymers and High Performance Materials, University of Southern Mississippi, Hattiesburg, MS, pp. 44–79.
72. Gordon, G. V., Tabler, R. L., Perz, S. V., Stasser, J. L., Owen, M. J., Tonge, J. S. *Book of Abstracts, 215th ACS National Meeting*, Dallas, TX, 1998.
73. Ekeland, R., Tonge, J. S., Gordon, G. V. http://www.dowcorning.com/content/publish-edlit/30-1139-01.pdf, accessed October 10, 2015.
74. Disapio, A., Fridd, P. *Int. J. Cosmet. Sci.* 1988, *10*(2), 75–89.
75. Yahagi, K. *J. Soc. Cosmet. Chem.* 1993, *43*, 275–284.
76. Nanavati, S., Hami, A. *J. Soc. Cosmet. Chem.* 1994, *45*, 135–148.
77. Li, W., Amos, J., Jordan, S., Theis, A., Davis, C. *J. Cosmet. Sci.* 2006, *57*, 178–180.
78. Isabelle Van Reeth, In: *Handbook of Cosmetic Science and Technology*, 4th edn., Barel, A. O., Paye, M., Maibach, H. I. (Eds.), CRC Press, Boca Raton, FL, 2014, pp. 321–330.
79. Klykken, P., Servinski, M., Thomas, X. http://www.dowcorning.com/content/publish-edlit/52-1068-01.pdf, accessed January 12, 2016.
80. Aliyar, H., Huber, R., Liles, D., Loubert, G., Schalau, G., Toth, S. US20140371317, 2014.
81. Cauvin, S., Le Meur, M., Liles, D., Thomas, X. WO2014019841, 2014.
82. Friedrich, A., Biehl, P., Zimmerman, S., Trambitas, A., Meyer, J. *SOFW J.* 2013, *139*(12), 40–47.
83. Roos, M., Giessler-Blank, S. http://www.construction-chemicals.com/product/construction-chemicals/Documents/Keeping-moisture-at-bay.pdf, accessed January 12, 2016.
84. Be, A., Liles, D. T., Wilhelmi, F. G. P. US5919296, 1999.
85. McAuliffe, A., S., Sarrazin, M.-J., Selley, D. S., Stammer, A. US8354480, 2013.
86. Harwell, J. H., Scamehorn, J. F. In: *Mixed Surfactant Systems*, Ogino, K., Abe, M. (Eds.), Marcel Dekker, New York, 1993, pp. 263–281.
87. Griffin, W. C. *J. Soc. Cosmet. Chem.* 1949, *I*(5), 311–326.
88. Griffin, W. C. *J. Soc. Cosmet. Chem.* 1954, *5*, 249–256.
89. Lewandowska, K., Staszewska, D. U., Bohdaneck, M. *Eur. Polym. J.* 2001, *37*, 25–32.
90. Serrallach, J. A., Jones, G. *Ind. Eng. Chem.* 1931, *23*, 1016–1019.
91. Bunge, D. J. US4954554, 1990.
92. Kondo, T., Sawatari, C., Manley, R. St. J., Gray, D. G. *Macromolecules* 1994, *27*, 210–215.
93. Liu, Y., Rastello, J., Sutton-Poungthana, L. J. US8470925, 2013.
94. Cauvin, S, Morgane, L. M., Liles, D. T., Thomas, X. J.-P., Vincent, A.-M. US20150190516A, 2015.

95. Lecomte, J.-P., Stammer, A., Campeol, F., Thibaut, M. US8445560, 2013.
96. Wanka, G., Hoffmann, H., Ulbricht, W. *Macromolecules* 1994, *27*, 4145–4159.
97. Nagarajan, R., Barry, M., Ruckenstein, E. *Langmuir* 1986, *2*, 210–215.
98. Alexandridis, P., Olsson, U., Lindman, B. *Macromolecules* 1995, *28*, 7700–7710.
99. Holmqvist, P., Alexandridis, P., Lindman, B. *Macromolecules* 1997, *30*, 6788–6797.
100. Ivanova, R., Lindman B., Alexandridis, P. *Langmuir* 2000, *16*, 3660–3675.
101. Ivanova, R., Lindman, B., Alexandridis, P. *Langmuir* 2000, *16*, 9058–9069.
102. Evans, S. M., Liles, D. T., Mohamed, M. A. US8877293, 2014.
103. Hill, R. M. (Ed.). *Silicone Surfactants*. Marcel Dekker, New York, 1999.
104. Gee, R. P., Keil, J. W. US4122029, 1978.
105. Keil, J. W. US4265878, 1981.
106. Keil, J. W. US4268499, 1981.
107. Starch, M. S. US4311695, 1982.
108. Zotto, A. A. US4988504, 1991.
109. Sela, Y., Magdassi, S., Garti, N. *Colloid Polym. Sci.* 1994, *272*, 684–691.
110. Mehta, S. C., Somasundaran, P. *Langmuir* 2008, *24*, 4558–4563.
111. Mehta, S. C., Somasundaran, P., Kulkarni, R. *J. Colloid Interface Sci.* 2009, *33*, 635–640.
112. Rosen, M. J. *J. Am. Oil Chem. Sci.* 1972, *49*, 293–297.
113. Gradzielski, M., Hoffmann, H., Robisch, P., Ulbricht, W., Grüning, B. *Tenside Surf. Deterg.* 1990, *27*, 366–379.
114. Hoffmann, H., Ulbricht, W. In: *Siloxane Surfactants*, Hill, R. M. (Ed.), Marcel Dekker, New York, 1999, pp. 97–136.
115. Schwarz, E. G., Reid, W. G. *Ind. Eng. Chem.* 1964, *56*(9), 26–31.
116. Kanellopoulos, A. G., Owen, M. J. *J. Colloid Interface Sci.* 1971, *35*, 120–125.
117. Bailey, D. L., Peterson, I. H., Reid, W. G. *Chem. Phys. Appl. Surface Active Subst. Proc. 4th Int. Congr.* 1967, Brussels, Vol. 1, pp. 173–182.
118. Maki, H., Saeki, S., Ikeda, I., Komori, S. *J. Am. Oil Chem. Soc.* 1969, *46*, 635–638.
119. Grüning, B., Koerner, G. *Tenside Surf. Deterg.* 1989, *26*, 312–317.
120. Binks, B. P. In: *Modern Aspects of Emulsion Science*, Binks, B. P. (Ed.), The Royal Society of Chemistry, Cambridge, U.K., 1998, pp. 26–33.
121. Kabalnov, A. S. In: *Modern Aspects of Emulsion Science*, Binks, B. P. (Ed.), The Royal Society of Chemistry, Cambridge, U.K., 1998, pp. 205–260.
122. Gordon, G. V., Liles, D. T. US9303125, 2016.
123. Liu, Y., Friberg, S. E. *J. Colloid Interface Sci.* 2009, *340*, 261–268.
124. Mills, P., Snabre, P. *Europhys. Lett.* 1994, *25*, 651–656.
125. Knoche, M., Tamura, H., Bukovac, M. J. *J. Agric. Food Chem.* 1991, *39*, 202–206.
126. Snow, S. A., Fenton, W. N., Owen, M. J. *Langmuir* 1990, *6*, 385–391.
127. Fenton, W. N., Keil, J. W. US4125470, 1978.
128. Kennedy, R. D, Lenoble, B. L. J., Liles, D. T, Liu, Y., Raynaud, E. WO2016014619, 2016.
129. Gee, R. P., Vincent, J. M. US5891954, 1999.
130. Vincent, J. M., Liu, Y., Liles, D. T. US6652867, 2003.
131. Wang, A., Jiang, L., Mao, G., Liu, Y. *J. Colloid Interface Sci.* 2001, *242*, 337–345.
132. Wang, A., Jiang, L., Mao, G., Liu, Y. *J. Colloid Interface Sci.* 2002, *256*, 331–340.
133. Dahms, G., Zombeck, A. *Cosmet. Toilet.* 1995, *110*, 91–100.
134. Gum, M. L. US4782095, 1988.
135. Shigeta, A., Kikuta, Y. US4906458, 1990.
136. Gruening, B., Hameyer, P., Weitemeyer, C. *Tenside Surf. Deterg.* 1992, *29*, 78–83.
137. Rentsch, S. F. US5387417, 1995.
138. Dickinson, E. *J. Dairy Sci.* 1997, *80*, 2607–2619.
139. Dimitrova, T. D., Saulnier, L., Verhelst, V., Van Reeth, I. In: *Polymeric Delivery of Therapeutics*, Morgan, S. et al. (Eds.), ACS Symposium Series, Washington, DC, 2010, Chapter 13.

140. Anseth, J. W., Bialek, A., Hill, R. M., Fuller, G. G. *Langmuir* 2003, *19*, 6349–6356.
141. Sakai, K., Ikeda, R., Sharma, S. C., Shrestha, R. G., Ohtani, N., Yoshioka, M., Sakai, H., Abe, M., Sakamoto, K. *Langmuir* 2010, *26*, 5349–5354.
142. Gašperlin, M., Šmid-Korbar, J. *Pharm. J. Slov.* 1992, *43*, 3–10.
143. Bancroft, W. D. *J. Phys. Chem.* 1913, *17*, 501–519.
144. https://en.wikipedia.org/wiki/Bancroft_rule, accessed on October 14, 2015.
145. Binks, B. P. *Langmuir* 1993, *9*, 25–28.
146. Friberg, S. E., Lapczynska, I. *Progr. Colloid Polym. Sci.* 1975, *56*, 16–20.
147. Friberg, S. E., Corkery, R. W., Blute, I. A. *J. Chem. Eng. Data* 2011, *56*, 4282–4290.
148. Binks, B. P. In: *Modern Aspects of Emulsion Science*, Binks, B. P. (Ed.), The Royal Society of Chemistry, Cambridge, U.K., 1998, pp. 2–8.
149. Davis, H. T. *Colloids Surf. A Physicochem. Eng. Asp.* 1994, *91*, 9–24.
150. Davies, J. T. Gas/liquid and liquid/liquid interface, *Proceedings of Second International Congress Surface Activity*, 1957, London, Vol. I, pp. 426–438.
151. Winsor, P. A. *Solvent Properties of Amphiphilic Compounds*, Butterworth, London, U.K., 1954, p. 207.
152. Beerbower, A., Hill, M. In: *McCutcheon's Detergents and Emulsifiers*, Allured Publishing Company, Ridgewood, NJ, 1971.
153. Salager, J.-L. *Progr. Colloid Polym. Sci.* 1996, *100*, 137–142.
154. Becher, P. *Emulsions: Theory and Practice*, 3rd edn., Oxford University Press, New York, 2001, pp. 338–372.
155. O'Lenick, A. J. Jr., Parkinson, J. K. *Cosmet. Toilet.* 1996, *111*, 37–44.
156. Kabalnov, A., Wennerstrom, H. *Langmuir* 1996, *12*, 276–292.
157. Rouviere, J., Razakarison, J. L., Marignan, J., Brun, B. *J. Colloid Interface Sci.* 1989, *133*, 293–301.
158. Messier, A., Schorsch, G., Rouviere, J., Tenebre, L. *Prog. Colloid Polym. Sci.* 1989, *79*, 249–256.
159. Binks, B. P., Dong, J. *Colloids Surf. A Physicochem. Eng. Asp.* 1998, *132*, 289–301.
160. Liu, Y. In: *Encyclopedia of Surface and Colloid Science*, 2nd edn., Somasundaran, P. (Ed.), Taylor & Francis, New York, 2013, DOI: 10.1081/E-ESCS-120049478.
161. Mitchell, D. J., Tiddy, G. J. T., Waring, L., Bostock, T., McDonald, M. P. *J. Chem. Soc. Faraday Trans. 1*, 1983, *79*, 975–1000.
162. Kunieda, H., Fukui, Y., Uchiyama, H., Solans, C. *Langmuir* 1996, *12*, 2136–2140.
163. Leal-Calderon, F., Schmitt, V., Bibette, J. *Emulsion Science, Basic Principles*, 2nd edn., Springer, Berlin, Germany, 2007, pp. 6–10.
164. Hall, S., Cooke, M., Pacek, A. W., Kowalski, A. J., Rothman, D. *Can. J. Chem. Eng.* 2011, *89*, 1040–1050.
165. Schultz, S., Wagner, G., Urban, K., Ulrich, J. *Chem. Eng. Technol.* 2004, *27*, 361–368.
166. Walstra, P. In: *Encyclopedia of Emulsion Technology*, Vol. 1, Basic Theory, Belcher, P. (Ed.), Marcel Dekker, New York, 1983, pp. 57–128.
167. Walstra, P., Smulders, P. E. A. In: *Modern Aspects of Emulsion Science*, Binks, B. P. (Ed.), The Royal Society of Chemistry, Cambridge, U.K., 1998, pp. 56–99.
168. Brösel, S., Schubert, H. *Chem. Eng. Process. Process Intensification* 1999, *38*, 533–540.
169. Lobo, L., Svereika, A. *J. Colloid Interface Sci.* 2003, *261*, 498–507.
170. Brooks, B. W., Richmond, H. N., Zerfa, M. In: *Modern Aspects of Emulsion Science*, Binks, B. P. (Ed.), The Royal Society of Chemistry, Cambridge, U.K., 1998, pp. 175–204.
171. Shinoda, K., Saito, H. *J. Colloid Interface Sci.* 1968, *26*, 70–74.
172. Dickinson, E. *J. Colloid Interface Sci.* 1981, *84*, 284–287.
173. Salager, J.-L. In: *Encyclopedia of Emulsion Technology*, Becher, P. (Ed.), Dekker, New York, 1988, pp. 79–134.

174. Brooks, B. W., Richmond, H. N. *Chem. Eng. Sci.* 1994, *49*, 1843–1853.
175. Silva, F., Pena, A., Minana-Perez, M., Salager, J.-L. *Colloids Surf. A Physicochem. Eng. Asp.* 1998, *132*, 221–227.
176. Groeneweg, F., Agterof, W. G. M., Jaeger, P., Janssen, J. J. M., Wieringa, J. A., Klahn, J. K. *Chem. Eng. Res. Des.* 1998, *76*, 55–63.
177. Klahn, J. K., Janssen, J. J. M., Vaessen, G. E. J., de Swart, R., Agterof, W. G. M. *Colloids Surf. A Phys. Chem. Eng. Asp.* 2002, *210*, 167–181.
178. Tyrode, E., Mira, I., Zambrano, N., Marquez, L., Rondon-Gonzalez, M., Salager, J.-L. *Ind. Eng. Chem. Res.* 2003, *42*, 4311–4318.
179. Sajjadi, S., Zerfa, M., Brooks, B. W. *Langmuir* 2000, *16*, 10015–10019.
180. Sajjadi, S., Zerfa, M., Brooks, B. W. *Chem. Eng. Sci.* 2002, *57*, 663–675.
181. Sajjadi, S., Jahanzad, F., Yianneskis, M., Brooks, B. W. *Ind. Eng. Chem. Res.* 2003, *42*, 3571–3577.
182. Sajjadi, S., Zerfa, M., Brooks, B. W. *Colloids Surf. A* 2003, *218*, 241–254.
183. Sajjadi, S., Jahanzad, F., Yianneskis, M. *Colloid Surf. A* 2004, *240*, 149–155.
184. Mira, I., Zambrano, N., Tyrode, E., Márquez, L., Peña, A. A., Pizzino, A., Salager, J. L. *Ind. Eng. Chem. Res.* 2003, *42*, 57–61.
185. Salager, J.-L. In: *Emulsions and Emulsion Stability*, 2nd edn., Sjoblom, J. (Ed.), *Surfactant Science Series*, Vol. 132, Taylor & Francis, London, U.K., 2006, pp. 185–226.
186. Rondon-Gonzalez, M., Sadtler, V., Choplin, L., Salager, J.-L. *Colloids Surf. A Physicochem. Eng. Asp.* 2006, *288*, 151–157.
187. Rondon-Gonzalez, M., Madariaga, L. F., Sadtler, V., Choplin, L., Marquez, L., Salager, J.-L. *Ind. Eng. Chem. Res.* 2007, *46*, 3595–3601.
188. Shinoda, K., Arai, H. *J. Phys. Chem.* 1964, *68*, 3485–3490.
189. Shinoda, K., Friberg, S. E. *Emulsions and Solubilization.* Wiley, New York, 1986.
190. Leaver, M. S., Olsson, U., Wennerstrom, H., Strey, R., Wuerz, U. *J. Chem. Soc. Faraday Trans.* 1995, *91*, 4269–4274.
191. Stubenrauch, C. (Ed.). *Microemulsions: Background, New Concepts, Applications, Perspectives.* Wiley, New York, 2009.
192. Pillai, V., Kanicky, J. R., Shah, D. O. In: *Handbook of Microemulsion Science and Technology*, Kumar, P., Mittal, K. L. (Eds.), Marcel Dekker, New York, 1999, pp. 743–753.
193. Morales, D., Gutierrez, J. M., Garcia-Celma, M. J., Solans, Y. C. *Langmuir* 2003, *19*, 7196–7200.
194. Sole, I., Maestro, A., Pey, C. M., Gonzalez, C., Solans, C., Gutierrez, J. M. *Colloids Surf. A Physicochem. Eng. Asp.* 2006, *288*, 138–143.
195. Sole, I., Maestro, A., Gonzalez, C., Solans, C., Gutierrez, J. M. *J. Colloid Interface Sci.* 2008, *327*, 433–439.
196. Roger, K., Cabane, B., Olsson, U. *Langmuir* 2010, *26*, 3860–3867.
197. Sottmann, T., Stubenrauch, C. In: *Microemulsions: Background, New Concepts, Applications, Perspectives.* Stubenrauch, C. (Ed.), John Wiley & Sons, New York, 2009, pp. 23–31.
198. Minana-Perez, M., Gutron, C., Zundel, C., Anderez, J. M., Salager, J. L. *J. Dispers. Sci. Technol.* 1999, *20*, 893–905.
199. Taisne, L., Cabane, B. *Langmuir* 1998, *14*, 4744–4752.
200. Rao, J., McClements, D. J. *J. Agric. Food Chem.* 2010, *58*, 7059–7066.
201. Shim, A. K., Tabler, R., Tascarella, D. US 7385001, 2008.
202. Liu, Y., Carter, E. L., Gordon, G. V., Feng, Q. J., Friberg, S. E. *Colloids Surf. A Physicochem. Eng. Asp.* 2012, *399*, 25–34.
203. Liu, Y., Friberg, S. E. *Langmuir* 2010, *26*, 15786–15793.

204. Solans, C., Esquena, J., Azemar, N. *Curr. Opin. Colloid Interface Sci.* 2003, *8*, 156–163.
205. Cameron, N. R. *Polymer* 2005, *46*, 1439–1449.
206. Gibas, R. A., Liu, Y., Pan, D.-L., Rastello, J. T., Weber, C. A. US20140243727, 2014.
207. Barnes, K., Gordon, G., Liu, Y. WO2010065712, 2010.
208. Liu, Y. US6737444, 2004.
209. Lifshitz, I. M., Slezov, V. V. *Sov. Phys. JETP*, 1959, *8*, 331.
210. Wagner, Z. *Electrochem.* 1961, *35*, 581.
211. Weers, J. G. In: *Modern Aspects of Emulsion Science*, Binks, B. P. (Ed.), The Royal Society of Chemistry, Cambridge, U.K., 1998, pp. 292–327.
212. Wilke, C. R., Chang, P. *Am. Inst. Chem. Eng. J.* 1955, *1*, 264–270.
213. Taylor, P. *Adv. Colloid Interface Sci.* 1998, *75*, 107–163.
214. Feng, Q., Hickerson, R., Starch, M., Van Reeth, I. US20070190012, 2007, and EP1740148B1, 2010.
215. Hill, R. M., He, M., Lin, Z., Davis, H. T., Scriven, L. E. *Langmuir* 1993, *9*, 2789–2798.
216. Wilson, M. E. US4877654, 1989.
217. Higuchi, W. I., Misra, J. *J. Pharm. Sci.* 1962, *51*, 459–466.
218. Fisher, P. D., Gee, R. P. US6103001, 2000.
219. Daoust, J. A., Lecomte, J.-P. H., Liles, D. T., Liu, Y., Roggow, T. A. II. WO2014159598, 2014.
220. Wolf, L., Hoffmann, H., Talmon, Y., Teshigawara, T., Watanabe, K. *Soft Matter* 2010, *6*, 5367–5374.
221. Wolf, L. Hoffmann, H., Watanabe, K., Okamoto, T. *Phys. Chem. Chem. Phys.* 2011, *13*, 3248–3256.
222. Castellino, V., Cheng, Y.-L., Acosta, E. *J. Colloid Interface Sci.* 2011, *353*, 196–205.
223. Hill, R. M. In: *Silicone Surfactants*, Hill, R. M. (Ed.), Marcel Dekker, New York, 1999, p. 28.
224. Bialek, A. I., Hill, R. M., Kadlec, D. A., Van Dort, H. M. US6498197, 2002.
225. O'Rourke, B., Stevenson, A. A. GB2450727, 2009.
226. Burger, W., Mayer, H., Schrock, R., Woltje, H., Deubzer, B., Lautenschlager, H. US6182759, 2001.
227. Hill, R. M., Soeung, M., Gonzalez-Roldan, M. US20150068755, 2015.
228. Lakatos, I., Toth, J., Lakatos-Szabo, J., Kosztin, B., Palasthy, Gy. *Progr. Mining Oilfield Chem.* 2003, *5*, 213–232.
229. Lakatos, I., Lakatos-Szabo, J., Bodi, T., Vago, A., Szentes, G. Reservoir conformance enhancement in gas fields using microemulsions. *Geosci. Eng.* 2012, *1*, 187–204.
230. Halloran, D. J., Hill, R. M., Wrolson, B. M., Zimmerman, B. L. US6479583, 2002.
231. Brook, M. A., Whinton, M., Gonzaga, F., Li, N. *Chem. Commun.* 2011, *47*, 8874–8876.
232. Cooper, K. G., Marshall, H. S. B. US3303048, 1967.
233. Burrill, P. M., Kohnstamm, W. S. US4098701, 1978.
234. Ona, I., Ozaki, M., Taki, Y. US4359545, 1982.
235. Abrutyn, E. S. In: *Conditioning Agents for Hair and Skin*, Cosmetic Science and Technology Series, Vol. 21, Schueller, R., Romanowski, P. (Eds.), Marcel Dekker, New York, 1999, pp. 167–200.
236. Ando, Y., Hirai, M. US8329816, 2011.
237. Teshigawara, T., Watanabe, K., Araki, H. WO2010110047, 2012.
238. Feng, Q., Johnson, B., Liu, Y. WO2014058887, 2014.
239. Gee, R. P. *Colloids Surf. A Physicochem. Eng. Asp.* 1997, *137*, 91–101.
240. Gräbner, D., Xin, L., Hoffmann, H., Drechsler, M., Schneider, O. *J. Colloid Interface Sci.* 2010, *350*, 516–522.
241. Sharma, S. C., Tsuchiya, K., Sakai, K., Sakai, H., Abe, M., Komura, S., Sakamoto, K., Miyahara, R. *Langmuir* 2008, *24*, 7658–7662.
242. Hoffmann, H., Sturmer, A., *Tenside Surf. Deterg.* 1993, *30*, 335–341.

243. Clark, N. A., Hurd, A. J., Ackerson, B. J. *Nature* 1979, *281*, 57–60.
244. Pusey, P. N., van Megen, W. *Nature* 1986, *320*, 340–342.
245. Mortensen, K., Brown, W., Norden, B. *Phys. Rev. Lett.* 1992, *68*, 2340–2343.
246. Christov, N. C., Denkov, N. D., Kralchevsky, P. A., Broze, G., Mehreteab, A. *Langmuir* 2002, *18*, 7880–7886.
247. Uddin, Md. H., Morales, D., Kunieda, H. *J. Colloid Interface Sci.* 2005, *285*, 373–381.
248. Frank, C., Strey, R., Schmidt, C., Stubenrauch, C. *J. Colloid Interface Sci.* 2007, *312*, 76–86.

3 Ionic Polymerization of Silicones in Aqueous Media

Ronald P. Gee

CONTENTS

3.1 INTRODUCTION

Silicone emulsions produced by ionic polymerization of siloxane monomers in water have now been used for many years in commerce. Emulsions produced by such processes provide any, or a combination, of the following characteristics: improved production efficiency, reduced manufacturing cost, increased stability, and provision of the unique characteristics or properties of silicone to aqueous products. Those emulsions produced by the process of emulsion polymerization generally have a particle size less than about 200 nm. This is the "critical particle size for creaming" for polydimethylsiloxane particles in aqueous emulsion. Such emulsions will not exhibit creaming and would not need remixing after storage and before use. Increased stability against shearing forces during industrial use in various applications has been found. As a result, such emulsions have been used extensively in such applications as protection of freshly printed ink on paper in the web printing process and in the treatment of fabric in textile mills to enhance the soft feel of textiles. Emulsions produced by this method can have extremely small particle size (nanoemulsion) such as 25 nm. These appear transparent or translucent and can be added to clear, water-based products to add the characteristics of silicone polymers without loss of customer product clarity. These nanoemulsions have been extensively used in the production of clear conditioning shampoo.

3.2 HISTORICAL DEVELOPMENT

A unique method to produce fine-particle-size emulsions of polydimethylsiloxane in water was first revealed in a patent by Hyde and Wehrly in 1959 [1]. The process typically involved pre-emulsifying D_4 dimethylcyclosiloxane in water containing a cationic surfactant such as octadecyltrimethylammonium chloride and then heating and catalyzing the process by the addition of sodium hydroxide. During emulsion polymerization (EP), the particle size underwent a significant decrease in diameter, and a high percentage (approximately 85%) of the D_4 was converted to polydimethylsiloxane. The remainder of the siloxane was a mixture of cyclosiloxanes consisting of D_4, D_5, and D_6. The actual catalyst was considered to be alkyltrimethylammonium hydroxide formed in situ by ion exchange, that is, a surface-active catalyst. At that time, emulsion-particle-size measurement was very difficult and generally limited to optical microscopy at 1000× magnification. Particle diameters less than 500 nm could not be easily quantified due to the small size and high velocity of movement (Brownian motion). The extremely small size could be detected by the physical appearance of the emulsion exhibited by the loss of "whiteness" toward a more translucent to, in some cases, transparent nature with a blue coloration.

Emulsion polymerization of silicones to produce emulsions of fine particle size has now been utilized for over 50 years. In general, the initial patent by Hyde found surface-active quaternary ammonium hydroxides to be suitable for anionic polymerization of D_4 in cationic emulsion. Cationic polymerization using strong acids with anionic, or even nonionic, surfactants was also cited. In 1966, Findlay and Weyenberg patented the use of alkybenzenesulfonic acids to polymerize cyclosiloxanes in water [2]. Weyenberg published a work that included the use of trialkoxysilanes in place of cyclosiloxanes to form aqueous suspensions of silsesquioxanes having very small particle size (less than about 40 nm) [3]. Various alkoxysilanes have been copolymerized with cyclosiloxanes in EP to produce copolymers, such as aminosilicones, in emulsion, which have been utilized extensively in the hair care product market. When certain process conditions are controlled, as explained by Gee [4], it is not necessary to pre-emulsify the cyclosiloxane before polymerization.

3.3 OVERVIEW OF PROCESS TYPES

There are a number of different aqueous polymerization processes and usable starting "monomer" materials. The starting silicone "monomer" may be cyclosiloxane, oligomer siloxanediol, dialkoxydimethylsiloxane, diacetoxydimethylsiloxane, or dichlorodimethylsiloxane. The latter three materials are little used due to the undesirable generation of alcohols, or acid, which results in high concentrations of chloride or acetate salts in the emulsion. The most frequently used monomer is octamethylcyclotetrasiloxane (D_4). Possible loci where polymerization could occur are: inside the monomer droplet, at the surface of the droplet, and in the aqueous phase. The mechanisms of these processes appear to be dependent on three primary factors:

1. Aqueous solubility of the starting monomer or oligomer
2. Structure and aqueous solubility of the surface-active catalyst (surfcat)
3. Total surface area of the silicone droplets being polymerized

The term "surfcat" is preferred over the term "inisurf," as the activating species of the cyclosiloxane EP is a catalyst and is not consumed. In contrast, inisurf implies an initiator that is consumed, as in free radical EP.

The types of aqueous silicone ionic polymerizations can be described and categorized relative to each other based on these three key attributes. Five processes are described in Table 3.1: precipitation polymerization (PP), suspension polymerization (SP), internal suspension polymerization (ISP), emulsion polymerization (EP), and mini-emulsion polymerization (mEP). Only three are generally studied and used in industry: SP, EP, and mEP.

The PP process type occurs when both monomer and surfcat are soluble in water. As the monomer polymerizes, the oligomer quickly reaches a molecular weight where it is no longer water soluble and precipitates. These oligomers then aggregate to form polymer particles stabilized by the surfcat and other surfactants present.

SP occurs when the monomer is completely insoluble in water and the surfcat is watersoluble. Characteristically, the monomer is an oligomeric siloxane diol, $HO(MeSiO)_nH$, having sufficient degree of polymerization (n) to be incapable of diffusion into/through water. The dimethylsiloxane oligomer diol typically used has n approximately equal to 35 and a viscosity of 80–100 centistokes. The oligomer is emulsified in water with an anionic surfactant derived from a strong acid. It is subjected to high shearing forces to produce an emulsion of small particle size (100–800 nm). The emulsion is then catalyzed with a strong acid to begin the step-growth polymerization by silanol–silanol condensation. After an appropriate period of time, the reaction is stopped at the desired molecular weight by neutralizing the catalyst with a base, such as sodium hydroxide or triethanolamine. No change in particle size occurs and the reaction occurs at the particle surface.

ISP is like the SP process using a water-insoluble "monomer," except that the surfcat is not water soluble and the condensation reaction occurs inside the

TABLE 3.1

Correlation of Ionic Polymerization Process Types

Monomer aqueous solubility	Soluble	Insoluble	Insoluble	Slight
Surfcat aqueous solubility	Soluble	Soluble	Insoluble	Soluble
Initial droplet diameter	NA	Small	Large	Small
Process type	PP	SP	ISP	EP and mEP

oligomer-diol particle. There are other types of ISPs that do not contain a surfcat and polymerization in the particle occurs by other chemistries. An example is the physical emulsification of polymers having SiH and $SiCH=CH_2$ functionality where polymerization is achieved by a hydrosilation reaction of these groups. These are outside the scope of this chapter.

The process is EP when the monomer is only slightly soluble in water and the surfcat is water soluble. Silicone EP is similar to free radical EP that polymerizes unsaturated organic monomers to high molecular weight organic polymers in water. However, the silicone EP process is an ionic polymerization of a silicone monomer in water. Although numerous studies have been reported [5–15], some aspects of the EP process are still not fully understood. The monomer exists initially as a second phase dispersed in water where the initial droplet size is dependent on shearing forces and the interfacial tension. Generally, if the cyclosiloxane is not pre-emulsified, the finished emulsion product will have some unemulsified "oil" consisting of polymers and cyclics. However, when EP conditions and formulations are carefully controlled by certain parameters [4], it is not necessary to pre-emulsify the cyclosiloxane. The process involves the reaction of the monomer either at the droplet surface or via the diffusion of the monomer into the aqueous phase, followed by reactions to produce siloxane polymer (see Section 3.7). There is a gradual disappearance of the initial droplets and formation of new polymer particles. Generally, the observed result is an emulsion having a considerably smaller particle size. Final particle size over a wide range is possible (10–1000 nm). In a similar manner, to free radical emulsion polymerization, there are three identifiable process "intervals" that can occur. Interval 1 is a particle formation period. Interval 2 generally begins when the surfactant concentration in the aqueous phase becomes less than the critical micelle concentration (cmc), and micelles are no longer present to supply surfactant to stabilize newly forming particles. Interval 2 is a period of increasing particle size and generally constant particle number. Increasing particle size is typically due to intake of polymer and/or monomer, but may also be due to particle–particle coalescence if the particles are not sufficiently stabilized. Interval 2 ends when initial monomer droplets are consumed. Interval 3 is essentially a period of polymerization in the particles.

The process of mEP is essentially a subset of EP where the initial droplet size is sufficiently small, and the total interfacial surface area great enough that the surface concentration of the surfactant (the Gibb's excess, Γ) is less than Γ_{cmc}. (At the cmc, the interface is saturated with surfactant, and the aqueous-phase surfactant solution concentration is at the cmc.) Therefore, no micelles are present in the aqueous phase. It is then believed that polymerization of monomer to polymer is forced to occur only in the mechanically formed initial particles.

3.4 PRECIPITATION POLYMERIZATION

There are few water-soluble dimethylsiloxane materials suitable as monomers. Dimethylsiloxane oligomer diols, where n equals 1–4, are soluble in water but are unstable and readily undergo condensation. They are thus not available commercially. However, dimethylsilanediol may be generated in situ by hydrolysis of dichlorodimethylsilane added, or slowly fed, to the water phase. The use of dichlorodimethylsilane to form an emulsion of silicone polymer in water was patented in 1984 [16]. Dichlorodimethylsilane reacts rapidly (exothermic) with water to form highly water-soluble dimethylsilanediol. This then undergoes step-growth condensation in the water phase to form oligomers that progress to form polymer and/or cyclize to form a mixture of cyclosiloxanes. An illustrative study by Palaprat and Ganachaud [17] provides details of the process. It was observed that, with continued reaction, the cyclosiloxanes formed also reacted to form polymer. It is noteworthy that the random condensation of dimethylsilanediol and other oligomers in the water phase would form oligomer diols having n of 4, 5, and 6, which can readily undergo cyclization. This could easily account for the formation of D_4, D_5, and D_6, without the need to invoke the concept of a "backbiting" reaction of the polymer upon itself to form them.

In a similar manner, diacetoxydimethylsilane reacts with water to produce dimethylsilanediol in the water phase and proceeds by PP to form a polydimethylsiloxane emulsion. In the use of either silane as a monomer, the particle size will be dependent on the amount and structure of surfactant and the salt content produced by the process.

3.5   SUSPENSION POLYMERIZATION

Suspension polymerizations of polysiloxane diols, having n greater than approximately 10–20 and typically about 35, may be conducted in anionic surfactant-stabilized emulsions catalyzed by a strongly acidic surfcat. Cationic surfactant-stabilized emulsions catalyzed by a strong base surfcat, formed in situ, can also be conducted.

Studies of the process and mechanism of the acid-catalyzed anionic emulsion were published by Saam and Huebner [18]. The emulsion of this SP study, at 21°C, contained a dimethylsiloxane oligomer diol of n equal to 35, dodecylbenzenesulfonic acid (DBSA), and water. Particle size ranged from 250 to 420 nm. The overall polymerization was considered to be a reversible reaction of silanol–silanol condensation with the elimination of H_2O, resulting in the step-growth of the polymer. At constant temperature, the rate constant was found to be a function of DBSA concentration, and particle size, that is, total interfacial surface area (A). The rate constant increased linearly with increasing A, supporting the mechanism of polymerization occurring at the surface of the particles. The activation energy was determined to be 18 kJ/mol (4.3 kcal/mol). The process was found to be second order in silanol. Equilibrium molecular weight was shown to be inversely related to reaction temperature. It appears that this is due to the increasing vapor pressure of water in the silicone particle (interface) with increasing temperature. The increasing concentration of water molecules, which provided hydroxyl groups to terminate the polymer, consequently reduced equilibrium molecular weight of the polymer. The starting oligomer diol of this study had a molecular weight (M_w) of 5000 g/mol and a polydispersity (M_w/M_n) of 2.9.

A study of a base catalyzed cationic emulsion SP process and mechanism was published in 2000 [19]. A dimethylsiloxane oligomer diol having a molecular weight (M_w) of 6000 g/mol and a narrow M_w/M_n of 1.1–1.5 was used. The particle size was approximately 250 nm, polymerization temperature 40°C, and the surfcat was benzyldimethyldodecylammonium hydroxide. It was observed that D_4, D_5, and D_6 cyclosiloxanes formed early in the process, and it was attributed to intramolecular redistribution (backbiting) reactions. Equilibrium (90%) between polymers and cyclics was observed after 40 hours reaction. During this time, the polymer molecular weight (M_w) increased to 60,000 g/mol and polydispersity, M_w/M_n, broadened to 2.0. With continued reaction up to 50 hours, the polydispersity continued to increase to approximately 3.5. The base-catalyzed cationic emulsion, SP, was determined to be second order in silanol with an activation energy of 34 kJ/mol.

The SP process clearly occurs at the particle surface and therefore small particle size is preferred to maximize reaction rate. The activation energy is considerably less than the EP of cyclosiloxanes (59 kJ/mol), and SP is preferably conducted at lower temperatures to minimize generation of cyclics.

3.6   INTERNAL SUSPENSION POLYMERIZATION

ISP occurs in an oil-in-water emulsion and produces a water-in-silicone-in-water multiple emulsion during polymerization. The surfcat is not water soluble and is dispersed in the silicone phase as aggregates and, to a small degree, adsorbed at the silicone–water interface. Necessarily, a hydrophilic surfactant is present to stabilize the large emulsion particles of the oligomeric siloxane diol. The same starting "monomer" as in SP is most suitable (n = 35). The surfcat is typically a sulfonic acid containing a longer hydrocarbon, such as dinonylnaphthalene sulfonic acid (DNNSA) or didodecylbenzenesulfonic acid (DDBSA), in contrast to dodecylbenzenesulfonic acid (DBSA), typically used in SP. The patented process [20] utilizes emulsions having large particle size (greater than 1 μm). The ISP process mechanism is that of step-growth polymerization by silanol–silanol condensation inside the particle. Larger particles have less interfacial surface area and thus take less surfcat from the internal silicone phase by adsorption at the silicone–water interface. The water molecules produced by the condensation reaction apparently form water particles in the silicone phase. (If the condensation reaction is observed in the unemulsified oligomer diol, the light-scattering colloidal effect can be seen due to the formation of water particles.) It is believed that the locus of the reaction is the silicone phase. In contrast to SP, where reaction rate *decreases* with increasing particle size, the reaction rate in ISP *increases* with increasing particle size. To illustrate, the rate of reaction in ISP using DNNSA in an emulsion having a very large 135 μm particle size was two times faster than that observed in an emulsion having 3 μm particle size [20]. Additionally, the effect of surfcat structure on the polymerization rate was shown by comparing DBSA with DDBSA in emulsions having one micron particle size. After 200 minutes reaction at 23°C, the polymer viscosities had increased from an initial 75 centipoises to 35,000 and 235,000 centipoises, respectively [20]. Thus, the difference in reaction locus of the condensation reaction: the silicone–water interface in SP versus inside the silicone in ISP. This process is useful where it is desirable to produce high molecular weight silicone in large particle size emulsions, starting with a low molecular weight siloxanediol oligomer (easier handling and lower cost)."

3.7 EMULSION POLYMERIZATION

To arrive at a full and current understanding of the mechanism of EP, it is necessary to provide a description of past research and proposed mechanisms, and then the most current findings, just published.

3.7.1 PAST PUBLISHED MECHANISMS

The early work of Hyde [1] and Weyenberg [2] on silicone EP of D_4 had led to a generally held mechanism of reaction at the droplet surface to form water-soluble oligomer siloxane diols, which then diffuse through the aqueous phase into surf-cat micelles, undergo silanol–silanol condensation producing insoluble polymer, and form new particles from the micelles. The original droplets decrease in size and essentially disappear. Consequently, it was believed that the initial mechanical emulsification of D_4 to a small droplet size was important to increase the interfacial surface area to maximize the polymerization rate. Research published by Weyenberg [3] describes the acid-catalyzed EP of pre-emulsified D_4 resulting in an emulsion of polymer particles having a diameter much smaller than the initial cyclosiloxane particles. Pre-emulsification of the D_4 to increase interfacial surface area and higher reaction temperature was utilized to increase the polymerization rate. (The EP of D_3 was also reported where it was not necessary to be pre-emulsified, and EP proceeded readily at 25°C.) Additionally, it was reported that trialkoxysilanes were used in place of cyclosiloxanes in this EP process to produce silsesquioxanes instead of polydimethylsiloxanes. The study of trialkoxysilane EP revealed the formation of small silanol-ended intermediates that condensed to form silsesquioxane colloidal particles (thus, better described as a PP process). No siloxane diol intermediates were reported in the EP of D_4. However, it was suggested that the cyclosiloxane EP mechanism must involve at least two reactions: opening of the cyclosiloxane, and polymer growth by silanol condensation of silanol-ended oligomer intermediates. It was emphasized that it was unknown whether the opening of the cyclosiloxane occurs before or after the transfer of the monomer into the aqueous phase. Even though no position was stated on this key issue, apparently, the hypothesis of the reaction beginning at the cyclosiloxane–water interface was favored, partly due to the very low aqueous solubility of cyclosiloxanes versus the large solubility of siloxane oligomer diols.

In 1982, Zhang et al. published a series on kinetic studies of the cyclosiloxane EP process [5–7]. They concluded that the EP process begins with the opening of the cyclosiloxane on the particle surface to form dimethylsiloxane oligomer silanolates that undergo addition polymerization with cyclosiloxane at the interface. Eventually, cyclosiloxane in the droplet becomes depleted and the condensation of the silanol-ended polymer molecules predominates. Variations of this general hypothesis continue to be recognized by other researchers. Still, confusing and conflicting conclusions are published. In one publication [8], a reaction scheme was given, but it was noted as "incomplete" and that "it is necessary to obtain more details about the localization of the different reactions."

In a subsequent publication [9], a schematic diagram was given to describe the EP process reactions, *"supposed"* to be occurring at the monomer droplet surface. Then the conclusion section claimed that this hypothesis *"seems now well-established."* A third publication [10] *"assumed"* that all reactions were confined to the original cyclosiloxane particles. A conclusion of the same publication, that oligomers of DP greater than 30 withdraw from the interface into the particle interior, strongly conflicts with the findings of Saam and Huebner [18] where the condensation of oligomers of 35 DP proceeds readily to a very high DP, and that the polymerization rate is proportional to the interfacial surface area (i.e., a function of particle diameter). In general, it may thus be said that cyclosiloxane EP is definitely a complex process and that the mechanism is not yet fully understood.

3.7.2 EXPERIMENTAL RESULTS OF PAST PUBLICATIONS

To clarify the mechanism of cyclosiloxane EP, a number of key publications were reviewed and the experimental findings only were compiled (i.e., mechanism proposals were disregarded). These compiled findings were then considered *collectively* to identify a unified mechanism proposal that could provide direction for future research studies to confirm or disprove that hypothesis. The following is a summary of the experimental findings:

Zhang et al. [5–7]—Cyclosiloxane polymerization rates were found to be linear up to approximately 60% conversion, and decrease in the order: $D_3 > D_4 > D_5$. Cyclosiloxane diffusion rates into the aqueous phase were found to decrease in the order: $D_3 > D_4 > D_5$. Cyclosiloxane diffusion rate appears related to polymerization rate. The molecular weight of the polymer being formed was relatively constant until approximately 70% conversion and the polymer dispersity (1.1–1.5) narrow. Beyond this point, polymer molecular weight increases rapidly, the number of polymer molecules rapidly decreases, and dispersity increases to about 2.0. Conversion reached an equilibrium maximum of approximately 93% polymer. Finished emulsion polymer particles were much smaller than the initial coarse dispersion of cyclosiloxane. Therefore, some form of mass transfer occurs in the EP process.

Hémery et al. [8–10]—Three related studies were published by this group. The first study observed EP of D_4 exhibited formation of polymer, D_5, and D_6 from the very beginning of the reaction. The same characteristics of polymer formation, as found by Zhang, were confirmed. A significant experimental finding was that small dimethylsiloxanediols, having n of 1–8 and mostly 2–6, were present as intermediates some time during the EP. These derivatives could have been formed in the aqueous phase.

Gee and coworkers [4,11]—Two relevant patents issued to Gee described a method to produce fine particle size emulsions by EP without the need to pre-emulsify the cyclosiloxane. Insights were provided into the effect of reaction temperature on particle size. It was shown that initial droplet size does not determine polymer particle size in EP.

Barrère et al. [12]—This publication examined both the EP method patented by Gee [4], called the "batch" process by Barrère, and a method of slowly feeding D_4 to the aqueous phase containing surfcat and nonionic surfactant. Unfortunately, the nonionic surfactant selected to study the "batch" process was not in accordance with the principles of the patent. As a result, the selected nonionic surfactant did not cause the size of particles formed to be sufficiently large to minimize the interfacial surface area and stabilize the particles.

Using the "starved-feed" process, the particle size formed was initially quite small and progressively increased during the feed period. Polydispersity of particle size was, and remained, low/narrow. This indicated EP beginning in Interval 1 and rapidly transitioning into Interval 2 when the process became surfactant starved due to very small particles.

Mohoric and Sebenik [13]—Mohoric and Sebenik observed the transport of monomers, in some manner, from the initial droplet into the aqueous phase and new, much smaller, polymer particles formed. Polymer particles formed were relatively constant in size during the EP. (Some variation of particle size was observed, but was probably due to limitations of the DLS instrumentation and the CONTIN software program used.)

Jiang et al. [14]—A mini-emulsion EP process (mEP) was studied. A series of pre-emulsions with varied initial particle sizes was emulsion polymerized at 60°C. Polymerization rate was observed to increase with decreasing pre-emulsion particle size. Figure 3.1 shows the effect of initial particle size on reaction rate.

The increasing polymerization rate observed under conditions of decreasing initial particle size was much greater than a predicted rate based on the reaction at the interface (calculated based on the relationship that the total surface area is inversely proportional to particle diameter, and by the projection of rate from run 5). Additionally,

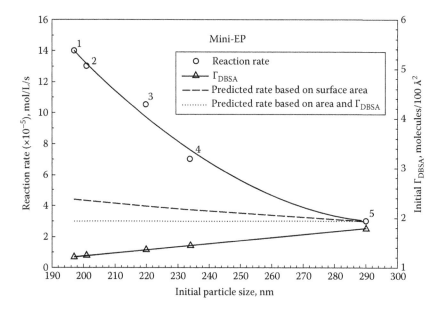

FIGURE 3.1 Effect of initial particle size on reaction rate. Runs 1–5, $[DBSA]_{aq.ph.}$ = 0.0262 $mol/L_{aq.ph.}$ (Plotted based on data taken from Jiang, S. et al., *Polymer*, 51, 4087, 2010, Table 1 and Figure 2.)

the interfacial surface concentration of DBSA surfcat (Γ_{DBSA}) also *decreases* with decreasing particle size and effectively nullifies the effect of increasing surface area. The predicted rate change, due to changing area and Γ_{DBSA}, would then be zero in this series. Thus, increasing reaction rate does not support the hypothesis of the reaction at the particle–water interface. The observed rapidly increasing polymerization rate must then be due to some other effect. An alternative hypothesis is that D_4 is polymerized in the aqueous phase by DBSA under conditions of insufficient diffusion rate of D_4 from the particles into the water phase, that is, D_4 mass transfer limitations.

A second series of experiments was conducted using portions of the same preemulsion, which had half the amount of DBSA and a particle size of 220 nm, each with progressively increased additional DBSA added. Figure 3.2 shows the effect of DBSA concentration on reaction rate. Each DBSA concentration is expressed as mol per liter aqueous phase, calculated from the data of Jiang, which was given in mol per liter emulsion.

Surface tension measurements show that runs 6–12 had initial particles less than saturated with DBSA at the interface, run 13 was saturated with DBSA, and runs 14–16 more than saturated with DBSA. The reaction rates of runs 6–9 progressively increased, runs 9–13 gradually decreased, and runs 13–16 were essentially constant. Runs 6–8 show that the rate increased as a linear function of DBSA concentration, particle size assumed to remain unchanged during the polymerization, and the process appears to be mEP. Runs 14–16 had micelles present and able to produce new

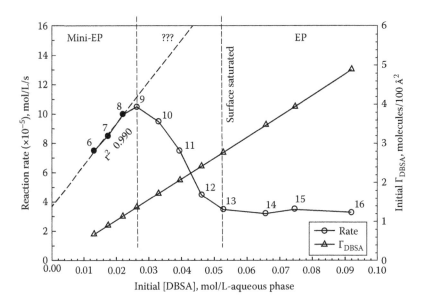

FIGURE 3.2 Effect of concentration of DBSA on reaction rate. Runs 6–16, initial particle size: 220 nm, with initial DBSA concentration recalculated from moles per liter emulsion to moles per liter aqueous phase. The "???" in the graph indicates that the mechanism type is uncertain in this range. (Plotted based on data taken from Jiang, S. et al., *Polymer*, 51, 4087, 2010, Table 1 and Figure 3.)

polymer particles during reaction. The process of runs 13–16 was EP and would be expected to have a final particle size less than the initial 220 nm. Unfortunately, no particle size and surface tension measurements were provided during and after the EP. Runs 9–13 (Figure 3.2, the range labeled "???") appear to indicate a transition from mEP to EP. The mechanism in this range is uncertain due to the lack of particle size data during reaction. See Section 3.8 for further discussion.

3.7.3 A New Mechanism: Coagulative Homogeneous Nucleation

The collective summary of the observed results of cyclosiloxane EP clearly indicates the possibility of EP proceeding in the aqueous phase. The process ultimately is a heterogeneous two-phase reaction. Phase-transfer catalysis has been eliminated [6]. However, no discussion or research has been published with respect to the broader principles of kinetics that have been developed for such reactions. Cox provides a concise overview of the principles of two-phase liquid–liquid reaction kinetics and specifically presents a description of the kinetics of aqueous–organic reactions [22]. This approach provides a means to distinguish between reactions that occur at the interface versus those that occur in the aqueous phase. He shows that the time required for an organic liquid to diffuse through the interface into water is about 10 seconds, and for a reaction to take place at the interface, the reaction half-life must be less than this. Most significantly, if the reaction half-life is considerably greater than this, it occurs in the aqueous phase. Hémery [8] reports a rate constant (k_1) for D_4 EP of 0.085/hour at 80°C and therefore the calculated half-life ($T_{1/2} = \ln 2/k_1$) would be approximately 30,000 seconds. Mohoric [13] provides information to enable calculation of the rate constant to be 0.01270/minute and a half-life of 3275 seconds. Jiang [14] provides information from run 1 to enable calculation of the rate constant to be 0.01543/min and a half-life of 2700 seconds. Such large reaction half-life values, combined with the summary of the observed findings, support the concept of an EP mechanism occurring in the aqueous phase. The following proposed mechanism for emulsion polymerization of cyclosiloxanes is offered to the scientific community for consideration and future research studies.

3.7.3.1 Proposed Mechanism Based on Experimental Results of Past Publications

Figure 3.3 shows the proposed mechanism of cyclosiloxane EP.

D_4 diffuses from the droplet into the aqueous phase. Reaction (Figure 3.3 and Table 3.2) begins by *initiation* in the aqueous phase, where D_4 molecules in solution encounter NOH surfcat molecules and form silanolate ions, which then undergo *Propagation* by direct addition of D_4. As it is consumed, D_4 is supplied to the aqueous phase by diffusion from the D_4 droplets. At some point, the polymer precipitates to form precursor particles. *Termination* may occur at any time by reaction of water with the active silanolate chain end. Precursor particles undergo coagulation to form stable mature particles. *Interfacial condensation* occurs at the particle–water interface of the newly formed particles by reaction of oligomer siloxanediols having various values of n, including large and small oligomers. This contributes to broadening of the polymer molecular weight distribution.

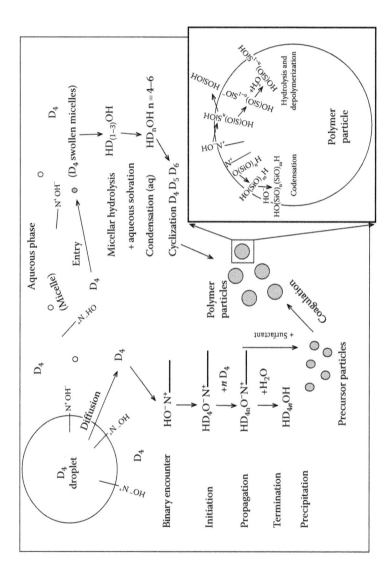

FIGURE 3.3   Schematic illustration of the newly proposed mechanism of cyclosiloxane EP.

TABLE 3.2

Equations of Proposed Cyclosiloxane EP Mechanism

$[N^+ OH^-]_{aqueous\ phase}$ > cmc of NOH; where: N^+ is the surface active quaternary ammonium ion

D_4/water interface

Diffusion: D_4 (droplet) $\rightarrow$ D_4 solution (aqueous phase)

Aqueous phase

A-1. Initiation: $D_4 + N^+OH^- \rightarrow HO(Si(CH_3)_2O)_3 Si(CH_3)_2O^- N^+$

A-2. Propagation (chain growth polymerization): $HO(Si(CH_3)_2O)_3 Si(CH_3)_2O^- N^+ + n\ D_4 \rightarrow$
   $HO(Si(CH_3)_2O)_{4n+3} Si(CH_3)_2O^- N^+$

A-3. Termination: $HO(Si(CH_3)_2O)_{4n+3} Si(CH_3)_2O^- N^+ + H_2O \rightarrow HO(Si(CH_3)_2O)_{4n+4} H + N^+ OH^-$

A-4. Micellar hydrolysis: D_4 (in micelles) $+ H_2O \rightarrow HO(Si(CH_3)_2O)_n H$ n = 1–4 (to aqueous phase)

A-5. Condensation: $2\ HO(Si(CH_3)_2O)_n H$, n =1–4 (random) $\rightarrow H_2O + HO(Si(CH_3)_2O)_n H$, n =3–6
   Cyclization: $HO(Si(CH_3)_2O)_n H$, n = 3–6 $\rightarrow D_3, D_4, D_5, D_6 + H_2O$
   Oligomerization: $\rightarrow HO(Si(CH_3)_2O)_m H$, m > ~8 $\rightarrow$ precipitation/particles form

Polymer particles

I-1. Interfacial hydrolysis and depolymerization:

$HO–(Si(CH_3)_2O)_n Si(CH_3)_2OH + N^+OH^- \rightarrow HO(Si(CH_3)_2O)_{n-1}Si(CH_3)_2O^- N^+ + HOSi(CH_3)_2OH$ (to A-5)

$HO(Si(CH_3)_2O)_{n-1} Si(CH_3)_2O^- N^+ + H_2O \rightarrow + HO(Si(CH_3)_2O)_{n-1} Si(CH_3)_2OH + HO^-$

I-2. Interfacial condensation (step-growth polymerization)

$HO(Si(CH_3)_2O)_{n-1} Si(CH_3)_2O^- N^+ + HO(Si(CH_3)_2O)_m H \rightarrow HO(Si(CH_3)_2O)_{n+m} H + HO^-$

Simultaneously, D_4 also diffuses into micelles and undergoes *micellar hydrolysis* by NOH surfcat in one or more reaction steps to form small oligomer diols (n = 1–3) that diffuse into the water phase. These small oligomer diols then undergo *condensation* and cyclize to form D_3, D_4, D_5, and D_6, or form larger oligomer diols that precipitate to form new particles or enter into existing particles. (This would also contribute to broadening of polymer molecular weight.) Polymer growth can then continue at the particle–water interface by interfacial condensation.

It is also proposed that *interfacial hydrolysis and depolymerization* occurs at the particle–water interface where polymer terminal silanol groups contact the water phase. The terminal Si in the interface is attacked by surfcat hydroxide ion, cleaving the end of the polymer chain to liberate a molecule of $HOSi(CH_3)_2OH$ into the aqueous phase. A molecule of water may then neutralize the silanolate end group, effectively shortening the polymer chain by one unit. Sequential repetition of these reactions results in decreasing polymer molecular weight and also the formation of cyclosiloxanes by aqueous condensation and cyclization of the $HOSi(CH_3)_2OH$ molecules. Alternatively, *interfacial condensation* occurs at the polymer particle surface by reaction of the polymeric silanolate (formed by interfacial hydrolysis) with the terminal silanol of another polymer chain to form one larger polymer chain, regenerating a catalyst hydroxide ion. The concentration of water vapor in the interface, which is a function of the reaction temperature, determines whether interfacial condensation or depolymerization predominates and governs the equilibrium molecular weight of polymer. Thus, it appears unnecessary to invoke "backbiting" as a cause of cyclics formation.

3.7.3.2 Determined Mechanism Based on Heterogeneous Reaction Kinetics

A new publication [21] reports an important study of base-catalyzed EP of dimethyl-cyclosiloxane in cationic emulsion utilizing the principles of two-phase liquid–liquid reaction kinetics [22]. These principles, introduced in Section 3.7.3, provide a method to objectively determine whether the locus of the reaction is at the surface of the cyclosiloxane droplet or in the aqueous phase. There are three types of two-phase aqueous–organic reactions: slow, moderate, and fast. Fast reactions have a half-life less than several seconds and occur at the organic droplet surface. Fast reactions are proportional to the total droplet interfacial surface area and are unaffected by the volume of aqueous phase. Reactions with a half-life substantially greater than about 10 seconds occur in the aqueous phase and are slow or moderate types. There are four distinguishing characteristics of slow and moderate two-phase aqueous–organic heterogeneous reactions:

- First order in reactant.
- Reaction half-life is greater than about 10 seconds.
- Rate is a function of the reactant aqueous solubility at the reaction temperature.
- Absolute reaction rate is proportional to the volume of aqueous phase (i.e., rate per aqueous phase volume is constant).

Moderate type reactions are sufficiently rapid that the reactant concentration in the aqueous phase is less than saturation, its aqueous solubility, due to mass transfer limitation (MTL) from the reactant droplets. Therefore, the reaction rate of moderate type reactions may be increased by decreasing the reactant droplet diameter (i.e., increasing interfacial area) to increase the mass transfer rate into the aqueous phase.

The reported results of this publication are as follows:

- Aqueous solubility of D_4, D_5, and D_6 over the range of 20°C–100°C was determined.
- Reaction rate is a linear function of cyclosiloxane aqueous solubility at the reaction temperature.
- EP conforms to slow/moderate two-phase heterogeneous liquid–liquid reaction characteristics.
- EP of cyclosiloxane occurs in the aqueous phase, but not in micelles.

It was concluded that dimethylyclosiloxane emulsion polymerization conforms to the kinetic characteristics of "slow" to "moderate" two-phase heterogeneous liquid–liquid reactions, where polymerization occurs in the aqueous phase. The rate of polymerization is thus a function of cyclosiloxane aqueous solubility ($c_{D_4}^0$), when not mass transfer limited:

$$\frac{d([D_4])}{dt} = k_1' f_s c_{D_4}^0 \text{ [CTAOH]}.$$

When the emulsion polymerization is mass transfer limited, the rate equation is:

$$\frac{d([D_4])}{dt} = k_1' f_s c_{D_4}^0 \ [CTAOH] \ \frac{(k_L a)}{(k_1 + k_L a)}$$

where a is the initial total droplet surface area. Increasing "a" by increasing the concentration of D_4 droplets or decreasing droplet size by greater emulsification only affects the reaction rate if the conditions of EP are mass transfer limited. The term f_s accounts for undefined factors, apparently due to the presence of surfactant and the process of EP, which are not a part of the basic two-phase heterogeneous reaction kinetics theory. Surfactant and the process of EP appear to enhance cyclosiloxane flux rate into the aqueous phase.

This study concluded that the locus of the reaction is clearly in the aqueous phase and the overall mechanism of cyclosiloxane EP is coagulative homogeneous nucleation. This determined mechanism is the same as that proposed in Section 3.7.3.1. Note that the proposed reactions in the polymer particle were not addressed in the publication, as the focus was on polymerization and particle formation.

3.8   MINI-EMULSION POLYMERIZATION

The mini-emulsion polymerizations of D_4 and F_3 are considered here. Details of a study of mEP of D_4 by Jiang [14] were already discussed in Section 3.7.2.

Barrère et al. [15]—This mEP study used F_3 (trifluoropropylmethylcyclotrisiloxane) in place of D_4 as monomer, didodecyldimethoxyammonium bromide (DDDAB), and very low amounts of NaOH as catalyst. Particle size was constant during the entire mEP, and the initial pre-emulsion particle size was sufficiently small to adsorb most surfactant on the particle surface and therefore no surfactant micelles were in the aqueous phase. F_3 is a cyclosiloxane with a very strained ring and therefore is highly reactive in comparison to D_4. Percent conversion to polymer was >95% within 15 hours at 40°C. Time for the completion of the polymerization "varied from less than a second to several days depending on experimental conditions." With continued reaction time, conversion to polymer gradually decreased to about 45%. (In bulk polymerization, percent conversion at equilibrium is characteristically only 10–20.)

The aqueous solubility of F_3 is unknown, but it would be expected to be much less than D_4. The factors of extremely low aqueous solubility, high monomer reactivity, and a possible reaction half-life less than 10 seconds indicate this mEP is a fast type heterogeneous liquid–liquid reaction and occurs at the F_3 particle surface. Whether the reaction, under conditions requiring a very long time, becomes a slow- or moderate-type reaction in the aqueous phase remains an open question.

If the proposed mechanism of Figure 3.3 is valid for dimethylcyclosiloxane EP, it is likely this scheme may also hold for mini-emulsion polymerization. A difference being that precipitated polymer would be absorbed into the initial D_4 particles due to insufficient surfcat/surfactant to stabilize potential new polymer pre-cursor particles. Based on the observations of Jiang [14], it appears there may be a period of overlap (Section 3.7.2) or uncertainty between EP or mEP, where new particle formation and polymer entry into initial D_4 particles occurs simultaneously. It seems that surf-cat/surfactant on the initial D_4 particle interface can be a source to stabilize new particles by desorption and diffusion to the polymer aggregates. Formation of new particles increases total interfacial surface area and, by adsorption, removes DBSA from the aqueous phase. It appears the variation of reaction rate in Figure 3.2 may be due to changes in the aqueous solution concentration of DBSA when it becomes less than the cmc. Thus, it may be that pre-emulsification for mEP may require an initial particle size considerably less than that needed to have the surface less than saturated with surfcat. This, of course, would also require the monomer to have some finite solubility in water. Consequently, mEP of D_4 would not be characterized as a process in which each particle functions as an isolated "reactor."

3.9   CONCLUDING REMARKS

The concept of EP as a two-phase, liquid–liquid heterogeneous reaction in the aqueous phase has been under study in our laboratories for a number of years. The same mechanism appears applicable in EP of cyclosiloxanes by acid catalysis in anionic emulsions. It is hoped to publish at least some of these results in the future. This review, along with the recent publication [21], establishes the mechanism and locus of emulsion polymerization. Investigation of this hypothesis by other researchers is encouraged to further prove its validity.

In conclusion, this chapter has presented an overview of the various colloidal chemistries and methods of ionic polymerization of dimethylcyclosiloxane in water to produce polydimethylsiloxane emulsions of commercial interest. It is hoped that the description of the relationship of the processes and their mechanisms will be of benefit to those wishing to develop aqueous emulsions useful to deliver the various and unique beneficial properties of silicone dispersions.

REFERENCES

1. Hyde, J., Wehrly, J. Polymerization of organopolysiloxanes in aqueous emulsion. U.S. Patent 2,891,920, 1959.
2. Findlay, D., Weyenberg, D. Method of polymerizing siloxanes and silcarbanes in emulsion by using a surface active sulfonic acid catalyst. U.S. Patent 3,294,725, 1966.
3. Weyenberg, D., Findlay, D., Cekada, J., Bey, A. *Journal of Polymer Science Part C* 1969, *27*, 27–34.
4. Gee, R. Method for making polysiloxane emulsions. U.S. Patent 6,316,541, 2001.
5. Zhang, X., Liu, X., Dai, D. *Gaofenzi Tongzun* 1982, *2*, 154–157.
6. Zhang, X., Yang, Y., Liu, S. *Polymer Communications* 1989, *4*, 266–270.
7. Zhang, X., Yang, Y., Liu, X. *Polymer Communications* 1982, *4*, 310–313.
8. DeGunzbourg, A., Favier, J., Hémery, P. *Polymer International* 1994, *35*, 179–188.
9. DeGunzbourg, A., Maisonnier, S., Favier, J., Maitre, C., Masure, M., Hémery, P. *Macromolecular Symposium* 1998, *132*, 359–370.
10. Barrère, M., Ganachaud, F., Bendejacq, D., Dourges, M., Maitre, C., Hémery, P. *Polymer* 2001, *42*, 7239–7246.
11. Gee, R. Anionic and cationic silicone emulsions. U.S. Patent 6,465,568, 2002.
12. Barrère, M., Capitao da Silva, S., Balic, R., Ganachaud, F. *Langmuir* 2002, *18*, 941–944.
13. Mohoric, I., Sebenik, U. *Polymer* 2011, *52*, 1234–1240.

14. Jiang, S., Qiu, T., Li, X. *Polymer* 2010, *51*, 4087–4094.
15. Barrère, M., Maitre, M., Dourges, M., Hémery, P. *Macromolecules* 2001, *34*, 7276–7280.
16. Terae, N., Abe, A. Method for the preparation of an aqueous emulsion of silicone. U.S. Patent 4,465,849, 1984.
17. Palaprat, G., Ganachaud, F. *ComptesRendusChemie* 2003, *6*, 1385–1392.
18. Saam, J., Huebner, D. *Journal of Polymer Science Part A: Polymer Chemistry* 1982, *20*, 3351–3368.
19. Barrère, M., Maitre, C., Ganachaud, F., Hémery, P. *Macromolecular Symposium* 2000, *151*, 359–364.
20. Gee, R., Wrolson, B. Emulsion containing silicone particles of large size. U.S. Patent 6,235,834, 2001.
21. Gee, R. *Colloids and Surfaces A: Physicochemical and Engineering Aspects* 2015, *481*, 297–306.
22. Cox, B.G. *Modern Liquid Phase Kinetics*. Oxford University Press, New York, 1994.

4 Silicone Water-Based Elastomers

Donald T. Liles

CONTENTS

4.1 INTRODUCTION

Silicone water-based elastomers, sometimes referred to as SWBE or silicone rubber latex or silicone latex, are aqueous dispersions or emulsions of silicone polymers that deliver silicone elastomers, usually in the form of films, upon removal of water [1–4]. Figure 4.1 shows a dried film of silicone water-based elastomer. These dispersions or emulsions can exist in several different forms, and they are used in a variety of applications, such as coatings, additives for coatings and inks, sealants, and binders for ceramics or other substances. In the area of coatings, silicone water-based elastomers are used as architectural coatings, air barrier coatings, fabric coatings, and release coatings. In all of these types of coatings, 100% of the binder is usually silicone and it is in the form of an elastomer, although there are some instances when silicone water-based elastomers are combined with organic polymer emulsions or dispersions to achieve particular properties resulting from the combined silicone-organic compositions.

Silicone elastomers result from cross-linking siloxane polymers, usually polydimethylsiloxane, PDMS [5]. Similarly, silicone water-based elastomers are derived from siloxane polymers such as PDMS and will have undergone or will undergo some type of cross-linking reaction during their preparation or during their application, respectively, in order to form silicone elastomers. Thus silicone water-based elastomers can be classified as precured emulsions or postcuring emulsions of silicone polymers, depending upon when cross-linking occurs.

Precured silicone emulsions consist of aqueous emulsions or dispersions of silicone elastomer particles that began as emulsions of reactive siloxane polymers. At some point after emulsification, the reactive silicone polymers undergo some type of cross-linking reaction within each emulsion particle to form silicone elastomer, and the final product becomes a dispersion of silicone rubber particles. Most often the pre-cross-linking reactions are carried out at ambient temperatures, but in some

FIGURE 4.1   Silicone water-based elastomer.

cases, heat can be applied to either speed up cross-linking reactions or to trigger a particular type of cross-linking reaction.

Removal of water from such a dispersion causes the silicone elastomeric particles to coalesce and form a coherent film. Water removal normally occurs at ambient temperatures, but in some cases elevated temperatures are used to speed up the drying process. Since the polymer chains within each particle are cross-linked, interparticle diffusion of polymer is nonexistent or limited, so the original integrity of the particles is maintained within the silicone elastomeric film indefinitely. Thus, silicone elastomeric films derived from precured silicone water-based elastomers could be construed as a collection of silicone rubber balls stuck together.

In postcuring silicone water-based elastomers, cross-linking occurs after the removal of water. These emulsions are available as either one-part or two-part system, and water removal takes place either under ambient conditions or with heat. In addition, some type of triggering mechanism to initiate cross-linking can be used, such as exposure to high energy radiation or to heat.

Two-part, postcuring silicone water-based elastomers have been commercially available for many decades. Such compositions are referred to as silicone release coatings or silicone paper coatings. They constitute a significant portion of the entire silicone release coating industry. Besides water-based silicone release coatings, there exist solventless silicone release coatings that contain no water or solvents. Both water-based and solventless silicone coatings are applied to paper and/or plastic films to provide release for a variety of substances. Examples of these coatings are release liners for pressure sensitive adhesives (PSA) and various polymers such as in prepregs, and release applications for foods such as in parchment paper. These silicone water-based elastomers are not a subject of this chapter, and the reader is directed to several treatises on release coatings for additional study [6,7].

This chapter describes silicone water-based elastomers with emphasis on precured emulsions. Also covered are their preparation, their properties, and some of their commercial applications.

4.2 PREPARATION

As silicone water-based elastomers, SWBE, are aqueous emulsions of silicone polymers that undergo cross-linking reactions to form silicone elastomers, their preparation begins with emulsions of reactive silicone polymers. Such emulsions of silicone polymers can, in principle, be based on any polydiorganosiloxane, the most common involve polydimethylsiloxane (PDMS). Figure 4.2 shows the structure of PDMS. The preparation of aqueous emulsions of PDMS is a vast subject in itself;

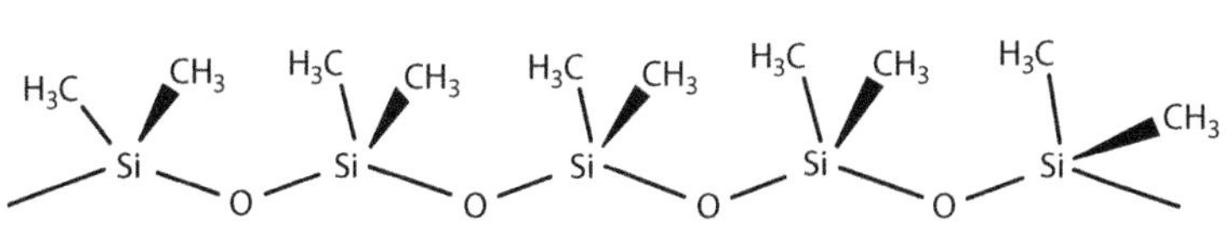

FIGURE 4.2   Structure of poly(dimethylsiloxane) (PDMS).

the most common routes to such emulsions are by emulsion polymerization of low molecular weight siloxane oligomers or polymers, or the direct emulsification of siloxane polymers. These two methods are briefly described, followed by discussions on cross-linking.

4.2.1 EMULSIONS OF PDMS

4.2.1.1 Emulsion Polymerization

In 1959, Hyde and Wherly reported the synthesis of high molecular weight PDMS aqueous emulsions by polymerizing cyclic PDMS oligomers such as D_4 or D_5 using strong mineral acids as polymerization catalysts [8]. (For an explanation of silicone "shorthand" nomenclature, i.e., D_4, see Chapter 1.) Although the compositions by the method of Hyde and Wherly were indeed aqueous emulsions of PDMS, they had shortcomings, such as low silicone content and poor stability, which prevented them from commercialization. A breakthrough came in 1966 when Findlay and Weyenberg reported that stable aqueous emulsions of high molecular weight PDMS could be made from cyclic dimethylsiloxane oligomers, water, and sulfonic acid catalysts that were surface active [9].

Dodecylbenzene sulfonic acid, DBSA, was the catalyst of choice and it remains the catalyst of choice to this day for preparing anionically stabilized, aqueous emulsions of high molecular weight PDMS [10]. In these emulsions, DBSA functions as both a polymerization catalyst and a surfactant. Hydroxyl-terminated linear siloxane oligomers can also be used in these emulsion polymerizations; in this case, the process is more accurately described as a suspension polymerization [11]. Emulsion polymerizations of siloxanes is the topic of Chapter 3.

The compositions prepared by the method of Findlay and Weyenberg consist of aqueous, anionically stabilized, oil-in-water (O/W) silicone emulsions of OH-terminated, high molecular weight siloxane polymers. Concentration of silicone can vary from about 35% by weight to about 70% by weight. Particle size of the emulsion can vary from about 200 to 1000 nm. As particle size becomes smaller, viscosity of these emulsions usually increases, so a trade-off between particle size and concentration becomes necessary. Thus, if a higher concentration is desired, the emulsion is prepared with a relatively larger particle size, and vice versa. Particle size can be controlled to a certain extent by varying conditions during the emulsion polymerization.

Molecular weight of the siloxane polymer of these emulsions is governed by the emulsion polymerization process, and it can be controlled to a fairly good extent by varying the conditions during polymerization. Since molecular weight has an impact on the mechanical properties of the final elastomer, emulsion polymers are designed to possess a certain target molecular weight or a certain molecular weight range. The optimum molecular weight of the siloxane polymer is most often determined by experiment, which usually involves preparing silicone water-based elastomer films and measuring their mechanical properties. It is not normally disclosed by the silicone water-based elastomer manufacturers, but suffice it to say that molecular weights above 10,000 Da are usually preferred.

4.2.1.2   Polymer Emulsification

It is also possible to prepare silicone water-based elastomers from emulsions of reactive siloxanes that have been obtained by direct emulsification of such polymers. Various methods for emulsifying siloxane polymers directly have been described in the patent literature [12–14]. Basically, the process involves subjecting mixtures of silicone polymer, surfactant, and water to high shear, followed by dilution with additional water.

In the case of direct polymer emulsification, molecular weight and hence viscosity of these polymers is usually lower compared to those polymers obtained by emulsion polymerization. This is because commercially viable processes for producing aqueous emulsions having all the necessary requirements for silicone water-based elastomers by direct emulsification of very high viscosity siloxane polymers have not matured enough at this time. Such lower molecular weight polymers in emulsion could be construed as being inferior to those polymers in emulsions obtained by emulsion polymerization. However, in many applications, performance of the final silicone water-based elastomers prepared from polymer emulsions by direct emulsification is adequate.

Particle size of polymer emulsions obtained by direct emulsification of siloxane polymers is usually larger than that of polymer emulsions obtained by emulsion polymerization. Since emulsion particle size in water-based coatings can have an impact on film quality and integrity (smaller is better), one might be led to believe that silicone water-based elastomers obtained via directly prepared polymer emulsions are inferior to those obtained via emulsion polymerization. In some cases, this is true, but in other cases it matters not as silicone water-based elastomers obtained via direct polymer emulsification are adequate for their intended applications.

A chief advantage of siloxane polymer emulsions prepared via direct emulsification over siloxane polymers obtained via emulsion polymerization is the practically unlimited choice of surfactant. Silicone water-based elastomers prepared via emulsion polymerization are restricted to anionic surfactants based on sulfate or sulfonate. In contrast, silicone water-based elastomers prepared via direct polymer emulsification can be made from all types of surfactants: anionic, cationic, nonionic, or amphoteric. In certain applications, choice of surfactant can have certain advantages. For example, it can be challenging to formulate ionic emulsions so that they are freeze/thaw stable, whereas nonionic emulsions are inherently freeze/thaw stable. It has also been suspected that the presence of sulfonate and sulfate in silicone water-based elastomeric films contributes to a decrease in weatherability and thermal stability properties compared to neat silicone elastomers.

Choice of reactive siloxanes for preparing silicone water-based elastomers via direct polymer emulsification is also much broader than it is with emulsion polymerization. While emulsion polymerization leads to siloxane polymers having primarily SiOH groups as the reactive component, direct polymer emulsification can in theory utilize any reactive functional group attached to siloxane polymers. The fact that many reactive functional groups used to vulcanize silicone polymers are reactive toward water limits this list substantially; however, the choices of reactive functional groups applicable for cross-linking silicone polymers obtained by direct emulsification are still more than what is applicable for the emulsion polymerization route.

4.2.2 CROSS-LINKING

Cross-linking transforms many polymers into elastomers and the same is true for emulsions of siloxane polymers. Cross-linking essentially connects polymer chains together via multiple linkages and such connections usually involve covalent chemical bonds, although not exclusively. Electrostatic bonds can also lead to cross-linking among polymers. Silicone water-based elastomers are invariably cross-linked via covalent chemical bonds either through direct siloxane chemical reactions (reactions directed at the silicon atom) or through organic functional groups attached to siloxane polymers.

As previously mentioned, cross-linking in silicone water-based elastomers can occur either prior to or after removal of water (or both). Most of the silicone water-based elastomers used commercially are cross-linked prior to removal of water and hence they are referred to as precured or pre-cross-linked emulsions. Postcuring silicone water-based elastomers, that is, those siloxane emulsions that cure or cross-link after removal of water exist, but the technology is considered to be in its infancy when compared to precured siloxane emulsions, with the exception of those two-part siloxane emulsions that are referred to as release coatings or paper coatings. Both precured and postcuring silicone water-based elastomers are discussed separately.

4.2.2.1 Precured Silicone Water-Based Elastomers

Axon is credited with making the first silicone water-based elastomer in 1962, referred to at the time as a self-curing silicone latex [15]. However, his discovery was neither patented nor disclosed in the open literature. Axon's latex consisted of an aqueous emulsion of hydroxyl terminated PDMS, a pre-hydrolyzed silane cross-linker such as methyltrimethoxysilane (MTM), and a tin (Sn) catalyst such as dibutyltindilaurate (DBTDL). This composition formed an elastomeric film of silicone rubber upon removal of water.

After Axon's discovery, steady progress on the science and technology of silicone water-based elastomers took place, but almost 20 years would pass before silicone water-based elastomers became a commercial reality. The discovery of stable, aqueous emulsions of high molecular weight siloxane polymers by Findlay and Weyenberg in 1966 paved the way toward viable silicone water-based elastomers. In one of their patent examples, these two investigators actually disclosed a silicone water-based elastomer when they emulsion copolymerized a silane, such as MTM, along with a cyclic oligomeric dimethylsiloxane, although they did not describe it as such [9]. The investigators reported that the emulsion formed an elastomeric film upon removal of water.

At about the same time frame, Cekada disclosed silicone emulsions that formed substantial silicone rubber films upon evaporation of water; such films consisted of emulsions of high molecular weight SiOH terminated PDMS, colloidal silsesquioxanes, and a Sn catalyst [16]. Colloidal silsesquioxanes are discussed in more detail in Chapter 7. In some of his examples, Cekada included a silane cross-linker, such as MTM, in the aqueous composition. At the time, Cekada et al. did not understand how their emulsions reacted to form silicone rubber films upon removal of water, but the prevailing theory was that crosslinking reactions occurred after removal of water.

It would be a little more than 10 years after these researchers' discoveries that commercially viable silicone latex products finally arrived at about 1980 [2]. Not long after that, Saam advanced a plausible mechanism for the chemistry involved in silicone water-based elastomers that was based on a precured phenomenon [1]. Saam proposed that cross-linking starts as soon as the silicone water-based elastomer components are combined, and this results in precured or pre-cross-linked particles of silicone polymer, or in other words, aqueous dispersions of silicone rubber. Removal of water from such aqueous dispersions results in coalescence of the particles to form coherent silicone elastomeric films. Gone was the notion that cross-linking reagents such as alkoxysilanes were preserved in the latex until removal of water, which could actually be years away from when the latex components were first combined.

Stein et al. studied similar emulsions using ^{29}Si NMR and concluded that cross-linking occurred prior to removal of water [17]. Thus Saam's precured explanation for describing how silicone water-based elastomers actually work has become the generally accepted mechanism behind their chemistry. These silicone water-based elastomers, having attributes that were suitable for certain commercial applications such as coatings, had finally come of age.

Such silicone water-based elastomer compositions consisted mainly of aqueous emulsions of high molecular weight PDMS reacted with a cross-linker and a Sn IV catalyst. Examples of Sn IV catalysts are dibutyltindilaurate or dimethyltindineodecanoate. The cross-linker can be practically any multifunctional silanol compound or precursor, such as a silane like $CH_3Si(OCH_3)_3$ or $Si(OCH_2CH_3)_4$. Usually the catalyst and the cross-linker are added to the PDMS emulsion and the emulsion is allowed to age for a certain experimentally predetermined period of time. During this period, which can be anywhere from several hours to several weeks, the cross-linking processes take place.

The cross-linking processes that occur in silicone water-based elastomers involve migration of catalyst and cross-linker from the aqueous phase into the polymer phase (polymer particles), followed by chemical reaction—condensation—with the silanol functional polymer chain ends, see, Figure 4.3. It is plausible that some hydrolysis and perhaps some condensation occurs with certain silanol precursors, such as alkoxysilanes, prior to their migration into the polymer particles. The extent of cross-linking can easily be monitored with respect to time by casting films from emulsions and measuring such properties as modulus, tensile strength, or swelling behavior in a solvent such as heptane. When fully cross-linked, silicone water-based elastomers will have a gel content of 80%–90% and a volume swell in heptane of 1000%–1500%.

As mentioned, the cross-linker for silicone water-based elastomers prepared with Sn (IV) catalysts can be practically any multifunctional silanol compound or a precursor thereof. In theory, any silane or siloxane having hydrolyzable groups and containing a functionality greater than two can function as a cross-linker for silicone water-based elastomers. The most common cross-linker is alkoxysilanes (Si–OR), but examples have been reported that use other functionalities attached to silicon: silicon hydride (Si–H), amino (Si–NR$_2$), amido (Si–NC(O)R), alkenoxy (Si–O=CR$_2$), acetoxy

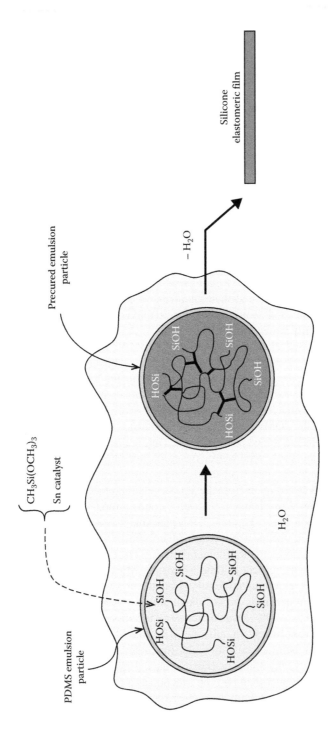

FIGURE 4.3 Silicone water-based elastomer preparation. Starting from silicone emulsion polymer, cross-linker and catalyst are added. After a sufficient ripening period, removal of water produces a film of silicone rubber. Catalyst and cross-linker are believed to migrate into polymer particles where they react with SiOH functional polymer chains. Cross-linker (depicted in the drawing as bold dark lines) can be practically any multifunctional silanol compound or precursor, such as alkoxysilane. Catalyst is normally dialkyltindicarboxylates such as dibutyltindilaurate or stannous carboxylic esters such as stannous octoate. Level of cross-linker is typically 0.5–1.0 percent based on polymer weight, while level of catalyst is 0.2–0.5 percent based on polymer weight. The precured particles are expected to contain residual silanol (SiOH) groups.

(Si–O$_2$CCH$_3$), aminoxy (Si–ONR$_2$), and oximo (Si–N=CR$_2$) [18–21]. In some of these cases, the leaving groups on silicon are reactive enough so a catalyst is not required.

In the presence of a Sn IV catalyst, cross-linked silicone emulsion polymers can also be formed in the absence of any of these types of cross-linkers if silica or a silsesquioxane (silicone resin) is present [16,22]. In this case, silicic acid, Si(OH)$_4$, or polysilicic acids (or SiOH functional silsesquioxane) are believed to be the cross-linker; it is known that small amounts of silicic acid or polysilicic acids are present in aqueous systems that contain silica [23].

Once formed and fully reacted, precured silicone water-based elastomer emulsions that contain a Sn IV catalyst and silica produce elastomers having diminished mechanical properties with respect to emulsion age. In other words, both tensile strength and elongation of elastomers from these precured emulsions are lower from aged emulsions compared to freshly prepared emulsions. The reason for this is believed to be due to the Sn IV catalyst, which is always active in the emulsion, and it contributes to possible polymer rearrangement and continued cross-linking of the siloxane polymers. Put another way, cross-link density of the silicone elastomer within the particles increases over time. In emulsions that contain silica, there is essentially an infinite supply of cross-linker, Si(OH)$_4$, available for continued cross-linking.

Cured, dry silicone water-based elastomers prepared with Sn IV catalysts also show instability toward water in certain circumstances. Stein et al. used stress relaxation techniques to study the effects of water on cured, dried silicone rubber films from silicone water-based elastomer emulsions and concluded that the polymeric network underwent siloxane bond rearrangements that were accelerated by the presence of Sn IV catalyst and water [24]. Indeed freshly prepared silicone rubber films from silicone water-based elastomer emulsions prepared with a Sn IV catalyst become weak and tacky upon boiling in water for several hours, which is indicative of the siloxane polymer experiencing reversion, that is, depolymerization. It is suspected that the shelf life phenomenon described here and the effects of water on cured, dried silicone elastomeric films from silicone water-based elastomers that contain Sn IV catalyst are related.

One method that overcomes both the shelf life property drift and poor resistance to water in silicone water-based elastomers is to substitute a Sn II catalyst for the Sn IV catalyst. Examples of Sn II catalysts are stannous octoate (stannous-*bis*-2-ethylhexanoate) or stannous oleate. An interesting property of the stannous catalysts is that they are susceptible to hydrolysis and hence deactivation by water, particularly at pH extremes. Stannous catalysts are also not organotin compounds as they contain no tin–carbon bonds. Hydrolysis of stannous catalysts forms the carboxylic acid (ligands) and eventually stannic oxide, SnO$_2$ [25]. Silicone water-based elastomers prepared with Sn II catalysts have significantly improved shelf lives over those emulsions prepared with Sn IV catalysts and the cured elastomeric films from such emulsions show no ill effects upon exposure to water [26].

4.2.2.2 Cross-Linking without Sn Catalysis

Tin and its derivatives have come under increased regulatory scrutiny, perhaps beginning with the activity centered on tributyltin oxide in 1999 [27]. Although organotin compounds such as dibutyltindilaurate are widely used in products in the developed world as stabilizers for polyvinylchloride and as catalysts for polyurethanes, there is natural concern that organotin compounds can enter and accumulate in the environment and in the food supply [28–30].

In any event, many end users and suppliers would rather not have tin in their products. Although it is very convenient to use Sn catalysts in silicone water-based elastomers, it is possible to prepare silicone water-based elastomers without Sn.

Trialkoxysilanes or tetraalkoxysilanes, when emulsion copolymerized with siloxane oligomers, leads to cross-linked or precured emulsions that form silicone elastomers upon removal of water without Sn catalysts [31–33]. Alkali silicates such as sodium silicate can also function as a cross-linker for hydroxyl functional siloxane emulsions. Saam reported a precured emulsion that was cross-linked with sodium silicate without a Sn catalyst [34]. However, the emulsion required a long gestation period of several months in order for it to produce an elastomer. This process could be shortened to several days with a dialkyltindicarboxylate catalyst. Siliconates such as sodium methylsiliconate ($MeSi(OH)_2ONa$) are also capable of cross-linking siloxane polymer emulsions (without a Sn catalyst) to form silicone elastomeric films upon removal of water [35]. In this case, the ripening period was on the order of 5–7 days.

Other methods for making precured emulsions of silicone water-based elastomers without Sn involve emulsifying siloxane polymers having functional groups that are capable of reacting to form cross-linked siloxane polymers. Free radicals generated either photochemically, thermally, or by high energy radiation can be used to cross-link emulsion copolymers of PDMS that contain vinyl or other unsaturated moieties [36]. Silicon hydrides are capable of reacting with unsaturated groups such as vinyl or allyl, in the presence of a hydrosilylation catalyst such as platinum [37]. These reactions can be utilized to cross-link siloxane polymers in emulsions to form silicone water-based elastomers [38].

A novel cross-linked silicone water-based elastomer that cures via Michael reactions is described [39]. Another example of a silicone water-based elastomer that cures via Michael reaction involving amino-functional siloxanes and diacrylate or polyacrylate cross-linkers is also described [40].

Finally, some of the reactive silanes or siloxanes previously described in Section 4.2.2.1, such as silicon hydride (Si–H), amino (Si–NR$_2$), amido (Si–NC(O)R), alkenoxy (Si–O=CR$_2$), acetoxy (Si–O$_2$CCH$_3$), aminoxy (Si–ONR$_2$), and oximo (Si–N=CR$_2$) are capable of cross-linking siloxane polymers in emulsion without the use of Sn catalysts. It should be stated that to this day, the Sn cured emulsions deliver superior mechanical strengths over all of the other silicone water-based elastomers.

4.2.2.3   Postcuring Silicone Water-Based Elastomers

Two-part, post-cross-linking, aqueous silicone emulsions represent a commercially significant field of silicone products used to provide silicone release coatings for industries, such as PSAs, labels, polymers, and foods. As stated before, these compositions are usually called silicone release coatings and are not referred to as silicone water-based elastomers, although technically they are. These silicone release coating emulsions are not a subject of this chapter and references to them have been given [6,7]. Aside from compositions that are considered aqueous silicone release coatings, there are two-part aqueous emulsions that fit in to the classification of silicone water-based elastomers, however. In these emulsions, the polymers used would be more in line with those polymers used in precured silicone water-based elastomers and they would perhaps contain other ingredients, such as reinforcing fillers. For example, in silicone release coatings, polymer molecular weights (in both aqueous and nonaqueous coatings) are significantly lower than those polymers used in silicone water-based elastomer.

Two-part, post-cross-linking silicone water-based elastomers are typically prepared from reactive polymers that cross-link via hydrosilylation involving SiH and Si–CH=CH$_2$ functional polymers and a platinum catalyst. In this manner, either the Pt catalyst is kept apart from the functional polymers, or the SiH functional polymer is kept apart from the Si–CH=CH$_2$ polymer and Pt catalyst until just before use.

One-part, post-cross-linking silicone water-based elastomers are even scarcer than two-part, post-cross-linking silicone water-based elastomers as they are technically much more difficult to design and formulate. In order for one-part, postcuring silicone water-based elastomer compositions to be successful, reactive functional groups must be preserved in emulsions for long periods of time, then they must be able to react upon removal of water. A post-cross-linking silicone water-based elastomer has been reported that cures by air oxidation [41]. Another example of a post-cross-linking silicone water-based elastomer is a composition based on PDMS having acryl functionality and cross-links upon exposure to ambient sunlight [42]. In any event, there is much room for continued development of one-part, post-cross-linking silicone water-based elastomers as their state of development and understanding is quite limited at this time. There could be significant advantages of these types of compositions, such as improved strength and improved weatherability.

4.2.3 Fillers

4.2.3.1 Reinforcing Fillers

Silicone rubber is inherently very weak and in order for it to have any appreciable mechanical properties, it requires reinforcing fillers, such as silica or silicone resins. The same is true for silicone water-based elastomers. Although it is possible to add solid forms of reinforcing fillers to silicone water-based elastomers, this method can have complications. In many cases, surfactant in silicone water-based elastomers can absorb onto the new surfaces resulting in a destabilization of the silicone elastomer particles. The preferred method for adding reinforcing fillers to silicone water-based elastomers is to first disperse them in water either with or without added surfactant, then add these dispersions to the silicone emulsion.

Aqueous dispersions of silica, colloidal silica, or silica sols, which are commercially available from a number of manufacturers, are the preferred form of silica for reinforcing silicone water-based elastomers. These silica sols are produced such that they contain usually spherical particles of silica having particle sizes that range from a low of 4–5 nm to a high of about 50 nm. As would be expected, the smaller particle size sols have lower concentrations, on the order of about 15%, while the larger particle size sols are more concentrated, up to about 50%. Colloidal silica has been reviewed by Iler and more recently by Roberts [23,43].

Fumed silica, also known as pyrogenic silica, is widely used to reinforce neat silicone elastomers, and it too can be used to reinforce silicone water-based elastomers. In this case, it is preferred to use aqueous, dispersed fumed silica such as Cabosperse® dispersions supplied by Cabot Corporation. These silicas differ from typical colloidal silicas in that they are structured (groups of primary spherical particles fused together) while most colloidal silicas are spherical, individual particles. Reinforcing silicone water-based elastomers with aqueous colloids or dispersions of silica is easy and consumes very little energy, as the silica colloids or dispersions are simply blended with the aqueous silicone precured emulsion using low shear. This contrasts with incorporating reinforcing filler into neat silicone polymer, whereby it is a very highly energy-intensive process that requires high shear.

The type and amount of silica for reinforcing silicone water-based elastomers will have a profound effect on mechanical properties of the cured, dried elastomeric films. This is discussed in more detail in Section 4.4.5.

Silicone resins can also be used to reinforce silicone water-based elastomers. As previously mentioned, colloidal silsesquioxanes as prepared by Cekada can be used for reinforcement [16]. Liu discloses a method for reinforcing silicone water-based emulsions with MQ resin [44].

Fibers can also be used in silicone water-based elastomers to achieve certain results. For example, a silicone water-based elastomer foam is described that uses glass fibers for reinforcement [45]. An electrically conducive silicone water-based elastomer is described that uses metal coated carbon fibers [46].

4.2.3.2 Nonreinforcing Fillers

Nonreinforcing fillers can also be used with silicone water-based elastomers. In this case, their primary use would be as extending fillers to either lower the cost or increase the solids content (or both) of the final composition. Calcium carbonate is perhaps the most widely used nonreinforcing filler for silicone water-based elastomers, but other fillers, such as ground mica, aluminia, talc, kalonite, calcined clay, carbon black, barytes, chalk, ground marble, zinc oxide, and aluminum trihydrate, can also be used. It should be realized that with some of these fillers, dispersants may be required to keep the filler particles from agglomerating after they have been combined with the precured silicone emulsions. This is normally not required when silica sols or aqueous dispersions of silica described here are used as they are anionically stabilized and hence they remain as discrete particles prior to removal of water.

4.2.3.3 Other Ingredients

Many water-based polymeric coatings, latexes, paints, and inks will contain a number of extra ingredients beyond the basic polymer dispersion or emulsion, which constitutes the coating binder. Such ingredients include, but are not limited to, fillers, pigments and their dispersants, coalescing aids, additional surfactants, wetting agents, flow control additives, thickeners and other rheology control additives, adhesion promoters, defoamers, additives for pH control, freeze/thaw stabilizers, UV stabilizers, mildewcides, and in-can preservatives. Each of these additives can be important to the overall success of the coating as it solves a particular problem or shortcoming with the coating.

Silicone water-based elastomeric coatings will also contain many of these additives, with perhaps the exception of UV stabilizers and coalescing aids. PDMS, being transparent to UV radiation, is well known to be resistant to the effects of UV radiation from the sun [47]. With its very low T_g, on the order of $-123°C$, PDMS needs no coalescing solvents to form films from aqueous dispersions. Perhaps some of the other additives listed here can perform a certain function in silicone water-based elastomers, but their use depends mainly upon the particular application. The most common additives used in silicone water-based elastomer coatings besides reinforcing filler and pigments are thickeners and other rheology control additives. In some cases, defoamers, adhesion promoters, and additives for pH control are also incorporated.

In most cases, silica colloids and dispersions require high pH in order for them to remain stable. Thus, silicone water-based elastomers usually require pH adjustments in order to accommodate these forms of silica. This is usually accomplished with alkalis such as sodium or potassium hydroxide or alkali carbonates. Ammonia or amines like AMP-95 can also be used to adjust pH [48,49]. pH is usually kept above 9 in these silicone water-based elastomers that contain silica.

Thickeners and other rheology control additives are also very important to silicone water-based elastomer coatings. Usually silicone water-based elastomer coatings are too low in viscosity to build adequate film thickness in architectural coatings. Thus, thickeners such as cellulosics, associative thickeners such as HEUR or HASE, are used to achieve adequate film build on substrates. The reader is referred to Heilen et al. for acquaintance to thickeners, rheology modifiers, and other additives for water-based coatings [50]. Once applied, the coating must not sag on vertical surfaces, and it should flow enough for brush or roller marks in the wet film to dissipate before drying. Proper rheology control additives usually address these issues in most coatings and the same can be said for silicone water-based elastomer coatings.

4.3 PARTICLE COALESCENCE AND FILM FORMATION

Removal of water from precured silicone water-based emulsions causes the silicone rubber particles to coalesce, whereby they stick together and form coherent films. Since the polymer chains within each particle are cross-linked, interparticle diffusion of polymer chains is either nonexistent or very limited. This leads to films in which the original integrity of the particles is maintained throughout the life of the coating. Figure 4.4 is a transmission electron micrograph of a typical silicone water-based elastomeric film. Thus, films from dried silicone water-based elastomer emulsions can be thought of as a bunch of silicone rubber balls stuck together. Particle size of elastomeric particles is consistent with particle size measurements of the aqueous dispersion obtained by laser light scattering.

Quite surprisingly, these films are capable of exhibiting significant strength for a silicone elastomer, and especially when one considers the elastomer a room temperature vulcanizing (RTV) silicone. Unfilled silicone water-based elastomeric films have a tensile strength on the order of only 30 psi (lb/in.^{2}) or 0.2 MPa, which is typical of unfilled, neat (non-water-based) silicone rubber. When properly reinforced, silicone water-based elastomeric films typically have a tensile strength of 450 psi or 3.1 MPa.

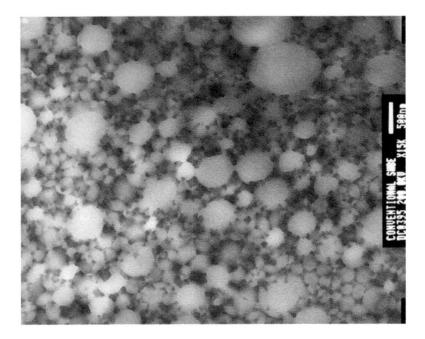

FIGURE 4.4 Scanning transmission electron micrograph of a cured, dried, silicone elastomer film. Image indicates that as polymer within each particle was cross-linked, the original integrity of the particles was maintained. Sample was Sn catalyzed and contained 15 parts per hundred resin (polymer) SiO_2 as colloidal silica of 4 nm size. Silica is not resolved in the image. White vertical scale bar represents 500 nm.

Silica is the reinforcing filler of choice for silicone elastomers and it is also the most desirable reinforcement for silicone water-based elastomers as well, see Section 4.2.3.1. Neat silicone elastomers utilize either precipitated silica or fumed (pyrogenic) silica for reinforcement; the preferred type of silica for silicone water-based elastomers is aqueous, colloidal silica.

Upon coalescence by removal of water from silicone water-based elastomers, the silicone elastomer particles are believed to be forced together by significant pressure developed by capillary forces that result from the drying process. Interparticle diffusion of polymer molecules is most likely insignificant because the polymer chains are cross-linked. It is believed that Van der Waals forces play a role in the integrity and strength of the resulting films. This is particularly so if one considers that the particles are deformed after removal of water. The deformed elastomer particles are like inflated balloons pressed together. Therefore, contact between the particles is significant, hence the adhesion. Indeed, attractive forces have been calculated for both deformable and nondeformable particles (during flocculation), and as expected, it is higher for deformable particles [51–54].

Although surfactant plays a role in particle interactions during coalescence, the presence of surfactant seems to have either no or little consequences on interparticle adhesion. The surfactant is insoluble in silicone elastomeric particles (cross-linked PDMS).

During dry down, surfactant most likely migrates to form domains between particles within the dried film.

In silicone water-based elastomers that are prepared from SiOH functional polymers, an alkoxysilane cross-linker and a Sn catalyst, the cross-linked siloxane polymers are believed to contain residual silanols, and these silanol groups are capable of undergoing silanol condensation reactions. Such reactions can be driven by evaporation of water since water is a byproduct of silanol condensation, but if an active catalyst is present, such as a Sn IV catalyst like dibutyltindilaurate, silanol condensation would be expected to be even more significant. Thus, it is suspected that upon removal of water from precured silicone water-based elastomers, silanol condensation between molecules from the opposite particles takes place that enhances interparticle adhesion in addition to that of normal Van der Waals attractive forces.

In addition to silanol condensation occurring between polymer molecules of separate precured silicone elastomer particles, colloidal silica, which is located at the silicone particle boundaries, and which also contains silanols, especially at its surface, is capable of participating in silanol condensation too. In other words, silanol condensation most likely occurs between polymer molecules of two separate particles and also between polymer and silica. Silanol condensation can also occur between silica particles. Thus, colloidal silica can act to "cement" silicone elastomer particles together upon removal of water from precured silicone water-based elastomers.

4.4 PROPERTIES OF SILICONE WATER-BASED ELASTOMERS

Silicone polymers and silicone elastomers are well known for their high performance. Silicone elastomers have excellent thermal stability, excellent low temperature flexibility, excellent weatherability, excellent permeability to gases (breathability), and excellent dielectric properties. Although silicone elastomers display inherently weak mechanical strengths, especially when compared to many other engineering polymers, when properly reinforced, mechanical strengths of silicone elastomers are adequate for many applications. Engineering polymers are those polymers used in many applications. Examples of engineering polymers are nylon, polyurethane, polystyrene, polymethylmethacrylate, etc.

Properties of silicone water-based elastomers, on the other hand, differ somewhat from neat (nonaqueous), silicone elastomers. Cured, dry, silicone water-based elastomers have poor dielectric properties, moderate high temperature stability, good to very good weatherability, excellent low temperature flexibility, and excellent breathability. Some of these properties are discussed separately here.

In summary, silicone water-based elastomers possess properties that can be significantly different from those of neat or nonaqueous silicone elastomers. The reason for this difference is attributed mainly to the surfactants, which are not removed and remain with the silicone elastomeric films. In some cases, the surfactants can break down and release substances that may catalyze depolymerization or other undesirable reactions of the siloxane polymers. Thus, one should not assume that a silicone water-based elastomer will behave exactly like a neat, nonaqueous silicone elastomer. The silicone water-based elastomer should be tested in its application prior to being selected for a particular use.

4.4.1 ADHESION

Adhesion of silicone water-based elastomers varies from excellent to poor, depending upon the type of substrate and the type of silicone water-based elastomer used. Nonreinforced silicone water-based elastomers have rather poor to marginal adhesion to most substrates. Adhesion is improved substantially when silicone water-based elastomers contain colloidal silica, in particular, colloidal silica of small particle size like 4–5 nm. Addition of some silane adhesion promoters increases adhesion even more. Adhesion of silicone water-based elastomers that contain colloidal silica is excellent to mineral surfaces, such as concrete, masonry, and glass. Adhesion to wood is good to fair, depending upon the type of wood and whether silane adhesion promoters are used. Adhesion to plastics, such as polystyrene, vinyl polymers, and polyethylene is poor. Adhesion to previously coated surfaces (adhesion to itself) is sometimes mediocre, but with the use of primers or adhesion promoters, very good adhesion can be achieved.

From the patent literature, amino-functional silanes and their reaction products with carboxylic acid anhydrides as well as amino-functional siloxanes appear to be the most common means for improving adhesion of silicone water-based elastomers [55–58]. The following amino-functional siloxanes are the preferred silanes for achieving good adhesion in silicone water-based elastomers:

$$(MeO)_3SiCH_2CH_2CH_2NHCH_2CH_2NH_2$$

$$(EtO)_3SiCH_2CH_2CH_2NH_2$$

where
    $Me = CH_3$
    $Et = CH_3CH_2$

Epoxy functional silanes, such as 3-glycidoxypropyltrimethoxysilane, are also used to enhance adhesion of silicone water-based elastomers; however, their performance is not as pronounced as the amino-functional silanes. Usually small amounts of these silanes are used, something on the order of 1% or 2% by weight.

4.4.2 THERMAL PROPERTIES

Although silicone rubber is well known for having excellent high temperature stability, this is not the case for silicone water-based elastomers. It is believed that residual surfactants in dried silicone water-based elastomer films decompose into acidic substances at elevated temperatures, and these substances have adverse effects on the silicone elastomer. This is particularly acute when sulfate or sulfonate surfactants are used. Figure 4.5 shows a TGA of a typical silicone water-based elastomeric film and a typical neat (nonaqueous) silicone elastomer.

Films from first generation silicone water-based elastomers were reported to have a maximum continuous rating of only 110°C, which is in sharp contrast to approximately 250°C for standard, heat-vulcanized silicone rubber and approximately 177°C for Sn catalyzed RTV silicone rubber [2]. Successive generations of silicone

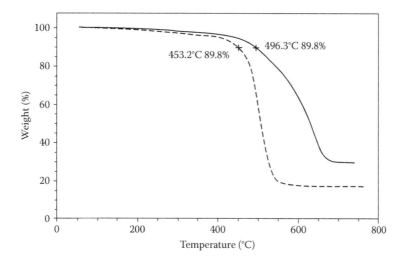

FIGURE 4.5 TGA (thermogravimetric analysis) of a silicone water-based elastomer (SWBE) and a conventional (nonaqueous) silicone elastomer. Dotted curve is SWBE while solid curve is conventional silicone rubber. SWBE contained 10 phr SiO_2 as 8 nm colloidal silica (Nalcoag® 1130). phr stands for parts per hundred resin (polymer) on a solids basis. Conventional silicone elastomer was peroxide cured and silica filled. ~10% weight loss occurred at 453.2°C for the SWBE and 496.3°C for the conventional silicone rubber. Measurements were carried out under N_2 atmosphere and at a heating rate of 10°C per minute using a TA Instruments Q-5000 thermogravimetric analyzer.

water-based elastomers have improved upon this, but they still are not comparable to nonaqueous silicone elastomers. An experimental silicone water-based elastomer has achieved a maximum continuous rating of 140°C.

Silicone elastomers are also well known for their low temperature properties as PDMS has a very low T_g, which is −123°C. PDMS will however crystallize at a higher temperature as its T_m is on the order of −54°C. This limits the use of most silicone elastomers to about −40°C. At low temperatures, silicone water-based elastomers behave the same as neat, nonaqueous silicone elastomers.

4.4.3 WEATHERABILITY

Silicone elastomers are well known for their excellent resistance to the effects of the weather. For example some RTV silicone sealants have been in outdoor service for over 50 years and still display properties that allow them to perform in their intended applications [59].

Although weatherability of silicone water-based elastomers is considered to be very good, it cannot be equated entirely to the weatherability of neat, nonaqueous silicone elastomers. Silicone water-based elastomers have not performed well as roof coatings, while nonaqueous silicone elastomers have worked well in this application. It is believed that surfactants, which remain in the cured, dried silicone elastomeric films, could affect silicone water-based elastomers' weatherability properties. The presence of an active Sn IV catalyst, such as dibutyltindilaurate, in some water-based elastomers could also be a factor. Silicone water-based elastomers have shown to perform very well under outdoor exposure and accelerated weather testing on vertical surfaces.

The combination of surfactants, an active Sn IV catalyst and the microstructure of precured silicone water-based elastomers may be acting in unison to influence the weatherability properties of cured, dried, silicone water-based elastomeric films. For example, the results from the study by Stein et al. on the effects of water on polymer integrity of silicone water-based elastomer films discussed in Section 4.2.2.1, may be related to the weatherability characteristics of the cured, dried films [24]. On the other hand, it is possible that surfactants and Sn catalyst are removed from the dried coating during the weathering process.

In any event, silicone water-based elastomers have weatherability properties that are quite adequate for some applications. For example, silicone water-based elasto-mers have found widespread use as architectural coatings for facades of commercial buildings. This topic is discussed in more detail in Section 4.5.1.

4.4.4 PERMEABILITY AND BREATHABILITY

PDMS is perhaps the most permeable polymer known and silicone elastomers based on PDMS have permeability to gases that are significantly higher than most other types of polymers. For example, silicone rubber based on PDMS is 25 times more permeable to oxygen than is natural rubber [5]. Permeability of silicone water-based elastomers is essentially the same of neat, conventional silicone elastomers. Water vapor permeability of a coating can be very important in certain applications, such as architectural coatings, whereby liquid water trapped behind a coating can exit via water vapor. If permeability of a coating is too low, the coating can form blisters due to trapped water. In applications that benefit from high permeability, silicones and silicone water-based elastomers are ideal. They both are impervious to liquid water, but they both allow water vapor to pass at relatively high rates.

Silicone water-based elastomers can also improve water vapor permeability of coatings based on organic binders. In this case, silicone water-based elastomers are combined with an organic water-based coating, such as acrylic latex, and the result-ing film is significantly more permeable than the original. Figure 4.6 shows the effect of silicone water-based elastomers on water vapor permeability of an acrylic latex.

4.4.5 MECHANICAL PROPERTIES

As previously mentioned, silicone elastomers based on PDMS possess inherently weak mechanical strengths, and they usually require reinforcement in order for them to have any utility in their myriad applications.

Mechanical strengths of silicone water-based cured, dried films are comparable to and sometimes can exceed that of neat, RTV silicone elastomers. Heat-vulcanized neat silicone elastomers (high consistency silicone rubber) have superior mechani-cal strength to those of both neat RTV silicone and silicone water-based elastomers.

Unreinforced silicone water-based elastomer films typically have a tensile strength that is on the order of 30 psi (0.21 MPa), which is the same for unreinforced, neat

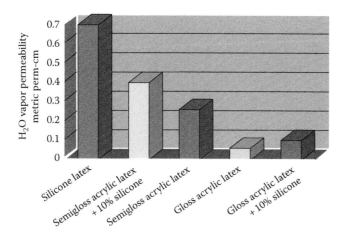

FIGURE 4.6 Water vapor permeability of silicone water-based elastomer and acrylic films. Silicone Latex was a pre-cured, Sn catalyzed, silicone water-based elastomer containing 60% silicone. Semigloss and gloss acrylic latex were commercial, fully formulated exterior latex paints. Semigloss and gloss acrylic latex with silicone were the same latex paints with 10% by weight of the silicone latex added. Metric perm-cm is $g(H_2O)/m^2/24$ h/mmHg/cm(thickness). Measurements determined using Payne permeability cups.

or nonaqueous silicone elastomers based on PDMS. When properly reinforced with colloidal silica, however, tensile strength of silicone water-based elastomer films is on the order of 450 psi (3.10 MPa).

Colloidal silsesquioxane resin and MQ resin are also useful for reinforcing silicone water-based elastomers [16]. Colloidal resins are discussed in more detail in Chapter 7.

Depending upon the type and level of reinforcing silica present, tensile strengths can reach as high as 750 psi (5.2 MPa), but more commonly these films are in the range of 300–450 psi (2.0–3.1 MPa). Figure 4.7 shows tensile properties of typical silicone water-based elastomeric films. Usually the type and level of reinforcing filler used in silicone water-based elastomers are determined (optimized) by experiment, which involves preparing a series of compositions having differing levels of reinforcing filler, casting films, and measuring their tensile properties.

Particle size of silica will have an impact on filler loading, with smaller particle size silicas requiring lower levels than larger particle size silicas to achieve the same tensile strength. It should be pointed out, however, that smaller size colloidal silica or silica dispersions are more dilute than larger size silica colloids or dispersions. Even though smaller particle size silicas (4–8 nm) are more effective reinforcing fillers than larger particle size silicas (20–50 nm), that is, their loading can be lower, they still lead to more dilute silicone water-based elastomer emulsions or dispersions compared to larger particle size colloids or dispersions. Typical silicone water-based elastomers contain 10–15 phr (parts per hundred resin or polymer) of small particle size silica on a solids basis. Figure 4.8 shows a dynamic mechanical analysis histogram for a typical Sn cured, silicone water-based elastomer, both filled and unfilled.

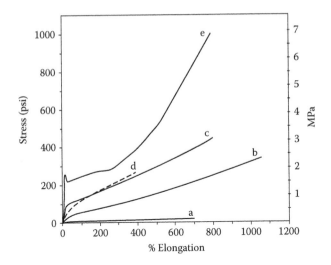

FIGURE 4.7 Tensile properties of silicone water-based elastomers: a is unfilled; b is filled with 18 phr aqueous, dispersed, fumed silica (Cabosperse® SC-2); c and d are filled with 15 phr 4 nm colloidal silica (Nalcoag® 1115); and e is filled with 40 phr 10 nm aqueous colloidal MeSiO$_{3/2}$ resin. Samples a, b, and c are Sn (II) catalyzed, sample d is uncatalyzed, and sample e is Sn (IV) catalyzed. phr stands for parts per hundred resin (polymer) on a solids basis.

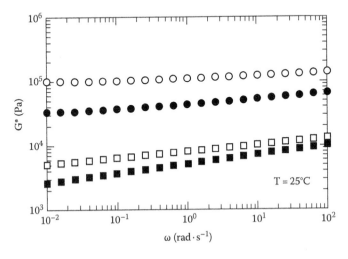

FIGURE 4.8 Dynamic mechanical analysis of silicone water-based elastomer. Solid symbols correspond to unfilled, Sn-cured silicone water-based elastomeric film. Open symbols correspond to Sn-cured, silicone water-based elastomer filled with 10 phr 4 nm colloidal SiO$_2$ (Nalco® 1115). phr stands for parts per hundred resin (polymer). Circles are G′ while squares are G″. Measurements obtained using an Ares® rheometer.

4.5 APPLICATIONS

Commercial applications of silicone water-based elastomers include coatings, coating additives, binders for ceramics and sealants, or caulks. Silicone water-based elastomers have been commercially available since about 1980. Construction coatings were the first products developed and these have evolved into specialized architectural coatings.

4.5.1 ARCHITECTURAL COATINGS

Architectural elastomeric coatings represent the largest single volume of use for silicone water-based elastomers. Such coatings have found significant opportunities in remediation of commercial buildings that have experienced problems with water penetration. The silicone water-based coatings have been designed to bridge cracks in building facades and provide protection from water ingress, even protection from wind-driven rain [60]. Silicone water-based elastomeric coatings are used on above-grade, exterior substrates, such as concrete block, fluted block, brick, stucco, synthetic stucco, poured concrete, precast concrete, EIFS (exterior insulation finish system) and previously coated masonry substrates. Tables 4.1 and 4.2 show properties of a typical silicone water-based elastomer architectural coating.

These silicone water-based elastomeric coatings have very good durability and maintain water protection properties for buildings for many years. They can be pigmented, they are one-component, and they are water-based. Figure 4.9 shows a typical application of silicone water-based elastomeric architectural coating. Although silicone water-based elastomeric coatings have found their niche in waterproofing commercial buildings, their use in residential construction is not significant, probably due to cost constraints of the latter.

TABLE 4.1
Properties of a Typical Silicone Water-Based Elastomeric Architectural Coating

Test	Property	Unit	Value
ASTM[a] D 2369	Solids content	Percent by weight	58.6
		Percent by volume	50.1
ASTM D 1475	Specific gravity	lb/gal (kg/L)	9.64 (1.155)
ASTM D 2196	Viscosity[b]	Centipoise (Pa · s)	40,000 (40)
ASTM D 1849	High temperature stability (constant viscosity)	Days	>28
EPA method 24	VOC (volatile organic content)	lb/gal (g/L)	<0.42 (<50)

[a] American Society for Testing Materials.
[b] Measured using Brookfield HAV, spindle #2, at 2 rpm.

TABLE 4.2
Properties of Cured Silicone Water-Based Elastomeric Architectural Coating

Test	Property	Unit	Value
ASTM[a] D 2240	Durometer hardness, shore A	Points	38
ASTM D 412	Tensile strength	psi (MPa)	>145 (1.00)
ASTM D 412	Elongation	Percent	600
ASTM D 1653	Permeance	English perms (ng/(m² · Pa · s))	43.2 (2480)
ASTM D 522	Room temp. Flex, 1/8″ mandrel		Pass
ASTM C 711	Low temp. Flex, 1/4″ mandrel		Pass
ASTM D 3274	Fungus resistance		No growth
ASTM D 6904	Wind driven rain[b]		Pass

[a] ASTM, American Society for Testing Materials.
[b] Measured on two coats of *Dow Corning*® Allguard silicone water-based elastomeric coating.

FIGURE 4.9 Typical commercial application for a silicone water-based elastomer architectural coating.

Silicone water-based elastomers are also being used as coatings to limit air leakage through building walls. The purpose of these coatings is to improve thermal efficiency of buildings by limiting transport of air through cracks or small openings in building walls. These coatings would be used in lieu of preformed polymer films such as spun-bonded polyethylene sheets to provide an extra layer of protection from both air and liquid water ingress or egress in building walls. These silicone water-based elastomer coatings are applied to building walls between the façade layer and a stud wall with sheathing. The coating is merely painted onto the wall and cures to form an elastomeric silicone film. This eliminates the need for fasteners, such as staples or adhesive tape, when preformed polymer sheeting is used. The "formed in place" silicone elastomeric film offers protection to the wall from the weather during the construction phase of the building and ultimately provides a barrier for air transport during the lifetime of the building.

A wood coating based on silicone water-based elastomer is described that consists of an aqueous emulsion of hydroxyl-functional PDMS, an amino-functional siloxane reacted with an anhydride, an epoxy-functional silane, a curing catalyst and a borate [61].

4.5.2   FABRIC COATINGS

Silicone water-based elastomers are also used in fabric coatings as solvent-borne silicone coatings are becoming less desirable. Although solventless silicone coatings have found widespread use, particularly in automotive airbag coatings, they have some limitations that certain silicone water-based elastomer coatings could address. For example, silicone water-based coating compositions can provide lower viscosity coatings which ultimately lead to coatings having lower coat-weights. They are one-part systems whereas the nonaqueous silicone coatings are usually two-part, and the silicone water-based coatings offer more flexibility in the choice of coating equipment. There are examples of airbag coatings in the patent literature that are based on silicone water-based elastomers [62,63].

A fabric finish consisting of silicone water-based elastomer has been described [64]. The finish reportedly reduces shrinkage of wool textiles and feels very soft and pleasant.

An interesting use for silicone water-based elastomeric coatings is for aquatic weed control in lakes or ponds [65]. A fabric coated with silicone water-based elastomer is submerged over the aquatic growth, whereby it blocks sunlight and the plants die. An experimental benthic-barrier cloth installed in a freshwater lake in Michigan, USA, performed for more than 20 years.

4.5.3   COATING ADDITIVES

In addition to being used as binders in coatings, silicone water-based elastomers are also used as additives in coatings and inks. In this case, only a small amount of the silicone emulsion is used, usually on the order of 0.5%–5%. Benefits from the silicone additive include slip, antimar, improved abrasion resistance, and waterproofing. In many cases, silicone orients predominantly at the surface of the dried coating, which impacts the coefficient of friction of the coating by making it lower.

A coating composition based on a silicone water-based elastomer that is enhanced with the reaction product of amino-functional silane plus anhydride is described [66]. The composition provides good release and lubricity to rubber items. Similar compositions have also been reported that are designed to provide adhesion to surfaces, excellent release, and good surface lubricity. These compositions are made from a silicone water-based elastomer, a special amino-functional siloxane, a polyurethane resin, and in one case, polymer particles [67,68].

Silicone water-based elastomers can also be added to other aqueous coatings based on organic polymers such as acrylates and polyurethanes to modify the mechanical properties of such coatings [69]. Being elastomeric, the silicone particles can provide toughening to the dried polymeric films. In addition, they can enhance breathability of the polymer film and also provide improved hydrophobicity. In these cases, the level of silicone in the coating is greater than 10% of the polymer binder.

4.5.4 SEALANTS AND CAULKS

Water-based silicone sealants have been prepared using silicone water-based elastomers. In this case, the solids content of the aqueous compositions is increased substantially with the addition of various fillers, such as silica and calcium carbonate [70]. In spite of their positive attributes, such as easy cleanup with water, easy handling, and low odor, these first-generation water-based silicone sealants were not very successful as they suffered from poor adhesion and mediocre durability. Perhaps these shortcomings can be overcome so that water-based silicone sealants will again be available.

4.6 SUMMARY

Silicone water-based elastomers, their preparation, their properties, and their major applications have been described. Silicone water-based elastomers are aqueous compositions that form silicone elastomers or silicone rubber upon removal of water at either ambient temperature or at elevated temperatures.

Silicone water-based elastomers consist of aqueous emulsions or dispersions of silicone polymer particles which either have undergone cross-linking reactions in the aqueous state or will undergo cross-linking reactions after they have been applied (dried). Thus, silicone water-based elastomers can be classified as either precured dispersions or postcuring emulsions, whenever cross-linking actually occurs. The majority of silicone water-based elastomers are of the precured type as the post-cross-linking emulsions have not been developed to any appreciable extent. Precured silicone water-based elastomers are prepared by crosslinking aqueous emulsions of high molecular weight PDMS. A number of routes exist for accomplishing this, but the composition delivering the best mechanical properties involves catalysis using tin compounds. Removal of water causes silicone elastomer particles to coalesce, whereby the particles adhere to each other producing coherent silicone elastomeric films. Silicone water-based elastomers are usually reinforced with colloidal silica, which leads to mechanical strengths on the order of 450 psi (3.1 MPa), which is comparable to many non-aqueous, RTV silicone elastomers.

Silicone water-based elastomers were commercialized around 1980, and they have found use mainly as coatings and coating additives. Architectural coatings are the predominant applications for silicone water-based elastomers.

ACKNOWLEDGMENTS

The author thanks H. A. Freeman and S. A. Wackerle for obtaining the electron micrograph of Figure 4.1. The author thanks Dr. Steven Swier for obtaining TGA data and Dr. Glenn Gordon for obtaining DMA data of silicone water-based elastomers. The author is indebted to Dr. Tatiana Dimitrova for very fruitful discussions about particle coalescence and interactions. Finally the author acknowledges the Dow Corning Corporation for support during the writing of this manuscript.

REFERENCES

1. Saam, J. C., Graiver, D., Baile, M., *Rubber Chem. Technol.*, 1981, *45*, 976.
2. Smith, S. B., *Proceedings of the Water-Borne and Higher-Solids Symposium*, 1983, Storey, R., Thames, S. (eds.), School of Polymers and High Performance Materials, University of Southern Mississippi, Hattiesburg, MS, pp. 67–81.
3. Liles, D. T., Lefler, H. V. III, *Proceedings of the 18th Water-Borne, Higher-Solids, and Powder Coatings Symposium*, 1991, Storey, R., Thames, S. (eds.), School of Polymers and High Performance Materials, University of Southern Mississippi, Hattiesburg, MS, pp. 161–173.
4. Liles, D. T., *Polymeric Materials Encyclopedia*, Vol. 10, 1996, Salamone, J. C. (ed.), CRC Press, Boca Raton, FL, pp. 7694–7696.
5. Polmanteer, K. E., *Rubber Chem. Technol.*, 1988, *61(3)*, 470–502.
6. Tracton, A. A., *Coating Technology Handbook*, 3rd edn., 2005, CRC Press, Taylor & Francis Group LLC, Boca Raton, FL.
7. Fairley, M., *Encyclopedia of Label Technology*, 2014, Tarsus Exhibitions and Publishing, Ltd., London.
8. Hyde, F., Wehrly, J., US2891920, 1959.
9. Findlay, D., Weyenberg, D., US3294725, 1966.
10. Weyenberg, D., Findlay, D., Cekada, J., Bey, A., *J. Polym. Sci. C*, 1969, *27*, 27–34.
11. Saam, J. C., Huebner, D. J., *J Polym. Sci.*, 1982, *20*, 335.
12. Narula, D., US4788001, 1988.
13. Tanaka, M., Okada, F., Oba, T. et al., US4814376, 1989.
14. Derian, P. J., Feder, M., Paillet, J. P. et al., US5763505, 1989.
15. Axon, G., Dow Corning Corporation Internal Technical Report, 1962.
16. Cekada, J., US3355406, 1967.
17. Stein, J., Lenoard, T., Smith, J., *J. Appl. Polym. Sci.*, 1993, *47*, 667–675.
18. Liles, D. T., US5089537, 1992.
19. Feder, M., US5140061, 1992.
20. Feder, M., US5145901, 1992.
21. Berg, D. T., Joffre, E. J., US5674937, 1997.
22. Johnson, R., Saam, J. C., Schmidt, C., US221688, 1980.
23. Iler, R. K., *The Chemistry of Silica: Solubility, Polymerization, Colloid and Surface Properties and Biochemistry of Silica*, 1979, John Wiley & Sons, New York, pp. 4–5.
24. Stein, J., Leonard, T., *Polym. Degrad. Stab.*, 1990, *28*, 311.
25. Donaldson, J. D., *Progress Inorg. Chem.*, 1967, *8*, 287–356. Inorganic Chemicals and Reactions.
26. Liles, D. T., *Proceedings of the 21st Water-Borne, Higher-Solids, and Powder Coatings Symposium*, 1991, Storey, R., Thames, S. (eds.), School of Polymers and High Performance Materials, University of Southern Mississippi, Hattiesburg, MS, pp. 161–173.
27. Benson, R., Concise International Chemical Assessment Document 14, 1999, World Health Organization, Geneva, Switzerland. Wissenschaftliche Verlagsgesellschaft mbH, Stuttgart, Germany.
28. Nath, M., *Tin Chemistry: Fundamentals, Frontiers and Applications*, 2008, Davies, A. G., Gielen, M., Pannell, K. et al. (eds.), John Wiley & Sons, Ltd., Chichester, U.K., pp. 413–429.
29. Craig, P. J., Eng, G., Jenkins, R. O., *Organometallic Compounds in the Environment*, 2nd edn., 2003, Craig, P. J. (ed.), John Wiley & Sons, Ltd., Chichester, U.K.
30. Byrd, J. T., Andrae, M. O., *Science*, 1982, *218*, 565.
31. Huebner, D. J., Saam, J. C., US4568718, 1986.
32. Huebner, D. J., Saam, J. C., US4584341, 1986.
33. Bauman, T. M., Huebner, D. J., US4618645, 1986.

34. Saam, J. C., US4244849, 1981.
35. Feder, M., Frances, J. M., US5004771, 1991.
36. Saam, J. C., Wegener, R. L., US4273634, 1981.
37. Marciniec, B., *Hydrosilylation*, Springer, Dordrecht, the Netherlands, 2009.
38. Willing, D., US4248751, 1981.
39. Joffre, E. J., Liles, D. T., Murray, D. L. et al., US5708070, 1998.
40. Dauth, J., Deubzer, B., Schrock, R. et al., 6177511, 2001.
41. Frances, J., Leising, F., Reeb, R., EP 0578893, 1994.
42. Liles, D. T., Murray, D. L., US5480919, 1996.
43. Roberts, W. O., *Colloidal Silica, Fundamentals and Applications*, 2005, Bergna, H. O., Roberts, W. O. (eds.), CRC Press, Taylor & Francis Group LLC, Boca Raton, FL.
44. Liu, Y., US8012544, 2011.
45. Graiver, D., Kalinowski, R. E., US4572917, 1986.
46. Graiver, D., Kalinowski, R. E., US4545914, 1985.
47. Klosowski, J. M., Wolf, A. T., *Sealants in Construction*, 2nd edn., 2016, CRC Press, Taylor & Francis Group, Boca Raton, FL.
48. Doyle, A. S., *Additives for Water-Based Coatings*, 1990, Karsa, D. R. (ed.), The Royal Society of Chemistry, Cambridge, U.K., Special Publication No. 76, pp. 77–89.
49. Brutto, P. E., Peera, A., Troester, L. E., *Proceedings of the 42nd Annual International Waterborne, Higher-Solids, and Powder Coatings Symposium*, 2015, Storey, R. F., Rawlins, J. W. (eds.), pp. 467–479.
50. Wernfried, H., John, H., David, H. et al., *Additives for Waterborne Coatings*, European Coatings Tech Files, 2009, Vincentz Network GmbH, Hannover, Germany.
51. Danov, K. D., Denkov, N. D., Petsev, D. N. et al., *Langmuir*, 1993, *9*, 1731–1740.
52. Danov, K. D., Petsev, D. N., Denkov, N. D., *J. Chem. Phys.*, 1993, *99(9)*, 7179–7189.
53. Denkov, N. D., Petsev, D. N., Danov, K. D., *J. Colloid Interface Sci.*, 1995, *176*, 189–200.
54. Petsev, D. N., Denkov, N. D., Kralchevsky, P. A., *J. Colloid Interface Sci.*, 1995, *176*, 201–213.
55. Okada, F., Oba, T., Nakazato, M., US4496687, 1985.
56. Kondo, H., Koshii, T., US4535109, 1985.
57. Graiver, D., Kalinowski, R. E., US4710405, 1987.
58. Hill, M. P., Tselepis, A. J., Wolf, A. T., US5807921, 1998.
59. Kimberlain, J., Laureys, B., Harres, N., *Durability of Building and Construction Sealants and Adhesives*, Vol. 5, 2015, ASTM International Selected Technical Papers, Carbary, L. D., Wolf, A. T. (eds.), STP 1583-20140065, ASTI International, West Conshohocken, PA, pp. 106–122.
60. Lefler, H. V. III, *Science and Technology of Building Seals, Sealants, Glazing and Waterproofing*, Vol. 2, 1992, ASTM STP 1200, Klosowski, J. (ed.), ASTM, West Conshohocken, PA, pp. 260–267.
61. Matsumura, K., Yamamoto, A., US7658972, 2010.
62. Inoue, Y., Momii, K., US5254621, 1993.
63. Veiga, M. J., Manuel, J., Satin, R. J., US6239046, 2001.
64. Nakazato, M., Ohba, T., Tanaka, M. et al., US5102930, 1992.
65. Elias, M. G., Pullman, G. D., US4577996, 1986.
66. Inokuchi, Y. Kuwata, S., US6147157, 2000.
67. Inokuchi, Y., US7531241, 2009.
68. Inokuchi, Y., US7652091, 2010.
69. Lind, D., Meyers, D., Shope, M., US7767747, 2010.
70. Liles, D. T., Shephard, N., *Science and Technology of Building Seals, Sealants, Glazing and Waterproofing*, Vol. 2, 1992, ASTM STP 1200, Klosowski, J. (ed.), ASTM, West Conshohocken, PA, pp. 280–291.

5 Silicone Elastomeric Powders

Donald T. Liles, Isabelle Van Reeth, and Mari Wakita

CONTENTS

5.1   INTRODUCTION

Silicone elastomeric powders, also referred to as silicone powders, are small particles of silicone rubber (elastomer) that are usually on the order of 2–10 μm in size (see Figure 5.1). Silicone powders are used in a number of applications that fall into the major categories of personal care, plastics modification, and coatings modification. Silicone powders can be prepared by several different techniques, such as cryogenic grinding of silicone rubber, atomization of liquid silicone rubber, and an aqueous emulsion process. This chapter will deal with only powders made by the emulsion process. In the emulsion process, reactive, liquid, silicone polymers are first emulsified followed by cross-linking or vulcanization and, finally, removal of

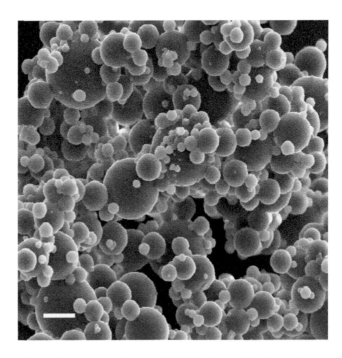

FIGURE 5.1   Scanning electron micrograph (SEM) of a typical silicone elastomeric powder. White scale bar represents 5 μm.

water [1,2]. Silicone powders were first prepared by Japanese researchers in the early to mid-1980s, and they became commercially available in the late 1980s [3–6].

Silicone elastomeric powders were first used for modifying thermosetting plastics, in particular epoxy molding compounds. Silicone powders have also been used to modify certain plastics for light diffusion. In addition, silicone elastomeric powders have found use in coatings whereby they reduce gloss in coatings and lead to coatings having a matte finish. The softness of silicone particles is also brought out in certain coatings that contain them as such coatings feel soft and pleasant to the touch and they can also have a textured appearance. Silicone elastomeric powders can also be used to modify polymeric surfaces to provide slip and lubrication and also to impart antiblocking behavior to polymers.

In personal care applications, silicone elastomer powders are used in skin creams and in shampoo compositions. The main benefits silicone powders bring to skin creams are their sensory appeal, their ability to absorb sebum, and their ability to mask the appearance of wrinkles on skin [7]. Silicone powders deliver a unique and very pleasant sensation to skin, hence such powders are popular among skin-care formulators. In addition, silicone powders have soft-focus properties that help to reduce the appearance of wrinkles on skin, especially around the eyes [7–9]. It should be understood that silicone powders on skin do not "cure" or lessen the severity of wrinkles, they merely mask or reduce the appearance of wrinkles. Silicone powders are also capable of absorbing sebum, an oily substance secreted by glands within the skin that makes it appear shiny [10]. Silicone elastomer powders also give conditioning to hair when incorporated into shampoos and/or conditioners. Finally, silicone elastomeric powders are capable of delivering active substances or emollients to skin, thanks to their capacity to absorb not only nonpolar, but also polar substances.

This chapter describes silicone elastomeric powders, their preparation, their properties, and some of their industrial as well as cosmetic uses.

5.2 PREPARATION

As stated previously, this chapter describes only silicone elastomeric powders that are prepared by an emulsion process, more specifically using aqueous emulsions. Preparation of silicone powders by the emulsion process involves three basic steps: first, form an aqueous emulsion of reactive silicone polymer particles; second, carry out a chemical cross-linking reaction of the polymers within such particles to form an elastomer; and third, remove water (see Figure 5.2).

Although silicone particles are prepared commercially by an aqueous emulsion process, it should be realized that a nonaqueous emulsion process should also be applicable for their preparation. The definition of an emulsion is a liquid dispersed in another liquid, with both liquids being immiscible in one another. Thus, it stands to reason that the three basic steps described above could be carried out in a liquid other than water. Since water is low cost and the know-how behind preparing aqueous silicone emulsions is well established among silicone producers, it is not surprising that aqueous emulsions are the method of choice for preparing silicone powders. There is a drawback to using water however, and that is the high energy demand for removing water from aqueous emulsions.

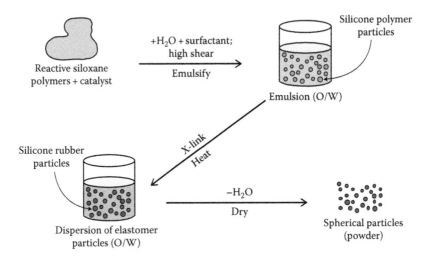

FIGURE 5.2 Scheme for preparing silicone elastomeric powder. Reactive siloxane polymers (functional polymers and cross-linker) and catalyst are combined with surfactant and H_2O. The mixture is emulsified via high shear to produce an emulsion. After cross-linking is complete, water is removed leaving spherical silicone elastomeric particles.

5.2.1 Choice of Silicone Polymers

Silicone polymers used to prepare silicone powders are primarily based on polydimethylsiloxane (PDMS), which is illustrated in Figure 5.3.

Molecular weight of siloxane polymers is not overly critical, although it is usually low enough such that silicone polymers can be easily emulsified. Perhaps, the main criteria for deciding upon the molecular weight of the starting silicone are the final properties of the elastomeric particles and also the ability to emulsify such silicone polymers, on both of which molecular weight has an impact. In most cases, molecular weight of the polymers is selected and optimized by experiment.

These polymers are functionalized with reactive groups that can participate in the cross-linking reactions, which transform siloxane polymer into silicone elastomer. The functional groups selected are dependent upon which type of chemical reaction is used to carry out the cross-linking step, and this is described in Section 5.2.2.

When selecting polymers for preparing silicone powders, starting polymers are chosen for the properties they impart to the final powders, as well as their availability and cost. It should be reiterated that the polymers selected need to be emulsified, and this factor can also influence the choice of polymers.

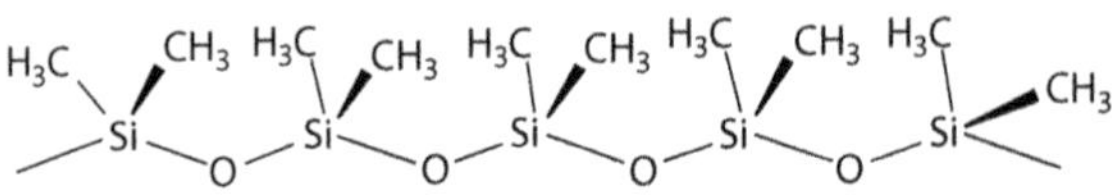

FIGURE 5.3 Structure of polydimethylsiloxane (PDMS).

5.2.2 CROSS-LINKING

Elastomers are usually formed when certain polymers are cross-linked. In the field of silicone technology, siloxane polymers can be cross-linked by a variety of methods to form silicone elastomers, which are commonly called silicone rubber [11]. When considering cross-linking reactive siloxane polymers in aqueous emulsions, thought must be given to the presence of water as water can interfere with or even prevent certain cross-linking reactions from occurring. In spite of the presence of water, there are several reactions that are either not affected or not significantly affected by water and thus, it is entirely possible to cross-link siloxane polymers in aqueous emulsions. Another factor to consider when selecting a cross-linking chemistry to prepare silicone elastomeric powders is the kinetics of the cross-linking reactions, or in other words, the speed of the reactions. For commercially prepared silicone powders, it is desirable to use cross-linking reactions that are fairly fast, as this leads to efficient use of chemical process equipment.

Several routes have emerged as being favored over the myriad reactions that are available to cross-link siloxane polymers. Perhaps the most popular reaction for producing silicone powders is platinum (Pt) catalyzed hydrosilylation (addition) of a silicon hydride to a vinyl substituted siloxane, Equation 5.1:

$$-\overset{|}{\underset{|}{Si}}-H \;+\; CH_2{=}CH{-}\overset{|}{\underset{|}{Si}}- \;\;\xrightarrow{\;\;Pt\;\;}\;\; -\overset{|}{\underset{|}{Si}}-CH_2CH_2-\overset{|}{\underset{|}{Si}}- \tag{5.1}$$

This reaction is not only very fast, but it is also unaffected substantially by the presence of water, and essentially no by-products are liberated. The SiH moiety is capable of reacting with water in the presence of Pt to form silanol (SiOH) plus hydrogen (H_2); however, this reaction occurs only to a minor extent when preparing silicone powders via aqueous emulsion (using the Pt catalyzed hydrosilylation reaction), and hence it is not significant. Even though generation of H_2 is not believed to be significant when preparing silicone powders by this method, one should always consider the *possibility* of H_2 formation and take the necessary precautions when designing process equipment for this application.

Hydrosilylation, of course, is widely used to transform a practically inexhaustible list of siloxane polymers into myriad useful products, with the basic reaction being addition of a silicon hydride to an unsaturated substituted (commonly vinyl) siloxane [12]. In silicone cross-linking technology via hydrosilylation, either the silicon hydride or the vinyl functional silicon can be in the form of siloxane polymers, oligomers, or silanes—the choice of which is usually governed by final properties of the cross-linked reaction product and availability of the starting materials. Most often, cross-linked siloxanes are prepared from mixtures of siloxane polymers or oligomers bearing either SiH or Si–CH=CH$_2$ functional groups that have been copolymerized with PDMS, and this is also typical of silicone powders. Pt catalyzed hydrosilylation is the method of choice for preparing silicone elastomeric powders used in cosmetics applications.

The level of functional groups relative to PDMS polymer is not overly critical; however, this ratio will influence cross-link density of the polymer network, and thus, it will certainly have an impact on final properties of the silicone elastomer powder.

For example, harder particles are prepared such that their cross-link densities are higher than those of softer particles; so harder particles would be designed to have a higher level of functional groups (SiH and $H_2C{=}CH{-}Si$) than softer particles. Usually, this level of functional groups, relative to PDMS in the polymer and molecular weight of the polymer, are optimized by experiment.

Placement of the functional groups within the polymer molecules is also not overly critical as either pendent groups on the polymer chains or groups in terminal positions, or both lead to silicone elastomer powders, provided not all of the reactive functional groups are in terminal positions. In this case, in which all of the reactive functional groups were in terminal positions, such a telechelic polymer composition would only lead to chain extension, and hence an elastomer would not be formed. In order to achieve cross-linking, at least some of the polymer molecules need to have more than two reactive functional groups per molecule. In linear polymer molecules in which the reactive functional groups are all pendent, some end capping or terminal group must be present. Usually, inert end capping groups such as Me_3SiO are present, but even reactive terminal groups such as SiOH or SiOR (alkoxy) can be used. Cyclic, functional siloxanes such as $(OSiMeH)_4$ or $(OSiMeCH{=}CH_2)_4$ are also capable of forming cross-linked networks with linear polymers, and therefore they too could be used to prepare silicone elastomeric powders.

In addition to hydrosilylation (addition) reactions for cross-linking, some condensation reactions can be used for preparing silicone elastomeric powders. Two of these reactions are shown here:

$$-\underset{|}{\overset{|}{Si}}-H + HO-\underset{|}{\overset{|}{Si}}- \xrightarrow{\ Sn\ } -\underset{|}{\overset{|}{Si}}-O-\underset{|}{\overset{|}{Si}}- + H_2 \qquad (5.2)$$

$$-\underset{|}{\overset{|}{Si}}-OR + HO-\underset{|}{\overset{|}{Si}}- \xrightarrow{\ Sn\ } -\underset{|}{\overset{|}{Si}}-O-\underset{|}{\overset{|}{Si}}- + ROH \qquad (5.3)$$

In reaction (5.2), SiH and SiOH are condensed usually in the presence of a tin (Sn) catalyst to form a siloxane (cross-link) with the liberation of H_2. The hydrogen formed is normally vented to the atmosphere, but in order to avoid hazardous mixtures of air and hydrogen, the reaction vessels are usually inerter with nitrogen.

In reaction (5.3), an alkoxy functional silicon compound, such as $MeSi(OCH_3)_3$ or $Si(OEt)_4$ (TEOS), is condensed with a silanol in the presence of a Sn catalyst to yield a siloxane cross-link with the liberation of an alcohol. Normally alcohol by-products are in such small quantities they do not pose a problem.

In certain silicone elastomeric powders, reactive functional groups are sometimes incorporated into the polymer network by the use of organofunctional silanes. Such functionalized silicone powders are normally used in applications when the particles are in close contact with organic polymers, such as in coatings or in organic resins. Although, practically any organofunctional silane could be used to functionalize silicone elastomeric powders, epoxy functional silanes are perhaps the most important.

5.2.3 Catalyst

The choice of cross-linking reactions for preparing silicone elastomeric powders dictates not only what choice of polymers to use, but what catalyst to use as well. The catalysts of choice for preparing silicone elastomeric powers are based on platinum (Pt) and tin (Sn).

5.2.3.1 Platinum Catalyst

The hydrosilylation reaction involving addition of a SiH moiety to an unsaturated hydrocarbon invariably calls for a Pt catalyst, although some catalysts based on other noble metals such as Rh could most likely be used. The form of Pt catalyst is chosen such that it is soluble in the silicone polymers, and also it is convenient to use. Speier's catalyst [13], H_2PtCl_6 in a solvent such as isopropanol, can be used, or Karlstad's catalyst [14], a Pt-vinylsiloxane complex, can also be used. Although the high cost of Pt is of concern, Pt catalysts are viable due to the extremely low concentrations used. Typical silicone elastomeric powders are prepared using 5–20 ppm (parts per million) Pt catalyst (calculated as elemental Pt based on polymer weight). A potential drawback to using Pt catalysts is that Pt is easily poisoned, in particular by many substances containing sulfur, phosphorous, or nitrogen. Thus, when using Pt as a catalyst, care must be taken to carry out silicone elastomeric powder syntheses in equipment that has been rigorously cleaned and is free from potential Pt poisons.

5.2.3.2 Tin Catalyst

Condensation reactions are also utilized to prepare silicone elastomeric powders and such reactions usually involve a catalyst based on Sn. Although these reactions are somewhat slower than the hydrosilylation reaction, they are fast enough to provide a viable route to preparing silicone powders. The favored reactions for preparing silicone powders via condensation involve reaction of either SiH or SiOR with SiOH with the formation of either H_2 or alcohol, respectively, as byproducts. The catalysts used to assist these reactions are Sn esters, such as stannous oleate, stannous octoate (stannous bis-2-ethylhexanoate), or organotin compounds, such as dibutyltindilaurate (DBTDL) or dimethyltindineodecanoate. Stannous esters have the advantage of being non-organotin compounds, and also, they are susceptible to eventual deactivation via hydrolysis [15], particularly in an aqueous environment, such as an emulsion. Although Sn catalysts are used at significantly higher levels than Pt catalysts, their levels are not considered excessive in most cases. Typical levels of Sn catalysts are on the order of 0.1%–0.5% by weight of polymer.

5.2.4 Surfactant

In most cases, emulsions used to prepare silicone elastomeric particles are stabilized by a surfactant. The choice of surfactant is not critical, although it must satisfy certain requirements in order to be useful for preparing silicone powders. Firstly, the surfactant must be effective at stabilizing the siloxane polymer particles and hence keep the particles from coalescing into the larger domains of silicone polymer during processing. Secondly, the surfactant must be compatible with the cross-linking reactions being performed, that is, they must not contain a Pt poison. Thirdly, the surfactant must not be overly toxic or contain toxic ingredients, especially if the intended use for the silicone powders is in cosmetics. As there are thousands of surfactants that are commercially available, and perhaps a large majority of them will suffice for preparing silicone elastomeric powders, the list can be shortened when considering the intended use of the silicone powders.

Surfactants are generally divided into the categories ionic and nonionic, depending upon if they possess a charge or not. Ionic surfactants are further divided into anionic, cationic, or amphoteric, depending upon whether the surfactant possesses a negative, positive, or both the charges, respectively. Although all of these types of surfactants have been used to prepare successful silicone elastomeric powders, the nonionic surfactants have emerged as being preferred.

It should be realized that the surfactant, or at least a portion of the surfactant, used to prepare the emulsion remains with the particles after removal of water and drying. Hence, the desire exists for choosing relatively "clean" surfactants or surfactants that are not only nontoxic, but are also nonirritating to skin. Various alcohol ethoxylates, with the exception of alkyl phenol ethoxylates, are generally the preferred type of surfactant used to prepare silicone elastomeric powders by the emulsion process. Although many alcohol phenol ethoxylates perform exceedingly well at stabilizing emulsions, they, as a general class of surfactants, have come under scrutiny due to regulation.

Examples of such alcohol ethoxylate surfactants useful for preparing silicone elastomeric powders include Brij® L23 (lauryl alcohol EO–23) and Genapol® UD 080 (unidecyl alcohol EO–8). It should be realized that numerous other surfactants can be used to prepare silicone elastomeric powders besides these surfactants, and combinations of surfactants can also be used if desired. The amount of surfactant used to prepare silicone elastomeric powders is also not very critical as the emulsion lifetime is very short. Usually, from 0.25% by weight to 5% by weight of surfactant based on silicone polymer is used.

It should be mentioned that choice of surfactant and amount can have an impact on particle size of the emulsion, and hence, particle size of finished silicone elastomeric particles. In general, more surfactant leads to smaller particle size emulsions and vice versa. However, other conditions during emulsion preparation, such as viscosity of oil or polymer to be emulsified, choice of high shear device, mixing speed, temperature, and pressure among other conditions, can also impact particle size of emulsions. Thus, varying the emulsification conditions (as described) along with the choice and level of surfactant can all be tailored to produce an emulsion of reactive siloxane polymers with a desired particle size that in turn produces silicone elastomeric powders having an acceptable or desired particle size distribution.

5.2.5 Process

The basic concept for preparing silicone elastomeric powders by the aqueous emulsion process involves three steps:

1. Emulsification
2. Cross-linking
3. Removal of water

The underlying principle involved when preparing these powders is to emulsify the composition before significant cross-linking occurs. If cross-linking proceeds to an appreciable extent prior to emulsification, a poor quality emulsion may result, or it may be impossible to emulsify the composition entirely. In the case of Pt catalysis, when the three reactive components (SiH, Si-vinyl, and Pt) are combined, reaction usually begins immediately, and in some cases, the composition can cross-link before it can be emulsified. Less reactive Si-vinyl siloxanes can be selected in this case, or a Pt inhibitor can be used. Inhibitors are used extensively in Pt-catalyzed silicone compositions, and they work by complexing with the Pt atom and preventing normal catalysis until they are removed, usually by elevated temperature. Typical Pt inhibitors are acetylenic diols, such as 3-methyl-2-butyn-2-ol or 3,5-dimethyl-1-hexyn-3-ol (Surfynol®61 surfactant), and esters of dibasic carboxylic acids, such as diethylfumarate.

The emulsion is formed by subjecting the mixture of reactive siloxanes (polymer + cross-linker), surfactant, and water to high shear. Any one of the typical high-shear devices used to prepare emulsions can be used for this operation, such as a homogenizer, a colloid mill, or a Microfluidizer® high-shear fluid processor. Once the emulsion is formed, cross-linking is allowed to proceed. This can occur at ambient temperature or it can be speeded up by elevated temperature.

Upon completion of the cross-linking reactions, the powders are harvested from the emulsion by removal of water. Water removal can be accomplished by several different techniques, such as filtering, centrifugation, evaporation, or spray drying. The latter has emerged as being the preferred method for removal of water. Once harvested and dried, the powders are packaged for delivery.

5.3   SPECIAL SILICONE ELASTOMERIC POWDERS

Over the many years since silicone elastomeric powders were first developed, silicone producers have sometimes modified their silicone powders to address certain shortcomings or to provide entirely new dimensions to the powders. One shortcoming, in particular, has been agglomeration. Silicone elastomeric powders can become agglomerated into larger size domains and breakup into primary particles can sometimes be challenging. Treating silicone powders with inorganic metal oxides or silicone resin can alleviate this tendency for particles to agglomerate. These methods, as well as some other modifications to silicone elastomeric powders, are discussed individually. It should be understood that some of these modified particles discussed are still in the experimental stage.

5.3.1   INORGANIC OXIDE MODIFICATION

It is possible to combine silicone elastomeric powders with small amounts of inorganic metal oxide particles, such as silica (SiO_2), alumina (Al_2O_3), or titania (TiO_2) [16]. Usually, the oxides are combined with silicone elastomeric powders by dry mixing. Silicone elastomeric powders can also be modified by combining aqueous dispersions of the particles and an aqueous metal oxide sol followed by removal of water [17]. Silica modification leads to silicone powder that is totally free flowing and free from agglomeration. Figure 5.4 is a micrograph of a silicone elastomeric powder to which has been added SiO_2. The presence of these oxides in silicone

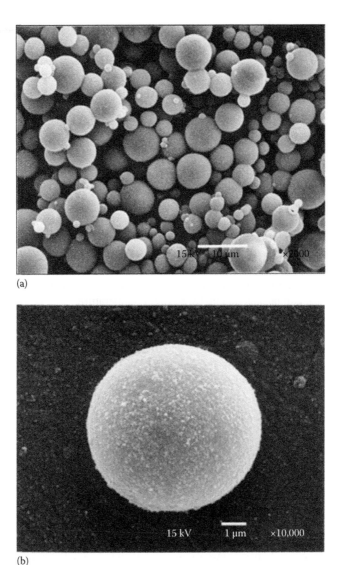

(a)

(b)

FIGURE 5.4 Scanning electron micrograph (SEM) of silicone elastomeric particles treated with SiO_2 particles. (a) Multiple particles; (b) single elastomeric particle with visible SiO_2 particles on surface.

elastomeric powders can change the sensory feel of the powders to some extent. Usually, the amount of oxides is kept to a minimum so as to not alter the sensory properties of the silicone powders significantly. Light scattering can also be affected by the presence of these oxides, and hence, such compositions can exhibit a different appearance on skin from that of the unmodified (pure) silicone powder.

Another method to incorporate metal oxides into silicone elastomeric powders is by a sol-gel technique. Metal alkoxides such as $Al(O\text{-}sec\text{-}Bu)_3$ or $Ti(OBu)_4$ are

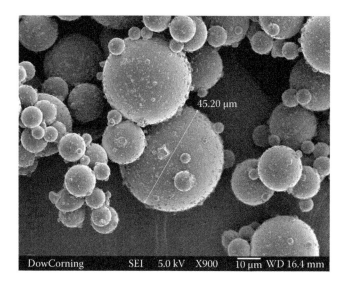

FIGURE 5.5 Scanning electron micrograph (SEM) of silicone elastomeric particles treated with In(Ot-Bu)$_3$. Smaller particles on surface of elastomeric particles are believed to be In$_2$O$_3$ as a result of hydrolysis of the alkoxide.

thoroughly mixed with a silicone powder and allowed to hydrolyze in air to form the corresponding oxide (Al$_2$O$_3$, TiO$_2$). Figure 5.5 shows silicone elastomer particles to which has been added In(O-t-Bu)$_3$. In some cases, the metal alkoxide is added to the reactive siloxanes prior to emulsification.

5.3.2 SILSESQUIOXANE MODIFICATION

It is also possible to modify silicone elastomeric powders by silsesquioxanes, such as methylsilsesquioxane (MeSiO$_{3/2}$). In this case, methylsilsesquioxane resin is deposited on the surface of silicone elastomer particles from reaction of MeSi(OCH$_3$)$_3$ in the presence of NH$_4$OH as a catalyst while the particles are still dispersed in the aqueous emulsion [18]. The main benefit of silsesquioxane modification is decreased agglomeration of the particles. Another use for silsesquioxane modification is for optical effects. Here, phenylsilsesquioxane (PhSiO$_{3/2}$) is formed around the outside of silicone elastomeric particles by reaction of a dispersion of silicone elastomeric particles with PhSi(OMe)$_3$ [19]. Since PhSiO$_{3/2}$ and PDMS have refractive indices significantly removed from one another, such powders exhibit optical effects that are different from unmodified silicone powders.

5.3.3 VERY LARGE PARTICLES

Most commercially prepared silicone elastomeric powders have a mean particle size that is on the order of 2–10 μm. This size leads to very good sensory properties on skin and it also provides good optical properties on skin. By varying the

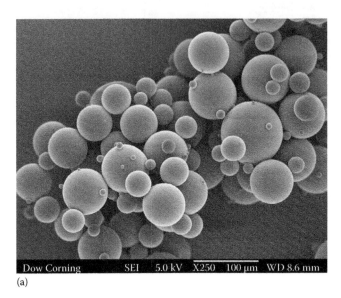

(a)

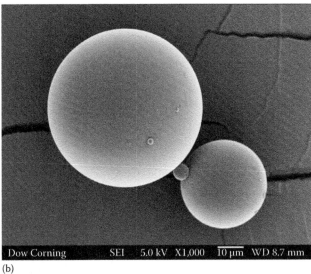

(b)

FIGURE 5.6 Scanning electron micrographs (SEM) of (a) large silicone elastomeric particles; (b) close-up view.

FIGURE 5.7 Optical micrograph of very large silicone elastomeric particles (silicone microspheres).

emulsification conditions, it is possible to arrive at particle sizes that are significantly larger than that of the typical 2–10 μm particles. Figure 5.6 shows an electron micrograph of 50–100 μm silicone elastomer particles.

Still larger particles—silicone microspheres—on the order of 300–500 μm are shown in Figure 5.7. Although these silicone microspheres are composed of the same polymer composition as typical 2–10 μm silicone elastomer particles, they feel completely different on skin. The 300 μm silicone microspheres feel somewhat gritty on skin; the feeling is not unpleasant, but it is certainly not the silky smooth feeling the smaller particles impart. A potential use for these large particles would be in a scrub composition for exfoliating dead skin cells from skin.

5.3.4 Metal Coated Particles

Using a chemical reduction technique involving SiH, it is possible to deposit certain metals onto the surface of silicone elastomeric powders [20]. Metals in the form of their salts that are lower than SiH on the electromotive series can be reduced by SiH. In this manner, Au, Ag, and Cu salts, when added to aqueous emulsions of silicone elastomeric particles in the presence of excess SiH, deposited the metal as nanoparticles on the surface of the silicone elastomeric particles. Figure 5.8 shows silicone elastomeric particles to which Au has been deposited. It was Michael Faraday who first reported the unusual colors of gold when it was subdivided into smaller and smaller particles [21]. Ag modified silicone elastomeric particles prepared in this manner exhibited antimicrobial activity.

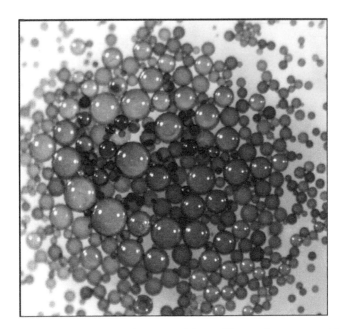

FIGURE 5.8 Optical micrograph of Au-coated large silicone elastomer particles. Largest particles are ~200 μm in diameter. Au nanoparticles on surface of elastomeric particles are not resolved.

5.3.5 Aqueous Suspension of Silicone Elastomeric Powders

In some cosmetics formulations, it is desirable to work with silicone elastomeric powders that disperse easily in water or are already dispersed in water. A typical aqueous suspension of silicone elastomeric particles is shown in Figure 5.9. Dispersions or suspensions of silicone elastomeric powders are prepared simply by omitting the water-removal step [22]. In reality, it is not quite that simple as the dispersion needs to remain stable for long periods of time. Thus, thought must be put into the design of the emulsion or dispersion in order to produce a suspension having the necessary stability. Factors like creaming, separation, and coagulation should be addressed for example. Another issue with the aqueous dispersion is microbial contamination. Since the suspension contains water, it is subject to contamination by microorganisms and it can harbor microbial growth. Thus, biocides or antimicrobial agents are usually added to the aqueous suspension of silicone particles in order to prevent microbial growth. The dry powder, on the other hand, is void of these issues because there is no water present and the powders are usually heat sterilized during the water-removal step.

5.3.6 Re-Dispersible Silicone Elastomeric Powders

Despite being prepared from an aqueous emulsion initially, silicone elastomeric powders can be quite difficult to redisperse in water, even in the presence of added surfactants. To overcome this problem, redispersible silicone powders were

FIGURE 5.9 Optical micrograph of aqueous dispersion of silicone elastomeric particles.

FIGURE 5.10 Redispersible silicone elastomeric powder. Left: conventional dry silicone elastomeric powder mixed with water. Right: dry silicone elastomeric powder treated with butylene glycol and mixed with water.

invented [23]. When small amounts of certain water-soluble substances such as glycols are combined with silicone elastomeric particles, the particles become very easy to redisperse. Figure 5.10 shows one of the redispersible powders that has been mixed with water, and for comparison, a conventional silicone powder mixed with water.

5.4 PROPERTIES

Silicone elastomeric powders consist of fine particles of spherically shaped silicone elastomer that have excellent thermal stability, and they also have very interesting sensory and lubrication properties.

5.4.1 PARTICLE SHAPE AND SIZE

The spherical shape of silicone elastomeric particles is a result of the emulsion process. Most emulsions have spherically shaped particles and emulsions of silicone elastomer powders are also spherical. Once the polymers within the emulsion particles have cross-linked, the particles are essentially nondeformable, so the integrity of the original emulsion particles is maintained throughout additional processing, which results in spherical particles of silicone rubber.

As stated previously, the particle size of silicone elastomeric powders is on the order of 2–10 µm. This value is for the mean particle size of collections of silicone powders. In other words, some samples of silicone powders will have a mean particle size near 2 µm, while mean particle size of other samples might be closer to 10 µm. Figure 5.11 shows a typical particle size distribution for silicone elastomeric powders. It should be understood that silicone elastomeric powders exist as distributions of particles, and for a stated particle size, approximately half of the particles will be below that value and half will be above it.

The emulsification process used to form the initial aqueous emulsion of silicone polymer particles is responsible for this distribution. The emulsification process used to prepare the initial emulsion can be adjusted to some degree so as to produce a broader or a more narrow range of particles. However, the high-shear devices used to prepare these emulsions most likely are not capable of producing a true monodisperse particle size emulsion, and hence, truly monodisperse silicone elastomeric powders.

It should be possible to prepare monodisperse silicone elastomeric powders using a technique called microfluidics. This process involves forcing one liquid through a

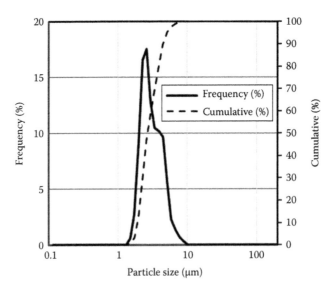

FIGURE 5.11 Particle size distribution of a typical silicone elastomeric powder. Volume percent $D_v0.5 = 5.71$ µm. Solid curve is volume percent while dashed curve is cumulative percent. Results obtained on a Malvern® Mastersizer S particle size instrument (version 2.81).

capillary orifice into another liquid such that the first liquid forms drops within the second liquid, with both liquids being immiscible in one another [24]. Usually, a surfactant is present, which stabilizes drops of the first liquid so they become dispersed in the second liquid. Particle size is affected by parameters such as orifice diameter, flow rate, viscosity, and interfacial tension among others, and as long as everything remains constant, a uniform droplet size is produced, which would translate into a monodisperse emulsion. The reader could easily visualize monodisperse silicone powders being prepared by this technique by using a curable silicone composition as the first liquid and the second liquid being water or some other immiscible liquid.

5.4.2 Mechanical Properties

It is possible to use a wide range of different polymer compositions for preparing cured networks that constitute silicone elastomeric powders, and this can lead to differences in mechanical properties of the silicone elastomeric particles, such as hardness and modulus. As these differences in mechanical properties can lead to different sensory experiences, silicone powder manufacturers supply their powders with a range of these mechanical properties. Perhaps the most meaningful property is hardness and usually this is measured as Shore A hardness or JISA hardness. Typical hardness of silicone elastomeric powders ranges from a low of around 20 to a high of around 70. Hardness is not actually measured on individual particles. Rather, it is measured on monolithic slabs of silicone elastomers that have been prepared with identical polymer compositions that are used to prepare the corresponding powders. Shear modulus of typical silicone elastomer powders is on the order of 0.5–0.6 MPa at 20°C and 1 Hz frequency (see Figure 5.12).

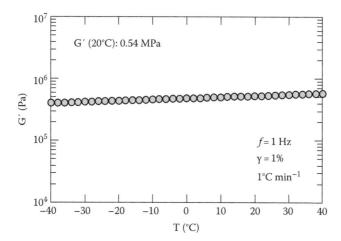

FIGURE 5.12 Dynamic shear modulus of a typical silicone elastomer particle composition. Modulus was determined from an elastomeric film cast from the identical elastomer composition used to prepare silicone elastomer particles. Measurements obtained on an ARES® controlled strain rheometer. f is frequency in Hz; γ is strain in percent.

5.4.3 Thermal Properties

Silicone rubber is well known for its excellent thermal properties, or in other words, its resistance to the deleterious effects of elevated temperature. Silicone elastomeric powders also possess excellent high temperature properties (see Figure 5.13). From the TGA (thermogravimetric analysis) data in Figure 5.13, it can be seen that a 10% weight loss corresponds to a temperature of approximately 437.5°C. Thus, silicone powders are useful in applications where high temperatures are prevalent. One such application is modification of thermoset epoxy resins, which is described in more detail in Section 5.5.1.

5.4.4 Sensory Properties

Silicone elastomeric powders, when applied to skin and rubbed, provide a very pleasant and highly agreeable sensation that is typically described as "soft and silky." Velvety is another descriptor of the sensory feel of silicone powders. The particles' spherical shape is believed to act like "ball bearings" and roll on the skin surface, while the low modulus of the silicone rubber particles is believed to provide a "cushion" feel to skin. These combined effects lead to the unique and extremely pleasant sensation when silicone elastomeric powders are applied to skin. Silicone elastomeric powders are very popular among skin care formulators due to the perception of younger, smoother skin. The threshold concentration for realizing these perceived benefits of silicone elastomeric powders in a skin cream is on the order of 3%–4% by weight [10].

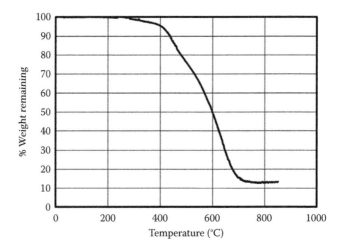

FIGURE 5.13 Thermogravimetric analysis (TGA) of silicone elastomeric powder. TGA run under nitrogen at 10°C min^{-1}. 10% weight loss occurred at 437.5°C.

5.5   APPLICATIONS OF SILICONE ELASTOMERIC POWDERS

Over approximately three decades, silicone elastomeric powders have been com-mercially available. They have found utility in a handful of applications, such as modifying thermoset and thermoplastic resins, modifying certain coatings, and uses in personal care including skin care and hair care. These topics are discussed individually.

5.5.1   THERMOSET RESINS

Modification of thermoset resins, in particular epoxy molding compounds, was the first commercial application for silicone elastomeric powders. Addition of silicone rubber powders to epoxy molding compounds allows for electronic devices encap-sulated with the molding compounds to withstand the harsh conditions of solder bath dipping without cracking or chipping. This is due to stress relief in the thermo-set plastic brought about by dispersed silicone elastomer particles [25]. The inher-ent thermal stability of the silicone elastomer powders makes them ideal for this application.

5.5.2   LIGHT DIFFUSION IN PLASTICS

Silicone elastomeric powders are used to modify certain plastics for light diffu-sion purposes. In this application, silicone powders are dispersed in thermoplastic resins which are formed into sheets that are used to provide a uniform source of light when it is projected through plastic sheets. Such sheets make up a com-ponent for TFT LCD TVs (thin film transistor liquid crystal display televisions) called a back light unit (BLU). The disparity in refractive indices between the silicone elastomeric powder and the thermoplastic resin results in significant light scattering within the plastic sheets such that the BLUs for these TVs perform as required.

5.5.3   COATINGS

Silicone elastomeric powders are used in certain coatings to modify the coatings' appearance. In some cases, silicone powders can impart a matte appearance to coatings by reducing gloss significantly. Figure 5.14 shows the effect silicone elas-tomeric powders have on the gloss of a typical water-based polyurethane coating. Precipitated silica is used extensively as a matting agent in coatings, and it is consid-erably lower in cost than silicone elastomer particles; however, it does not provide the burnish resistance to coatings as do silicone elastomer particles. Silicone powders can also lead to coatings that possess a soft and pleasant sensory feel, also unlike silica. Silicone powders can be used in both oil-based and water-based coatings. Dry silicone elastomeric powder or aqueous suspensions of the particles can be used in coatings, although aqueous suspensions are much easier to disperse in the coating compared to dry powders.

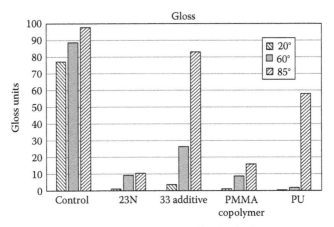

Matting agent at 10% active level

FIGURE 5.14 Effect of silicone and other particles on gloss of an acrylic latex coating. Samples consisted of Joncryl® ECO 2124 acrylic latex with 10% matting agent (active basis) added. Control is without matting agent. 23N is a dry silicone elastomeric powder; 33 additive is an aqueous dispersion of silicone elastomeric particles; PMMA copolymer is a polymethyl methacrylate-co-ethylene glycol dimethacrylate; PU is an aqueous polyurethane matting agent (Neorez® R-1010). Gloss measured at 20°, 60°, and 85°. Lower gloss values are better. Glossmeter used: ELCOMETER 408.

5.5.4 MODIFICATION OF SURFACES

Silicone elastomeric powders can be used to provide lubrication, release, antiblocking, and slip to polymer surfaces [26]. These applications appear to be related to the particle shape of silicone elastomeric powders and the inherent low surface energy of silicone polymers.

5.5.5 APPLICATIONS IN SKIN CARE

Due to their soft elastomeric nature, spherical shape, unique and pleasant sensory effect, ability to absorb oils, and optical effects, silicone elastomeric powders are used widely in skin care applications. Targeted benefits in skin care applications include silky feel, sebum absorption, matte finish, soft focus, wrinkle masking, and delivery of actives. These topics are discussed individually.

5.5.5.1 Absorption of Liquids and Delivery of Actives

Silicone elastomeric powders are cross-linked networks of siloxane polymers, and thus they are insoluble in all solvents. They are, however, swellable in certain solvents, and this is an important property in terms of their utility as potential agents for delivery of actives to skin. Due to their swelling behavior and also due to the particles' cumulative surface area, silicone elastomeric particles can take up moderate quantities of various oils and still remain dry to the touch. Therefore, silicone

TABLE 5.1

Absorption of Various Liquids by Different Silicone Elastomer Powders[a]

Liquid	9506 Powder[b]	9701 Cosmetic Powder[c]	EP-9215 Cosmetic Powder[d]
Octyl palmitate	3	1	0.5
Isopropyl myristate	5	1.4	0.8
Mineral oil	3	1.5	0.6
Castor oil	0.5	0.9	0.6
Sunflower oil	0.5	0.8	0.5
Isododecane	2	3	1.5
Cyclopentasiloxane	4	4.1	1.6
Water	1	1.1	0.5
Ethanol	1.3	1.1	0.4

[a] Values represent maximum weight in gram of liquid absorbed per gram of powder while composition still remains dry to the touch.

[b] 9506 Powder—INCI name: Dimethicone/Vinyl Dimethicone Crosspolymer. Low hardness particle (Shore A 30 hardness).

[c] 9701 Cosmetic Powder—INCI name: Dimethicone/Vinyl Dimethicone Crosspolymer (and) Silica. Low hardness particle treated with SiO_2 (Shore A 40 hardness).

[d] EP-9215 Cosmetic Powder—INCI name: Dimethicone/Vinyl Dimethicone Crosspolymer. Higher hardness particle (Shore A 60 hardness).

elastomeric powders can be used to carry certain oily substances or emollients to skin such as sunscreens, fatty esters, mineral oil, and glycerides.

As is true for most elastomeric networks, higher cross-link density leads to lower swelling. Thus, harder silicone elastomeric powders absorb less oil than softer particles as the former possess a higher cross-link density that the latter. Particle size can also affect silicone powders' ability to absorb liquids as some liquids are absorbed at the particles' surfaces. (Total surface area of particles varies inversely with particle size.) Table 5.1 lists various liquids that are absorbed by several different commercially available silicone elastomeric powders with the powders still remaining dry to the touch.

5.5.5.2 Absorption of Sebum

Sebum, the oily substance that accumulates on skin, consists of about 57.5% glycerides and free fatty acids, 26% wax esters, 12% squalene, 3% cholesterol esters, and 1.5% cholesterol [27]. Sebum is secreted by sebaceous glands which are found primarily on the scalp, face, shoulders, and upper chest of the human body. When sebum accumulates on skin, its presence can make skin appear shiny, which is undesirable, especially when it is on the face. When applied to skin, silicone elastomeric particles are capable of absorbing sebum and can lead to a reduction in the level of sebum on the surface of skin.

The level of sebum on skin can be quantitatively measured to determine how effective silicone powders are at controlling sebum accumulation on skin.

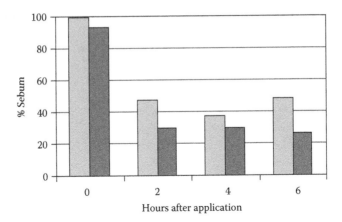

FIGURE 5.15 Level of sebum remaining on human skin after treatment with a formulation containing 23% by weight silicone elastomeric powder. Light bars (left) represent Dow Corning® 9506 and dark bars (right) represent Dow Corning® 9701. Both silicone powders are low hardness; 9701 has a surface treatment of SiO_2 particles. Values were obtained using a Courage + Khazaka Gmbh. Sebumeter® SM810 and are at a 99% confidence level.

Figure 5.15 shows the relative amounts of sebum remaining on skin after treatment with two kinds of silicone elastomeric particles. In some investigations, artificial sebum is used to determine performance of certain compositions for controlling sebum. Table 5.2 illustrates the level of (artificial) sebum absorption of different silicone elastomeric powders. For this particular property, the water-dispersible silicone elastomeric powder (*Dow Corning®* EP-9801 Hydro Cosmetic powder) offers additional possibilities to the cosmetic chemist as no oils are needed to disperse the powder. This reduces competition for absorption of sebum (from otherwise added oils) therefore allowing maximum efficiency from a very simple aqua gel-base formulation up to 6 hours [28].

TABLE 5.2
Artificial Sebum Absorption of Different E Powders[a]

Liquid	9506 Powder[c]	7901 Cosmetic Powder[d]	EP-9801 Hydro Cosmetic Powder[e]
Artificial sebum[b]	1.75	1.01	0.85

[a] Values represent maximum weight in grams of liquid absorbed per gram of powder with powder still remaining dry to the touch.
[b] Artificial sebum is a mixture of: triolein (CAS 122-32-7) 60%; oleic acid 20%; squalene 20%.
[c] 9506 Powder—INCI name: Dimethicone/Vinyl Dimethicone Crosspolymer: low hardness particle (Shore A 30 hardness).
[d] 9701 Cosmetic Powder—INCI name: Dimethicone/Vinyl Dimethicone Crosspolymer (and) Silica: low hardness particle treated with SiO_2 (Shore A 40 hardness).
[e] EP-9801 Hydro Cosmetic Powder—INCI name: Dimethicone/Vinyl Dimethicone Crosspolymer: higher hardness particle (Shore A 60 hardness).

Acne (*Acne vulgaris*) is a chronic disorder of the skin that is prevalent among adolescents, and its treatment usually involves reducing sebum production and controlling bacterial growth [27]. Although reducing the level of sebum on the surface of skin could be a beneficial strategy for treating acne, it is not known at this time if silicone elastomeric particles offer a therapeutic benefit for treating acne by reducing the level of sebum on the surface of skin. Silicone elastomeric particles having antimicrobial activity could possibly be of therapeutic benefit for acne or other disorders of the skin in which uncontrolled or undesirable bacterial activity is involved.

5.5.5.3 Optical Properties, Soft Focus, and Wrinkle Masking

In an aging population, cosmetic products that hide the effects of aging are highly desirable. Wrinkles and fine lines on skin trap light in crevices and cause the skin tone to look uneven. Pigments from foundations and makeup tend to collect in skin crevices and accentuate fine lines and wrinkles. Simply by covering up the imperfections in some cases creates an unnatural, caked-on appearance. There is a need for materials that can mask the signs of aging, while at the same time project the natural skin tone.

Compositions that coat with transparent materials which give sufficient differences in refractive indices could result in a "soft-focus" effect. The details of the principle and requirements for the soft-focus effect are described by Emmert [29]. In a simplified description, the soft-focus effect is apparent when the total reflection (TR) is minimal (<20% of total light), total transmission (TT) is maximal (>75% of total light), while diffused light reflection (DR) is maximal (80%) and diffused light transmission (DT) is maximal (>50%). When these criteria are met, the total light transmission makes the natural skin tone show through (like a back light) to give a natural glow, while the diffused transmission would even out the skin tone and hide imperfections. Suspension of silicone elastomeric particles have demonstrated soft-focus performance that is further increased by the presence of additional silica aerogel as illustrated in Figure 5.16.

The oil-in-water (O/W) formulation containing the combination of silicone elastomer suspension and the silica aerogel was further evaluated *in vivo* with VISIA CR (an optical imaging system for complexion analysis by Canfield Scientific). The material was applied around the crowfeet area of the face near the eyes where fine wrinkles are the most visible [9]. The pictures in Figure 5.17 illustrate an immediate effect on the reduction of the fine wrinkles.

5.5.6 APPLICATION IN HAIR CARE

Silicone elastomeric powder suspension has been evaluated in hair care products, such as shampoos, rinse off and leave on conditioners, resulting in improved hair smoothness and softness. These conditions are also described as a velvety feel when compared to a control. Thus, it is believed that silicone elastomeric powders provide conditioning to hair by bringing a lubricating effect to hair fibers, which allows them to move past one another very easily. Another use for silicone elastomeric particles

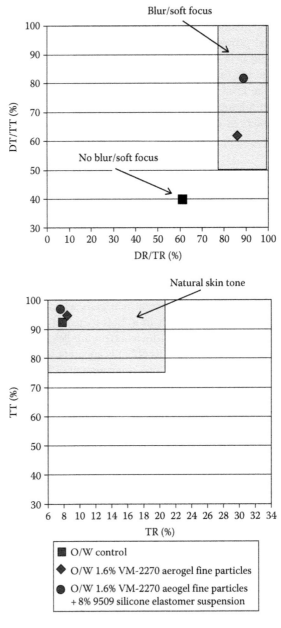

FIGURE 5.16 Soft focus results for an O/W formulation containing elastomeric powder suspension alone and in combination with silica aerogel compared to a formulation control. Formulation respectively contains no silicone powders, 8% by weight of Dow Corning® 9509 silicone elastomer suspension, and 8% by weight of 9509 plus 1.6% by weight of Dow Corning® VM-2270 aerogel fine particles. Values were obtained using a UV Vis spectrophotometer with integrating sphere. DT, TT, DR, and TR stand for diffused transmittance, total transmittance, diffused reflection, and total reflection, respectively.

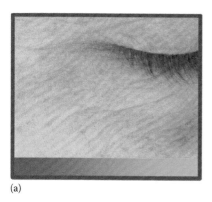
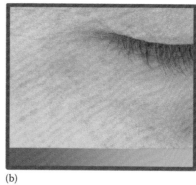

(a) (b)

FIGURE 5.17 Photographs showing neat skin before (a) and 5 min after (b) treatment of crow's feet wrinkles (facial wrinkles radiating from under or outer corner of eyes) with an oil-in-water (O/W) composition containing 8% by weight of Dow Corning® 9509 silicone elastomer suspension and 1.6% by weight of Dow Corning® VM-2270 aerogel fine particles.

in hair care is for reducing the amount of sebum on hair and the scalp [30]. Although hair care is one area where silicone elastomeric powders have been found to be useful in personal care applications, it is not nearly as significant as silicone elastomeric powders' use in skin care. Perhaps this is because numerous other silicone compositions also condition hair very well and many of such compositions are significantly lower in cost.

5.6 CONCLUSIONS

Silicone elastomeric powders, their preparation, properties, and their uses have been described. Silicone elastomeric powders are fine spherical particles of silicone rubber that are based on PDMS that have particle sizes on the order of 2–10 μm. Silicone elastomeric powders are generally prepared by an aqueous emulsion process whereby reactive siloxane polymers are emulsified and allowed to cross-link. Removal of water leaves spherical silicone elastomeric particles or powder. Silicone powders are used to modify certain plastics and also certain coatings. Silicone elastomeric powders generally have properties of silicone rubber, some of which include high temperature stability and low modulus (softness). They also have unusual sensory properties which make them useful in certain cosmetics formulations. In cosmetics, they are used for their extraordinary and pleasant sensory properties, oil and more specifically sebum-absorption capabilities, and soft-focus property, which helps to mask wrinkles on skin.

ACKNOWLEDGMENTS

This chapter is dedicated to the authors' dear friend and colleague, Yoshitsugu Morita, who was one of the original inventors of silicone elastomeric powders. The authors express their appreciation to Jennifer Stasser for obtaining the micrographs of Figures 5.1 and 5.9, to Dr. Glenn Gordon for obtaining the rheological data shown in Figure 5.12, to Timothy Roggow III for obtaining the optical micrograph in Figure 5.8, to Deborah Baily for obtaining the electron micrographs of Figures 5.6 and 5.7, to Rose Bao for providing data in Table 5.2 and the picture in Figure 5.10, to Anne Marie Vincent for providing data in Table 5.1 and in Figures 5.15 and 5.16, and to Bertrand Lenoble for providing the data in Figure 5.14. The authors thank the Dow Corning Corporation for supporting their work in the field of silicone elastomer particles and also for support during the writing of this manuscript.

REFERENCES

1. Liles, D. T., Morita, Y., Kobayashi, K., *Polym. Prepr. (Am. Chem. Soc., Div. Polym. Chem.)*, 2001, *42*(1), 240–241.
2. Liles, D. T., Morita, Y., Kobayashi, K., *Polym. News*, 2002, *27*, 406–411.
3. Hanada, T., Morita, Y., US 4,594,134, 1986.
4. Shimizu, K., Hamada, M., US 4,742,142, 1988.
5. Yoshida, K., Shimizu, K., Hamada, M., US 4,743,670, 1988.
6. Oba, T., Mihama, T., Futatsumori, K., US 4,761,454, 1988.
7. Liles, D. T., Lin, F., *Polym. Prepr. (Am. Chem. Soc., Div Polym. Chem.)*, 2008, *49*(2), 1087–1088.
8. Liles, D. T., Lin, F., *Polymeric Delivery of Therapeutics*, ACS Symposium Series 1053, 2010, Morgan, S. E., Lockhead, R. Y. (eds.), American Chemical Society, Washington, DC, pp. 207–219.
9. Vincent, A.-M., Tomalia, M. K., Tonet, G., Canfield, L., *Paris 2014 28th Congress*, Paris, France, October 27–30, 2014, pp. 189–194.
10. Vervier, I., Courel, B., *Cosmet. Toiletries Mag.*, November 2006, *121/11*, 65–74.
11. Polmanteer, K. E., *Rubber Chem. Technol.*, 1988, *61*(3), 470–502.
12. Marciniec, B., *Hydrosilylation*, 2009, Springer, Dordrecht, the Netherlands, p. 8.
13. Speier, J. L., Webster, J. A., Barnes, G. H., *J. Am. Chem. Soc.*, 1957, *79*, 974–979.
14. Karstedt, B. D., US 3,775,452, 1973.
15. Donaldson, J. D., *Prog. Inorg. Chem.*, 1967, *8*, 287–356.
16. Morita, Y., Yokoyama, N., US 5,387,624, 1995.
17. Morita, Y., Sakuma, A., US 5,948,469, 1999.
18. Inokuchi, Y., Kuwata, S., US 5,538,793, 1996.
19. Inokuchi, Y., US 6,376,078, 2002.
20. Liles, D. T., US 8,715,828, 2014.
21. Faraday, M., *Phil. Trans. R. Soc. London*, 1857, *147*, 145–181.
22. Kuwata, S., Nakazato, M., Inokuchi, Y., US 5,871,761, 1999.
23. Morita, Y., Kobayashi, K., US 7,399,803, 2008.
24. Weitz, D. A. et al., *Mater. Today*, 2008, *11*(4) 18–27.
25. Block, H., US 4,853,434, 1989.
26. Blackwood, W. R., Habermahl, J., Liles, D. T., US 6,534,126, 2003.
27. Wilkinson, J. B., Moore, R. J. (eds.), *Harry's Cosmeticology*, 7th edn., 1982, Chemical Publishing Co., New York, pp. 16–18.
28. Van Reeth, I., Bao, R. X., *SOFW-J.*, 2014, *140*(8), 20–24.
29. Emmert, R., *Cosmet. Toiletries Mag.*, 1996, *111*, 57.
30. Anderson, G. T., Milow, C. A., Luo, X., US 7,491,382, 2009.

6 Janus Emulsions
Some Fundamentals

Stig E. Friberg

CONTENTS

6.1 INTRODUCTION

Commercial preparations of multiple emulsions have demonstrated difficulties to achieve satisfactory storage stability. The traditional colloid stability approach becomes too complex when several oils of small mutual solubility are involved. The multiple interfaces each require different surfactants, and when the surfactants inevitably mix, their effect is reduced.

However, recent research indicates a significant stabilization action of interfacial thermodynamics and points out that the interfacial free energy actually contributes to the stabilization of multiple oil drops.

This chapter will outline the relevance of interfacial thermodynamics and oil volume ratios for multiple emulsions. The foregoing analysis uses a water–silicone–vegetable oil Janus emulsion as a model system. The purpose of using a silicone oil and a vegetable oil in the model is twofold. First, it is convenient: silicone oil and vegetable oil are mutually insoluble, and the values of the pertinent interfacial tensions in the system present an opportunity for investigation into the fundamentals of multiple emulsions. Second, in commercial applications, silicone oils are often combined with other oils to bring together the advantages of both in one formulation. Cosmetic and personal care products provide prominent examples [1,2]. The result of the fundamental research can be applied to commercially useful systems [3].

Recent studies based on Janus emulsions of water–silicone–vegetable oil have added to the fundamentals of emulsion science, revealing a thermodynamic emulsion stabilizing factor for the first time, as mentioned. Earlier, thermodynamics attracted scant attention; single oil emulsions did not offer a constructive role for it. The positive interfacial free energy played a dominant role in the total emulsion free energy, making such systems decidedly unstable. With this fact as a basis, the research naturally focused on the *kinetics* of destabilization, and improved "emulsion stability" was defined as a slower destabilization *rate*. This relative stability has, until recently [4], also been the focus of research into multiple emulsions, but the approach was markedly changed with the introduction of the microfluidics method [5]. Actually, the new preparation method did not only change the emulsion science, it indicated a large number of potentially new and advanced applications [6,7]. Of these may be mentioned the use of totally engulfing techniques for living cells [8] and for microorganisms in general [9], the obvious use for drug delivery systems [10], and synthesis of asymmetric solid particles [11] and even self-propelled particles [12].

With these new emulsions and their applications, thermodynamics now became relevant in emulsion science because the microfluidics method of emulsion preparation, with virtually no agitation, gave suitable systems to explore the role of equilibrium interfacial thermodynamics for the drop configuration. For the first time, experimental results could be compared to the equilibrium drop configuration in a meaningful manner [7,13,14], revealing that interfacial thermodynamics actually play a decisive role in the formation and properties of multiple emulsions.

In addition, another recent event significantly contrived to re-direct the research focus toward thermodynamics. Unexpectedly, Janus emulsions with silicone oil were also prepared by the traditional high-energy emulsification process [15–17]. There is less obvious—if any—correlation of such systems with equilibrium thermodynamics, giving additional impetus to correlating experimental results for the topology of aqueous multi-oil for systems [18,19] with equilibrium configuration: an analysis that has reached a highly advanced state with the recent contribution by Shardt et al. [20]. Moreover, the discovery also indicated the potential for batch preparation of nano-particles with complex and uniform structures [21], thereby potentially removing the volume constraints of the microfluidics method. All these developments show an area in rapid development, supported by related discoveries of less traditional Janus structures [22,23].

In a few words, these recent developments have ascertained a significant role of interfacial thermodynamics in emulsion science. Silicone oil emulsions are an excellent example to illustrate the effect of interfacial equilibria to favor different drop configurations: entirely separated drops, totally engulfed ones, or Janus drops. In addition to the equilibria at the contact line between the three liquids, the overall drop topology also depends on the relative volumes of the involved liquids. In the following sections, both factors will be evaluated after a brief introduction of interfacial thermodynamics and multiple oil emulsion configurations.

6.2 INTERFACIAL TENSIONS IN MULTIPLE EMULSIONS—AN OVERVIEW

There are but a few experiments reported on interfacial tensions in multiple emulsion drops, but a recent publication [14] about the combination of a hydrophobic silicone oil with a monomer, tripropyleneglycol diacrylate, and an aqueous solution

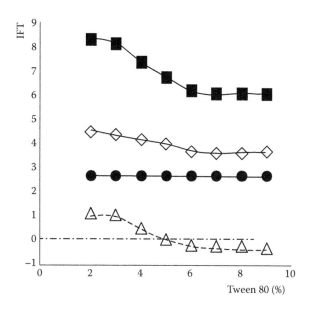

FIGURE 6.1 The interfacial tensions in the system: tripropyleneglycol diacrylate (TP), a silicone oil (SO), and an aqueous solution (Aq) of Tween 80 as a function of the surfactant concentration. ■ $\gamma_{SO/Aq}$, ◇ $\gamma_{TP/Aq}$, ● $\gamma_{SO/TP}$, △ $\gamma_{SO/Aq}-\gamma_{TP/Aq}-\gamma_{SO/TP}$. (From Pannacci, N. et al., *Phys. Rev. Lett.*, 101, 16450, 2008.)

of a surfactant, Tween 80, offers a helpful overview of the trends, Figure 6.1. The results are, in fact, an eminently useful illustration of how interfacial tensions may affect the configuration of multiple emulsion drops, and a brief review is instructive of the consequences.

The interfacial tensions between the aqueous solution and the oils, Figure 6.1, are reduced with increased concentration of the surfactant, while the number for the tension between the two oils is virtually constant. This fact has a substantial effect on the preferential configuration of emulsion drops, actually distinguishing between double emulsions and the Janus variety. At the lowest concentrations of surfactant, $[S] < 5.2\%$, the difference between the interfacial tensions, $\gamma_{SO/Aq}-\gamma_{TP/Aq}-\gamma_{SO/TP}$, is positive (the lowest curve in the figure). As a result, in this range of surfactant concentration, the total interfacial free energy of a double emulsion drop is lower than that of a Janus emulsion, that is, it is more stable thermodynamically. That difference may, at a first glance, be considered of minor importance, but is, in actual fact, substantial. An approximate number for a 5 μm drop would be of the order 10^{-11} J, 10 orders in excess of the number for kT, 4×10^{-21} J!

The ramification for the preparation of double emulsions is both wide and essential. At this level of surfactant, a double emulsion would spontaneously form and, in addition, the long-term storage stability concerns are limited to the TP/Aq interface as long as the interfacial tension difference in question is positive. The inner SO/TP interface is thermodynamically stable and the double emulsion would only have to be stabilized against coalescence of the total drops. That is, the TP/W interface is the only one of concern, as long as the surfactant in question is also useful against its coalescence. As a result, the engulfed drop *configuration* would be stable, as has recently been experimentally verified for a Janus emulsion during more than a month of creaming and coalescence [24,25]. Of course, the final emulsion destabilization to two separate oil layers is inevitable, but the configuration per se has been shown to be unexpectedly stable [24], and there is an obvious need for further experimental investigations to clarify the final destabilization of a Janus emulsion.

When the surfactant percentage is increased, Figure 6.1, to a range of 6%, the tension difference in question becomes negative; there is an interfacial equilibrium and now the thermodynamics favor a partially engulfed configuration, an equilibrium Janus drop, Figure 6.2.

In actual fact, the equilibrium of the tensions and angles, Figure 6.2a and Equation 6.1, favor the configuration of the Janus drop, Figure 6.2a and b.

$$\gamma_{SO/Aq} = \gamma_{TP/Aq} \cos\beta + \gamma_{SO/TP} \cos\delta \qquad (6.1)$$

Figure 6.2a offers an image of the angles β and δ that characterize the equilibrium at the contact line. Figure 6.2b shows the site of the contact line and, in addition, all the elements of the Janus drop, which will be utilized in later sections for a more detailed analysis. These are the local equilibria, which are temporary and not to be confused with the overall thermodynamics of a Janus drop. An emulsion of such drops is thermodynamically unstable and will spontaneously separate into distinct liquid layers with time, but the Janus drop equilibria define a local free energy minimum, which has been shown to affect the behavior of the emulsion.

Pending a quantitative analysis of the features in Figure 6.2b, the equilibrium topological consequences of the changes with the surfactant concentration in Figure 6.1 are instructive. Figure 6.3 presents a plain graphic of the changes with correct dimensions.

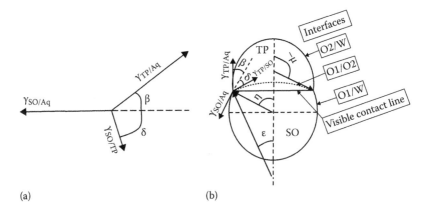

(a)   (b)

FIGURE 6.2 An illustration of the forces and angles of interfacial equilibrium at the contact line in a Janus drop (a) and all its pertinent factors (b).

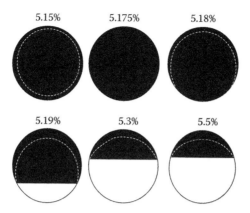

FIGURE 6.3 The equilibrium profiles of SO/TP Janus drops with a TP volume fraction of 0.1 in emulsions of different surfactant concentrations, Figure 6.1, in weight percent. The critical surfactant concentration shifts from double emulsion drops to the Janus variety, S_{crit} = 5.175%. The visible part of the SO sphere is white, while the black area shows the coverage of SO by TP. The dashed white line denotes the penetration of SO into the TP cap.

The images in Figure 6.3 bring to light the effect of interfacial tension equilibrium at the contact line. At surfactant concentrations, less than S_{crit}, 5.175%, even by an infinitely small amount, the TP covers the entire SO sphere and the dimension of the TP layer depends on its volume fraction. For the entire range $S < S_{crit}$, a variation of the surfactant concentration has no effect on the visible drop topology; a view of the drop shows only a TP surface. The only significant difference with surfactant concentration is a change in the total interfacial free energy. This constant appearance is in contrast to the behavior at $S > S_{crit}$. Now the TP covers virtually the entire SO sphere only for an infinitely small increase of surfactant concentration in excess of S_{crit}, but measurable increases give rise to the topologies shown in Figure 6.3. The reason for the reduced coverage is the enhanced values of both angles β and δ, Figure 6.1, as will be later shown in a quantitative manner. As a consequence of the augmented angles, the coverage of SO by TP and the penetration of TP into the SO cap are consistently reduced with increased surfactant concentration.

The images in Figure 6.3 are all for a TP volume fraction of 0.1 and the topology variation with a changed volume fraction is also of some interest. Figure 6.4 shows profiles of a series of drops at a surfactant percentage of 5.1% (a) and 5.2% (b), also with correct dimensions.

The figures give an illustration of the effect of the interfacial equilibrium for the drop appearance. For example, without the penetration of SO into TP, the visible contact line would be at surface coverage of 0.5, instead of the value 0.99 in the example. These aspects will be given a quantitative analysis in the following sections. In addition, the appearance for $V_{TP} = 0$ is not trivial; it will be examined in a different context.

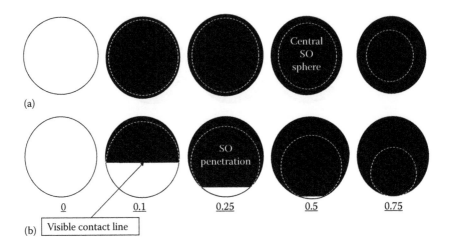

(a)

(b) | Visible contact line |

0   0.1   0.25   0.5   0.75

FIGURE 6.4 A comparison of the topology for SO/TP drops for surfactant concentration 5.1% (a) and 5.2% (b) by weight [14], with underlined numbers showing volume fractions of TP. Visible SO is white, while perceivable TP is black (a and b). The dashed line denotes the limit of SO penetration into the TP cap.

6.3 CORRELATION BETWEEN INTERFACIAL TENSION RATIOS AND PREFERRED TOPOLOGY

The preceding section gave some preliminary aspects of the correlation between the interfacial tensions and the preferred drop topology; the following analysis will quantify these aspects and extend the range of ratios between the tensions. In the analysis, the volumes marked as TP and SO in Figure 6.2 are given the more general denotations, O2 and O1 respectively. The quantitative evaluation is made general by defining the ratios as

$$\gamma_{O2/W} = c\gamma_{O1/W}$$

and

$$\gamma_{O1/O2} = k\gamma_{O1/W}, \tag{6.2}$$

but the range of c is restricted to $c < 1$ to avoid redundant expressions. The lowest value in the k range leads to $1 > k + c$, Equation 6.1, and the traditional interpretation is O2 spreading on O1 giving a double emulsion, O1/O2/W. However, this is a limited reading because of the general assumption that the aqueous solution in the emulsion is the continuous phase. For a case with V_{O2} significantly greater than V_W, the emulsion would be O1/W/O2, Section 6.5.1.

Increasing k, to give $k + c > 1$, results in partial coverage of O1 by O2 to form Janus drops, (O1 + O2)/W, according to Equation 6.2. This simple Janus topology is found for sufficient fraction of the aqueous phase. Experimental evidence [12] has shown a sufficiently small volume of the aqueous phase, actually to give extremely complex emulsions, for example, a triple Janus topology, (O1 + O2)/W/O2/O1 [12].

TABLE 6.1

Conditions for the Drop Configurations in Figure 6.1 for c = 0.9

Drop Configuration	k Range
O1/O2/W[a]	$k < 0.1$
Janus drops	$0.1 < k < 1.9$
Separate drops	$k > 1.9$

[a] See comments in text.

A final increase of k to $k > 1 + c$ generates an emulsion of separate drops of the two oils. Table 6.1 presents the k ranges for different topologies with $c = 0.9$.

The interfacial tension span is wide for Janus emulsions in Table 6.1, but recent research [16] shows a significant part of the span to incorporate unusual and unexpected phenomena. These are effects of the volume ratios of the oils, which are evaluated in the following assessment of the correlation between Janus drop topology, interfacial tensions, and relative volumes of the emulsion liquids.

6.4 EQUILIBRIUM AT THE CONTACT LINE

Balancing the forces in Figure 6.2a along and perpendicularly to the $\gamma_{SO/Aq}$ direction gives the angles β and δ, Figure 6.2

$$\beta = acos\left(1 + c^2 - k^2\right)/2c \qquad (6.3)$$

and

$$\delta = acos\left(1 - c^2 + k^2\right)/2k \qquad (6.4)$$

These equations control the conditions at the contact line, while the entire drop configuration, Figure 6.2b, also depends on the relative volumes of the two dispersed liquids. Unfortunately, the latter feature is not easily calculated from the given volume fractions. The expressions become prohibitively complex, and Guzowski et al. [7] opted to use a computer program to correlate volumes and topology. In the present contribution, the volumes are instead calculated from the geometrical features in Figure 6.2 and the correlations between volume ratios and topology are evaluated ex post facto [12–16]. The features in Figure 6.2b provide a series of relationships [10], Equations 6.5 and 6.6

$$\mu = \eta + \beta \qquad (6.5)$$

$$\varepsilon = \eta - \delta \qquad (6.6)$$

Furthermore, assuming $r_{O1/W} = 1$, the radii $r_{O2/W}$ and $r_{O1/O2}$ are attained

$$r_{O2/W} = \frac{\sin \eta}{\sin \mu} \tag{6.7}$$

$$r_{O1/O2} = \frac{\sin \eta}{\sin \varepsilon} \tag{6.8}$$

The volumes of O1 and O2, equal to the experimental volumes SO and TP, Figure 6.2, are calculated via pre-volumes, φ_{O1} and φ_{O2}, separated by the plane through the visible contact line.

$$\varphi_{O1} = \pi(1 + \cos \eta)^2 (2 + \cos \eta)/3 \tag{6.9}$$

$$\varphi_{O2} = \pi(r_{O2/W} - \cos \mu)^2 (3 - r_{O2/W} + \cos \mu)/3 \tag{6.10}$$

Calculating the V_{O1} and V_{O2} requires the volume $\varphi_{O1/O2}$, which is the volume of the cap between the contact line plane and the O1/O2 interface.

$$\varphi_{O1/O2} = \pi r_{O1/O2}^3 (1 - \cos \varepsilon)^2 (2 + \cos \varepsilon)/3 \tag{6.11}$$

The volumes are

$$V_{O1} = \varphi_{O1} + \varphi_{O1/O2} \tag{6.12}$$

and

$$V_{O2} = \varphi_{O1} - \varphi_{O1/O2} \tag{6.13}$$

These equations are geometrically straightforward, and may at first glance give the impression to offer only trivial features. However, there is one unexpected consequence, which was recently brought to light [11]. If $\beta = \eta$ in Equation 6.7, then $r_{O2/W} = \infty$ with far reaching consequences [11] for the understanding of the association between interfacial equilibrium and volume ratios, as will be illustrated in a numerical example.

6.5   NUMERICAL EXAMPLES OF THE COMBINED INTERFACIAL TENSION AND RELATIVE VOLUME EFFECT

The computations must, with necessity, limit the number of variables in each approach. In the following narrative, the primary calculations are using a model in which the space around the drop is infinite, that is, the only interaction by the continuous phase is with a single drop. The model for the drop per se offers two alternatives: the radius of one of the caps may be invariant, or its volume may be. The first case, a constant O1 radius, is primarily examined in the numerical calculations. The second case, a constant cap volume, is subsequently presented as a modification of the results from the first case. Seen from a physical point of view, the first choice is less realistic, because it entails a simultaneous addition of O2 and removal of O1, but it was chosen because of the computational simplicity.

6.5.1 CONSTANT O1 CAP RADIUS

A complete evaluation of the combined effect of the interfacial equilibrium and of the oil volume ratios would entail a space far in excess in a single chapter, and the calculations are restricted to a geometry in which the visible contact line is positioned at $\eta = 60°$, Figure 6.2b. As a consequence, the *visible* coverage of O1 by O2, defined by the location of the discernible contact line, Figure 6.2b, remains invariant at 0.25. Furthermore, as mentioned, the radius of the O1 cap was kept constant to facilitate the computations [11], while the constant volume of the O1 cap, more in line with the recent experimental results [26], will be evaluated in the subsequent section. Furthermore, the primary calculations were limited to the case of $c = 0.9$, but k spanned the entire range $0.1 \leq k \leq 1.9$ in order to illustrate some unusual effects. However, the relationship between k and the variables in drop geometry, Figure 6.2b, is not easily envisioned , and the ensuing account will instead be based on the more easily visualized value of β. Figure 6.5 shows the angles β and δ for the Janus drop range of k with $c = 0.9$ and $\eta = 60°$, Figure 6.1.

The value of β, Figure 6.5, is monotonically increasing, which makes it a useful variable to use in order to illustrate the change in drop topology with varied interfacial tensions. The effect of $\gamma_{O2/w}$ on the topology is only gradual, while $\gamma_{O1/O2}$ may

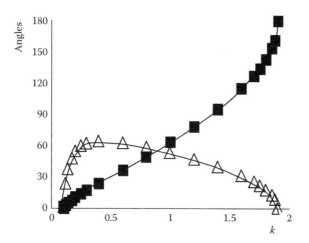

FIGURE 6.5   The angles β (■) and δ ($\triangle$) for the example in the text.

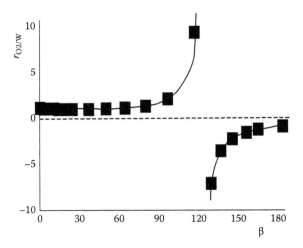

FIGURE 6.6 The radius $r_{O2/W}$ versus the angle β for $c = 0.9$ and $\eta = 60°$.

have a radical outcome, as will be demonstrated by the numerical example, in which c was held constant while k was varied. Figure 6.6 illustrates $r_{O2/W}$ versus angle β.

The radius is fairly constant for small β's, but increases rapidly when approaching 120°, at which point it reaches infinity. Angle β in excess of this value sees negative numbers for the radius in accordance with earlier results [10,11]. The sign reversal at $\beta = 120°$ in the present example is, in fact, a radical change of the $r_{O1/W}$ curvature from concave to convex. As a consequence, the (O1 + O2)/W emulsion undergoes a selective inversion from (O1 + O2)/W to (O1 + W)/O2 [11], Figure 6.7.

The images 4 and 5 in Figure 6.7 illustrate the fact that a Janus drop inversion is theoretically predicted. The related experimental evidence [14] unequivocally reveals an inversion from an aqueous continuous phase to a silicone oil one by reduced aqueous phase volume fractions. However, the emulsification was with high energy and the increased oil volume was for the oil with the greater interfacial tension toward the aqueous solution. Hence, the resulting emulsions were complex, for example, triple Janus emulsions (VO + SO)/W/VO/SO [14]. The area of selective inversion undeniably presents a virgin area of further research; keeping in mind that the change between images 4 and 5, Figure 6.7, cannot take place by increasing the volume of O2 only, being attainable first by reducing the volume of the aqueous solution.

In addition to demonstrating the potential for selective one-oil inversion, the results in Figures 6.6 and 6.7 also bring to light a more central aspect: the interfacial tension equilibrium at the contact line is decisive for the limit to O2 coverage of O1 reached by adding O2 to an (O1 + O2)/W emulsion. For a constant c, each value of k (via the corresponding β) decides the maximum coverage, ψ, of O1 reachable by the addition of O2. Equation 6.5, combined with the correlation between the coverage ψ and the angle η through $\psi = (1-\cos \eta)/2$ offers the limit of coverage versus k, Figure 6.8.

The curve in Figure 6.8 gives the coverage of O1 by O2, when the added amount of O2 approaches infinity, that is, it's the maximum attainable *visible* coverage of

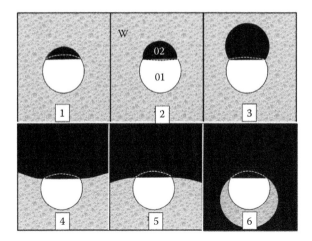

FIGURE 6.7 Images of Janus drops for $c = 0.9$ and $\eta = 120°$. The dashed line shows the central projection of the O1/O2 interface. Values for k and β are:

	k	β
1	0.14	5.02
2	0.3	17.1
3	1	63.3
4	1.6	114.6
5	1.7	126.9
6	1.875	161.4

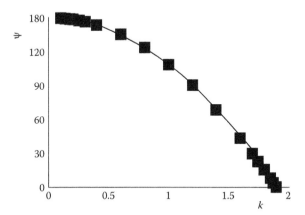

FIGURE 6.8 The coverage of O1 by O2, ψ, versus k for $c = 0.9$.

O1 by adding O2, the location of the contact line, Figures 6.2b, 6.3, 6.4, and 6.7. With $c = 0.9$, no more coverage of the O1 is conceivable, irrespective of the O2 amount added. However, the special case entailed large—but not very realistic—k numbers in order to illustrate the inversion in as graphic a manner as possible. For physical systems, for example, with the silicon oil combined with a vegetable

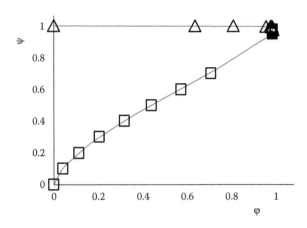

FIGURE 6.9 The fraction, ψ, of O1 surface covered by O2 versus the volume fraction, ϕ, of O2 ($\square$) as $V_{O2}/(V_{O1} + V_{O2})$ and water ($\triangle$) as $V_w/(V_{O1} + V_w)$. $c = 0.633$, $k = 0.45$.

oil and k numbers at the level of 0.4, Figure 6.1, the inversion will take place only at extremely high coverage of the SO sphere as in the following example.

A combination of $c = 0.6333$ and $k = 0.45$ from Figure 6.1 is used to illustrate the combined effects of interfacial equilibrium and relative volumes. At first, the correlation between the volume fractions, ϕ, and the coverage of the O1 drop, ψ, is instructive, Figure 6.9.

The curves in Figure 6.9 immediately point to one aspect of the O2 inversion, that is, it happens first, when the volume fraction of it, $V_{O2}/(V_{O2} + V_{O1})$, reaches 1 at extremely large coverage, ψ, of O1 (0.9731) by O2. In addition to this obvious feature, several aspects of the behavior at this point need comments. At first, the conditions in Figure 6.9 are valid only for a single Janus drop in infinite space; limiting the space to a spherical shell results in the two finite volumes gradually changing with η. In the second place, the amount of water is by definition initially infinite and the conditions during the actual inversion are not trivial. In short, is there a stage in the process when both water and O2 are simultaneously infinite? This would be the case, for $\beta + \eta = 0$, but even the smallest deviation from that condition means one of the volumes becomes finite. Hence, numerically it is not possible to find a point of equal and infinite volumes with the c and k numbers in Figure 6.9.

As highlighted, coverage in excess of the critical value is attained only by removing water from the system, making the formed water cap in the Janus drop, W + O1, smaller with increased coverage, image 6, Figure 6.9. As a minor but essential point, it should be noted that the coverage of O1 in excess of $\psi = 0.9731$ is still by O2, as pictured by the sixth image in Figure 6.7. The coverage of O1 in the range of O2 inversion is informative, Figure 6.10.

It is of interest to notice the fact that for $\psi = 1$, when O1 is completely covered by O2, the volumes of both O1 and W become zero. That is, there is no drop of any kind; the system consists entirely of the liquid O2!

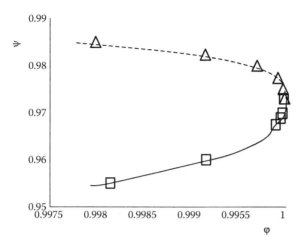

FIGURE 6.10 The coverage fraction of O1 by O2 in the range of the O2 inversion. □, (O1 + O2)/W emulsion; △, (O1 + W)/O2 emulsion.

These highlights are for a case in which the $r_{O1/W}$ is kept constant, because such a condition facilitates the computations, as mentioned. Physically, the approach means the increase in ϕ_{O2} is accompanied by a reduction of O1 to accommodate the calculated ψ. Other experimental results [23] would be better represented by an invariant O1 volume. This situation is examined in the next section.

6.5.2 CONSTANT O1 VOLUME

The results with $r_{O1/W} = 1$, Figure 6.9, show the volume of O1, as well as the volume of water, to reach zero for $\psi = 1$. As a contrast, the condition of $V_{O1} = 4\pi/3$ for the examined case would lead to rapidly increasing O1 radii to retain its volume invariant. Table 6.2 demonstrates the huge numbers for the O1 radius for almost complete

TABLE 6.2

The Radius of O1 for Large Coverage of O1 by O2; r_{O1} with No Coverage Equals 1

$1 - \psi$	r_{O1}	d	H	V_W
5E–08	1,021.32	0.9135	1.02E–3	7.62
5E–09	3,227.75	0.9129	3.23E–05	7.56
5E–10	10,205.1	0.9128	1.03E–05	7.55
4E–10	11,409.5	0.9128	9.19E–06	7.5
2.5E–10	14,431.8	0.9127	7.27E–06	7.55
5E–11	32,269.4	0.9127	3.28E–06	7.54
5E–12	102,043	0.9127	1.08E–06	7.54
5E–13	322,678	0.9127	3.83E–07	7.54

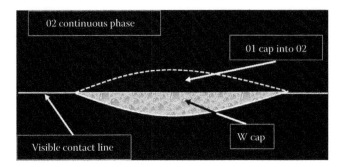

FIGURE 6.11   The topology of the O1 + W drop in the O2 continuous phase for $\psi \to 1$. The visible contact line is virtually straight, because of the large value of r_{O1}.

coverage of O1 by O2. As a comparison it may be noted that for $\psi < 0.9731, r_{O1} = 2.2$, a number that would not affect the coverage at the O2 inversion.

The numbers in Table 6.2 present some of the factors in the conditions close to complete coverage of O1 by O2. In spite of the tremendously large r_{O1}, the projected length of the visible contact line, d, is still small and combined with the minute values of the maximum distance between the r_{O1} and the visible contact line, means that the O1 volume is represented by the volume of the O1 cap penetrating into O2, Figure 6.11. In the same manner, an accurate volume of the W cap, V_W, is limited by the contour of the $r_W(= -r_{O2}$ in the calculations) and the visible contact line Figure 6.11.

V_W and V_{O1} are obviously both equal to zero for $\psi = 1$, and the system consists entirely of a one-phase O2 liquid, equal to the conditions for the earlier case for $r_{O1} = 1$. However, for $\psi \to 1$, the situation is entirely different; $V_W = 7.54$ and $V_{O1} = 4\pi/3$. For these conditions, V_W/V_{O1} approaches a constant value of 1.8 as compared to the corresponding value of 4.13 for the case of a constant r_{O1}.

With these sections, the main features of the drop topology have been correlated to the interfacial tensions and the volume ratio of the oils, based on the range of surfactant concentrations in Figure 6.1, giving interfacial tensions according to Equation 6.2. However, a comprehensive treatment requires an evaluation of the relative stability of the Janus topology versus the alternatives, for example, separate and engulfed drops, Table 6.1.

6.6 INTERFACIAL FREE ENERGY FOR DIFFERENT DROP TOPOLOGIES

The relevant comparison of the interfacial free energies of the Janus drops are with engulfed and separate drops. The engulfed drops are favored for surfactant concentrations less than S_{crit}, Figure 6.1, that is, a range, where Equation 6.1 applies, while the separate drops, although thermodynamically favored for $k > 1 + c$, also are relevant for final destabilization to three liquids of the Janus emulsion. The second process is the relevant choice for the present treatment.

The surface free energy is calculated based on Equations 6.5 through 6.13. The areas are

$$A_{XX/YY} = 2\pi r_{XX/YY} h_{XX/YY} \tag{6.14}$$

in which $h_{XX/YY}$ is the maximum distance from the visible contact line and the XX/YY interface, Figure 6.2b. The total interfacial free energy of a Janus drop is

$$IFE_{tot} = 2\pi \left(h_{SO/Aq} r_{SO/Aq} + c\, h_{VO/Aq} r_{VO/Aq} + k h_{SO/VO}\, r_{SO/VO} \right) \tag{6.15}$$

These equations are geometrically straightforward and may, at first glance, give the impression to offer only trivial features. However, there is one unexpected consequence, which was recently brought to light [16]. If $\eta + \beta = 180°$, in Equation 6.7, $r_{O2/W} = \infty$, that is, this η is the maximum that can be reached by adding VO to the drops. The aspect is illustrated by the conditions for the surfactant concentration of 0.66% in the aqueous phase, which gives $c = 0.9604$ and $k = 0.7827$, in turn giving $\beta = 0.8205$, $\delta = 1.114$, and a maximum η of 133°. Hence, by adding VO to a drop of SO/Aq, no addition of VO will cover more than a fraction of 0.739 of the SO surface. To attain greater coverage, the emulsion has to change from (SO + VO)/Aq to (SO + Aq)/VO [16].

The expressions for the total interfacial free energy of separate and entirely engulfed ones, SO/VO, are algebraically simple.

Engulfed drops:

$$IFE_{tot} = (4\pi)^{1/3} \left(3^{2/3} \right) \left(k \left(V_{SO} \right)^{2/3} + c \left(V_{SO} + V_{VO} \right)^{2/3} \right) \tag{6.16}$$

Separate drops:

$$IFE_{tot} = (4\pi)^{1/3} \left(3^{2/3} \right) \left(\left(V_{SO} \right)^{2/3} + c \left(V_{VO} \right)^{2/3} \right) \tag{6.17}$$

in which the volumes are for the oils in Janus drops with $r_{O1/W} = 1$.

The total interfacial free energy ratios between Janus drops and the two alternatives are given in Figure 6.12 for realistic c and k numbers.

The numbers in Figure 6.12 bring to light several essentials. At first, the interfacial free energy of the Janus drop is less than that of the alternative drops, the conclusion of which directly follows from Equation 6.2. Second, the interfacial free energy for separate drops equals that of the Janus drop for $\phi = 0$ and $\phi = 1$, which is trivial, because for these ϕ's, the drop consists of only one oil. However, so is the case for the engulfed drop at $\phi = 0$, but now the results show a most significantly lower interfacial free energy for the Janus drop. The explanation lies with the fact that $\phi = 0$ means the O2 has no volume, but it still covers O1, because the surface is infinitely thin and has no volume.

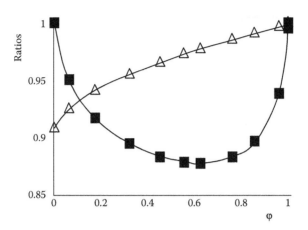

FIGURE 6.12 Ratios of total interfacial free energy between a Janus drop and separate drops (■) or an engulfed drop (△) versus the O1 volume fraction. $c = 0.6$, $k = 0.5$.

ACKNOWLEDGMENTS

The author is grateful to Ugelstad Laboratory for support, to Susan H. Friberg for editorial efforts, and to associate professor Lingling Ge at Yangzhou University, China, for generously providing typical interfacial tension numbers for a model system.

REFERENCES

1. Girboux, A.-L., Starch, M. Formulating with silicone and natural lipids. *Household Pers. Prod. Ind.* 2005, *42(12)*, 100, 102–104.
2. Girboux, A.L., Courbon, E. Enhancing the feel of vegetable oils with silicone. *Cosmet. Toiletries* 2008, *123(7)*, 65–74.
3. Lind, D., Meyers, D., Shope, M. US Patent 7767747, 2010; Kim, G., Sousa, A., Meyers, D., Libera, M. Nanoscale composition of biphasic polymer nanocolloids in aqueous suspension. *Microsc. Microanal.* 2008, *14*, 459–468.
4. Aserin, A. *Multiple Emulsions: Technology and Applications.* Hoboken, NJ: John Wiley & Sons, 2008.
5. Nisisako, T., Okushima, S., Torii, T. Controlled formulation of monodisperse double emulsions in a multiple-phase microfluidic system. *Soft Matter* 2005, *1*, 23–27.
6. Chen, C.H., Abate, A.R., Lee, D.Y. et al. Microfluidic assembly of magnetic hydrogel particles with uniformly anisotropic structure. *Adv. Mater. Sci.* 2009, *21*, 3201–3204.
7. Guzowski, J., Korczyk, P.M., Jakiela, S. et al. Structure and stability of multiple micro-droplets. *Soft Matter* 2012, *8*, 7269–7278.
8. Schmidt, J.J., Rowley, J., Kong, J. Hydrogels used for cell-based drug delivery. *J. Biomed. Mater. Res. A* 2008, *87A*, 1113–1122.
9. Choi, J.-H., Jung, J.-H., Rhee, J.W. et al. Generation of monodisperse alginate micro-beads and in situ encapsulation of cell in microfluidic device. *Biomed. Microdev.* 2007, *9*, 855–862.
10. Vladisavlevic, G.T., Shimitsu, M., Nakashima, T. Production of multiple emulsions for drug delivery systems by repeated SPG membrane homogenization: Influence of mean pore size, interfacial tension and continuous phase viscosity. *J. Membr. Sci.* 2006, *284*, 1–2.

11. Wang, J.T., Wang, J., Han, J.-J. Fabrication of advanced particles and particle-based materials assisted by droplet-based microfluidics. *Small* 2011, *7*, 1728–1754.

12. Golestanian, L., Liverpool, T.B., Ajdari, A. Designing phoretic micro- and nano-swimmers. *New J. Phys.* 2007, *9*, 126.

13. Neeson, M.J., Tabor, R.F., Grieser, F. et al. Compound sessile drops. *Soft Matter* 2012, *8*, 11042–11050.

14. Pannacci, N., Bruus, H., Bartolo, D. et al. Equilibrium and nonequilibrium states in microfluidic double emulsions. *Phys. Rev. Lett.* 2008, *101*, 16450.

15. Hasinovic, H., Friberg, S.E. A one-step process to a Janus emulsion. *J. Colloid Interface Sci.* 2010, *354*, 424–426.

16. Hasinovic, H., Friberg, S.E. One-step inversion process to a Janus emulsion with two mutually insoluble oils. *Langmuir* 2011, 27, 6584–6588.

17. Ge, L., Shao, W., Lu, S., Guo, R. Droplet topology control of Janus emulsion prepared in one-step high energy mixing. *Soft Matter* 2014, *10(25)*, 4498–4505.

18. Friberg, S.E., Kovach, I., Koetz, J. Equilibrium topology and partial inversion of Janus drops: A numerical analysis. *Chem. Phys. Chem.* 2013, *14*, 3772–3776.

19. Friberg, S.E. Selective emulsion inversion in an equilibrium Janus drop. 1. Unlimited space. *J. Colloid Interface Sci.* 2014, *416*, 167–171.

20. Shardt, O., Derksen, J.J., Mitra, S.K. Simulations of Janus droplets at equilibrium and in shear. *Phys. Fluids* 2014, *26*, 012104.

21. Ge, L., Lu, S., Guo, R. Janus emulsions formed with a polymerizable monomer, silicone oil and Tween 80 aqueous solution. *J. Colloid Interface Sci.* 2014, *423*, 108–112.

22. Bormashenko, E., Bormashenko, Y., Pogreb, R. et al. Janus droplets: Liquid marbles coated with dielectric/semiconductor particles. *Langmuir* 2011, *27*, 7–10.

23. Xu, J., Ma, A., Liu, T. et al. Janus-like Pickering emulsions and their controllable coalescence. *Chem. Commun.* 2013, *49*, 10871–10873.

24. Leonardi, G.R., Monteiro e Silva, S.A., Guimarães, C.M., Friberg, S.E. An unexpected stabilization factor during destabilization of a Janus emulsion. *Colloids Interfaces Commun.* 2015, *8*, 14–16.

25. Leonardi, G. et al. *Colloid Interface Sci.* 2015, *449*, 31–37.

26. Hasinovic, H., Friberg, S.E., Kovach, I. et al. Destabilization of a dual emulsion to form a Janus emulsion. *Colloid Polym. Sci.* 2014, *292*, 2319–2324.

7 Silica and Silsesquioxane Dispersions Obtained from Alkoxysilanes

Leon Marteaux and Donald T. Liles

CONTENTS

7.1   INTRODUCTION

Silica and certain silsesquioxane dispersions obtained from alkoxysilanes are examples of sol-gel processing. They are obtained from the hydrolysis and condensation of metal alkoxides when the water-to-alkoxide molar ratio (R value) is greater than 2, or a base (alkaline) catalyst is used [1]. In these conditions, spherical or disk-shaped particles are obtained. One typical example is the Stöber process, reported in 1968, for preparing monodispersed, spherical silica particles from tetraethylorthosilicate (TEOS) with an R value between 7 and 25 and ammonia concentrations from ~1 to 7 molar [2]. Spherical particles of silsesquioxanes such as methylsilsesquioxane ($CH_3SiO_{3/2}$) can be prepared in a similar manner starting from organosilanes like $CH_3Si(OCH_3)_3$ [3].

Colloidal silsesquioxanes, also referred to as silicone resin microemulsions, silsesquioxane microemulsions, or silsesquioxane dispersions, were discovered in the mid-1960s by Cekada and Weyenberg [4]. In this process, organosilanes undergo hydrolysis and condensation in water in the presence of a surface-active catalyst. The particles are spherical and range in size from about 10 nm to about 100 nm.

In the 1980s, Schmidt pioneered a new type of noncrystalline organic–inorganic hybrid material called Ormosil (for organically modified silica) or Ormocer® hybrid polymers [5]. These materials have the ability to present strong chemical bonds between organic and inorganic components. At the same period, Avnir et al. introduced organic dyes into oxide-gel nanocomposites [6]. In this case, there are no chemical bonds between organic and inorganic components, and the obtained materials are therefore called nanocomposites. The concept has been extended to biological and medical nanocomposites [7]. Improved biocompatibility between the inorganic matrix and enzymes, cells, and living tissues has been obtained by Livage et al. by the use of biopolymers like chitosan and alginate [8].

In parallel to these developments in sol-gel science, Beck et al. disclosed in the early 1990s processes for making mesoporous materials from alkoxides to significantly improve heterocatalysis yields [9]. The porous structures were obtained by first hydrolyzing TEOS in water in the presence of a template made up of surfactant micellar rods. Removal of surfactant by calcination or washing produces very high surface area silica particles having regularly spaced pores.

The missing link between surfactant-templated, mesoporous materials and hollow silica microspheres and core-shell microencapsules was discovered by Stucky et al. in 1996 [10]. The combined long-range oil-in-water (O/W) emulsion with short-range cooperative assembly of silica and surfactants at the O/W interface creates ordered composite mesostructured phases that were also macroscopically structured. This was the very first *in situ* method where an active ingredient was first mixed with shell precursors before emulsification to produce silica-based core-shell microcapsules. The subsequent removal of the volatile active ingredient produced silica hollow spheres.

7.2   SOL-GEL PROCESS

The sol-gel process has been the topic of countless publications and text books illustrating the specificity and the complexity of the process [11,12]. The sol-gel process can be defined as the colloidal route used to synthesize ceramics with an intermediate stage including a sol and/or a gel state [13].

Since the 1960s, most of the fundamental understanding has been gained to support the manufacture of silica gels, aerogels, glasses, silsesquioxanes, ceramics, entrapped and immobilized enzymes, membranes, and hybrid organic–inorganic materials in water-depleted conditions. As a result, the sol-gel process found more and more applications such as in protective and smart coatings, separative chromatography, catalysis, diagnostics, biotechnology, building waterproofing, optical lenses, restoration, and controlled release [14].

In order to obtain an acceptable toxicological profile, the usual choice is to start from tetraethylorthosilicate (TEOS), $Si(OCH_2CH_3)_4$, instead of tetramethoxysilane (TMOS), $Si(OCH_3)_4$, as a precursor, as the latter silane releases methanol. While TEOS is water insoluble, its hydrolysis product, orthosilicic acid, $Si(OH)_4$, is highly soluble in water. The total conversion of TEOS into silica, SiO_2, is sequentially obtained by hydrolysis (a) and condensation (b) as expressed in Equation 7.1:

$$\text{(a) } Si(OCH_2CH_3)_4 + 4H_2O \longrightarrow Si(OH)_4 + 4C_2H_5OH$$

$$\underline{\text{(b) } Si(OH_4) \longrightarrow SiO_2 + 2H_2O} \quad (7.1)$$

$$Si(OCH_2CH_3)_4 + 2H_2O \longrightarrow SiO_2 + 4C_2H_5OH$$

The use of this chemistry is delicate because the structure and porosity of the silica produced depends on many physical parameters such as temperature, pH, and ionic strength [11]. Due to its sequential nature, the hydrolysis and condensation of TEOS can follow multiple pathways [11]. From TEOS to silica, no less than 13 different intermediate species belonging to the categories of ethoxysilanes, ethoxysiloxanes, or ethoxyhydroxysiloxanes can be generated. Their structures and molecular weights determine the properties of the intermediates (SiO_2 sols) or final products (SiO_2 gels). Condensation of the orthosilicic acid leads to higher molecular weight species having reduced water solubility, eventually leading to the formation of precipitates. The water insoluble condensate is a negatively charged polyelectrolyte at pH above 4.5, the pK_a of silanol [15]. The complete reaction, similar to an emulsion polymerization process, involves at least one phase transfer of the precursor. Because the hydrolysis of TEOS generates ethanol, which is a good solvent for both TEOS and water; TEOS becomes increasingly soluble in the water/ethanol continuous phase during the hydrolysis process. The three-dimensional (3-D) structure of the resulting silica has been found to be dependent on a region in the ternary phase diagram in which the hydrolysis/condensation reaction is conducted.

pH also influences the relative rates of the hydrolysis, condensation, and dissolution reactions and has been summarized in Iler's graph [11]. It follows from this graph that the condensation rate in solution is at a maximum for pH ranging from 6 to 7 and at a minimum for pH ranging from 1.5 to 2. In alkaline media, the condensation rate of silicic acid falls sharply and at pH above 11, a reverse depolymerization or hydrolysis of Si–O–Si bonds takes place. It can be anticipated that this differentiation of the reaction kinetic can lead to differentiated silica structures [16].

A key aspect of this chemistry is the combined impact of pH and ionic strength on the spatial organization of the silica produced [11]. Conducting the hydrolysis and condensation at basic pH and low pI leads to the synthesis of the so-called "Stöber" silica [2]. Particles grow in size with a decrease in number leading to a colloidal silica sol of individual spherical nanoparticles. In acid solution or in the presence of flocculating salts, particles aggregate into 3-D fractal networks and form gels.

Finally silica starts to solubilize from pH 10 at room temperature [11]. In contrast, silica is stable at very acidic pH which is an asset compared to many organic materials.

7.3 MAIN DISPERSION TYPES

Main dispersion types of silica and organomodified silicone resin can be classified into silsesquioxane sols, silsesquioxane microspheres, hollow spheres, microspheres, core-shell, and polynuclear microcapsules (Figure 7.1).

7.3.1 COLLOIDAL SILSESQUIOXANES PARTICLES AND OTHER RESINS

Colloidal silsesquioxanes are aqueous dispersions of silsesquioxane resins ($RSiO_{3/2}$) that have particles of colloidal dimensions, less than 100 nm. Colloidal silsesquioxanes, also referred to as silicone resin microemulsions, silsesquioxane microemulsions or silsesquioxane dispersions, were discovered in the mid-1960s by Cekada and Weyenberg [4]. They represent a fascinating part of aqueous silicones, as colloidal silsesquioxanes can be prepared with a wide range of compositions, and therefore they offer a wide set of properties.

7.3.1.1 Preparation of Colloidal Silsesquioxanes

The Cekada process involves emulsion polymerization of alkoxy silanes such as $CH_3Si(OCH_3)_3$ or $C_6H_6Si(OCH_3)_3$ in water in the presence of a surface-active polymerization catalyst, the most common being dodecylbenzenesulfonic acid (DBSA). During this process, silane undergoes hydrolysis and condensation to form nanoparticles of silsesquioxane that are stabilized in part by the surface-active catalyst. The reaction proceeds at room temperature (RT) or optionally it can be heated. The reaction is terminated by neutralizing DBSA with a base such as NaOH or amines such as triethanolamine. Alcohol by-product is usually removed from the aqueous dispersion by distillation or vacuum stripping. The result is an aqueous dispersion of

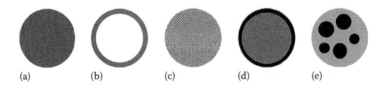

FIGURE 7.1 Schematic showing: silsesquioxane particles (a), hollow spheres (b), nano/microsphere (c), core-shell nano/microcapsules (d), polynuclear nano/microcapsule (e).

spherically shaped silsesquioxane nanoparticles from about 10 to 60 nm in diameter and from 5% to 15% active content.

Particle size of the silsesquioxane nanospheres can be controlled to some extent, mainly by adjusting concentration of the reactant silane(s). In this manner, lower concentrations of silane(s) lead to smaller particles. It should be realized that colloidal silsesquioxanes prepared by the Cekada process do not exhibit a monodisperse particle size distribution.

Like colloidal silica sols are processed, aqueous sols of silsesquioxanes can be concentrated by evaporation of water. However, the final concentration of these sols is only on the order of approximately 15% by weight because they take on a thick, gel-like consistency above this concentration. Silsesquioxane sols that have formed a gel-like state can usually be diluted with water, which will restore the system to the original free-flowing liquid state.

In addition to the strong acid DBSA, salts of long-chain carboxylic acids at pH > 7 can also be used in the preparation of colloidal silsesquioxanes. For example, stearic or oleic acid can be neutralized with NH_3, an amine, or NaOH and the salt becomes the surface-active catalyst [17].

Colloidal silsesquioxanes can also be prepared using cationic surface-active catalysts such as quaternary ammonium halides or hydroxides. In this case, the reaction mixture is neutralized with a carboxylic acid such as acetic acid to terminate polymerization [4].

Cekada and Weyenberg reported that trialkoxysilanes, $RSi(OMe)_3$, possessing a variety of different R groups could be used to prepare colloidal silsesquioxanes. R groups like methyl (Me), ethyl (Et), propyl (Pr), isopropyl (*i*-Pr), butyl (Bu), cyclobutyl (cyclo-Bu), amyl, cyclohexyl, phenyl (Ph), and trifluoropropyl could all be used to prepare the corresponding colloidal silsesquioxanes. In addition, mixed silanes could also be used to prepare colloidal silsesquioxanes. Furthermore, particles with layered compositions could be prepared by adding different silanes in stages during the course of the reaction. Cekada and Weyenberg also disclosed that organofunctional silanes having amino, glycidoxy, vinyl, acryl, methacryl, trifluoroproply, and mercaptopropyl functionality could be used to prepare colloidal silsesquioxanes.

Besides silsesquioxane (T), various other colloidal resins can be made using the basic Cekada process to arrive at compositions such as MQ resins, MT resins, TD resins, TDQ resins, and the like. (For an explanation of the M, D, T, and Q notations, please refer to Chapter 1.) The building blocks for these colloidal resins can be silanes or low molecular weight siloxanes. For example, Q can originate from TEOS, $Si(OEt)_4$, M can originate from hexamethyldisiloxane (MM), and D can originate from $Me_2Si(OMe)_2$. Thus colloidal MQ resin having a particle size of about 60 nm can be prepared by emulsion polymerization of a mixture of TEOS and MM in water in the presence of DBSA. Figure 7.2 shows a typical particle size distribution for colloidal MQ resin prepared from TEOS and MM.

It is also possible to prepare SiH functional colloidal resins by starting with SiH functional silanes such as $(EtO)_3SiH$ or $(MeO)_2SiMeH$. Low molecular weight siloxanes such as $Me_3SiOSiMeHOSiMe_3$ can also be a source of SiH when preparing these colloidal resins. Thus SiH functional colloidal resins can be constructed to have the SiH group present in the M, D, or T units. These SiH functional units can

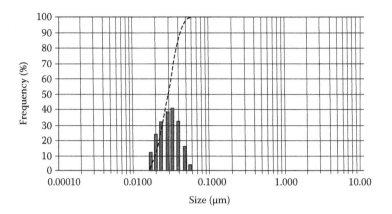

FIGURE 7.2 Particle size distribution of colloidal MQ resin. Molar M:Q ratio of resin is 0.9. Volume averaged median particle size is approximately 31 nm. Dotted line represents cumulative percent. Particle size measured on Microtrac UPA-150.

of course be copolymerized with other nonfunctional M, D, T, and Q units, so a very wide range of colloidal SiH functional resins can be formed. As SiH is capable of reacting with water, especially at pH extremes and in the presence of certain catalysts such as amines and metal salts, colloidal SiH resins lack long-term stability. Short-term stability (days to weeks) can be usually attained.

7.3.1.2   Metal Treated Silsesquioxane Particles

Starting from SiH functional colloidal silsesquioxane particles, metals can be precipitated in the colloid in their elemental state in the form of nanoparticles [18]. Although it is believed that these metal nanoparticles are deposited upon the surfaces of the colloidal siloxane particles; it has not been proven at this time. The process involves reduction of metal ion by SiH, so any metal salt that is capable of being reduced by SiH can be deposited onto colloidal silsesquioxane particles in this manner. Examples of such metals include gold, silver, copper, and platinum. In all of these cases, color changes occur upon the addition of the metal salt solution to SiH functional colloidal particles. Figure 7.3 shows a typical SiH-functional colloidal silsesquioxane to which has been added $AuCl_3$.

7.3.1.3   Properties of Colloidal Silsesquioxanes

Appearance of the colloidal silsesquioxanes can be clear to slightly hazy to almost totally opaque, depending upon particle size and concentration. For example, colloids of $MeSiO_{3/2}$ having a particle size of 10–15 nm and a solids content of 6%–10% are typically water clear while colloids of MQ resin having a particle size of about 40 nm and a solids content of 12% are slightly hazy. These colloidal silsesquioxanes can be prepared such that they contain an unusually low level of surfactant and yet remain stable. Colloidal silsesquioxanes of particle sizes even as low as 10 nm can be prepared having a surfactant concentration of 5% based on weight of dispersed phase. In sharp contrast, many microemulsions that have droplet sizes less than

(a)   (b)

FIGURE 7.3   (a) Colloidal T-D^H resin, particle size approximately 20 nm. (b) Same colloid to which has been added AuCl$_3$. Colloid contains approximately 6% silsesquioxane. Gold concentration is approximately 50 ppm as elemental Au based on silsesquioxane. Both colloids remained stable after 2 years storage at ambient conditions.

100 nm require significantly high levels of surfactant to maintain stability due to the very large surface area (cumulative) of small droplet emulsions. For example, some microemulsions having droplet sizes on the order of 50 nm or less will contain from 25% to 100% surfactant based on weight of oil or dispersed phase. The reason for low level of surfactant to stabilize colloidal silsesquioxane sols is believed to be due to the presence of silicate ion at the particle surface which contributes to stability. This would be envisioned to be similar to how colloidal silica is stabilized by silicate anion (SiO$^-$), which is balanced by a basic cation (Na$^+$, K$^+$, NH$_4{}^+$).

In any event, aqueous compositions containing low levels of surfactants can have certain advantages. For example, microemulsions usually do not make effective hydrophobing treatments or water proofers as the presence of excess surfactant often overwhelms any waterproofing or hydrophobing ability of the dispersed waterproofing substance. As it turns out aqueous, colloidal MQ resin which contains only 5 phr (parts per hundred of resin) surfactant works quite well as a hydrophobing composition (see Figure 7.4).

Removal of water from colloidal silsesquioxanes such as MeSiO$_{3/2}$ usually forms particles or powders and not coherent films. This behavior, which would be expected from a high T$_g$ and a highly cross-linked polymer, is quite similar to colloidal silica sols when they are evaporated to dryness. The dried resin powders can easily be micronized to further reduce their particle size, but most likely individual primary

FIGURE 7.4 Hydrophobing effect of colloidal MQ resin. Colloidal MQ resin having M:Q molar ratio of 0.9 and an active content of 5% by weight was applied to masonry substrate and allowed to dry under ambient conditions for 7 days. Water drops placed onto treated surface via a pipette. Image taken 1 hour after water drops were applied.

FIGURE 7.5 Scanning electron micrographs of dried colloidal methyl silsesquioxane, which to the unaided eye resembles a white powder. White scale bars represent 10 μm for top images and 100 nm for bottom images. Primary silsesquioxane particles are approximately 45 nm in diameter.

particles on the order of 10 nm cannot be achieved as the particles will agglomerate significantly. Figure 7.5 shows particles of dried colloidal $MeSiO_{3/2}$ at several magnifications, finally revealing primary particles on the order of 40–50 nm. Figure 7.6 shows atomic force microscope (AFM) images of dried colloidal $MeSiO_{3/2}$ having primary particles on the order of 10 nm.

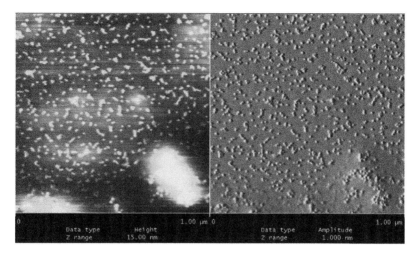

FIGURE 7.6 Atomic force micrographs of dried T–D colloidal silsesquioxane. Molar ratio of T:D is 9, image size is 1.0 μm. Primary silsesquioxane particles are approximately 10 nm in diameter.

7.3.2 SILSESQUIOXANE MICROSPHERES

In 1968, Stöber described a process for preparing spherical, monodisperse particles of silica from silicate esters such as TEOS or TMOS [2]. The process, which was based on earlier work by Kolbe [19], is commonly called the Stöber process and the silica particles are referred to as Stöber silica [2]. In this process, TEOS is added to a stirred alcohol–water solution containing ammonia as a catalyst. Hydrolysis of the silane occurs with concomitant condensation of formed silanols to produce silica particles having a monodisperse distribution of particle sizes. The silica particles have diameters between 50 and 2000 nm, and this is controlled to a certain extent by the type of silicate ester used, the type of alcohol used, and volume ratios of reagent silanes, alcohol, and water. Larger particles of silica, up to 1000 μm, can be prepared by a variation of the Stöber process using surfactants [20,21]. Porous silica particles have been prepared via a similar process using organosilanes such as methyltrimethoxysilane (MTM), $CH_3Si(OCH_3)_3$, along with TEOS followed by calcining the silica to "burn out" the organic content thus leaving voids in the silica particles [22].

In what might be considered an extension of the Stöber process, Kimura in 1985 patented a process for preparing spherical, monodisperse particles of methylsilsesquioxane (methyl-T resin) by reacting MTM with an aqueous solution of ammonia or an amine [23]. The reaction mixture can be performed with or without the addition of heat, but the preferred method is to heat the mixture for 1–2 hours at reflux, which is 70°C–90°C. A variation of this process is to react a silane such as MTM in an aqueous medium of a surface-active agent, an alkali metal hydroxide or ammonium hydroxide and water [24]. Particles of polysiloxane microspheres of narrow particle size in a range of 50 nm to about 10 μm can be produced. In some cases, a precipitated cake of agglomerated particles is formed and the individual particles are obtained by milling the particle mass. These spherical methyl silsesquioxane particles are claimed to be useful as anticaking agents for powder products, fillers for synthetic resins, and additives for cosmetics. The particles are also useful for wear resistance, antiblocking, and light-diffusing applications [3,23,24]. The particles are commercially available and are sold under the trade name of Tospearl® Microspheres.

7.3.3 Hollow Spheres

7.3.3.1 Hollow Glass Microspheres

The first process for producing hollow glass microspheres was patented in 1957 by Veatch [25]. It consisted of spray drying a slurry containing aqueous sodium silicate, boric acid, and urea as blowing agent to obtain, after grinding, solid particles smaller than 250 μm. The particles were then suspended in a heating zone between 540°C and 1610°C for 0.5–10 seconds to fuse the particle and liberate nitrogen, which expands the particle, thus forming hollow glass spheres. Yet these severe conditions would make it very difficult to encapsulate organic actives within the hollow glass microspheres.

7.3.3.2 Hollow Silica Microspheres (HSM) by the Sol-Gel Method

Hollow silica microspheres (HSM) became an active topic since the twenty-first century, and have been mainly documented by the academia, as shown in Figure 7.7. HSM can be obtained from different routes, using gas, solids, W/O emulsions, O/W Pickering emulsions, and surfactant micelles as the core materials to template them.

7.3.3.2.1 HSM Based on the Gas Cores Route

One of the first methods of making HSM has been described by Ravindra in 1991 [26]. The technique consists of creating, from spray nozzles, a continuous stream of uniform hollow drops of TEOS solution and exposing them to an ammoniated vapor to obtain HSM of controlled size and porosity. In 2001, Mann et al. introduced gas as

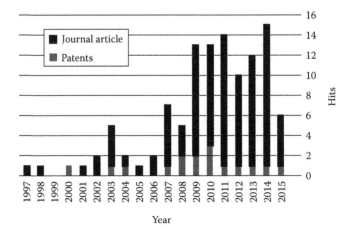

FIGURE 7.7 Number of publications found in SciFinder database using the keywords "hollow silica microspheres." Search conducted on March 2015.

the core material by vortex mixing a solution of cetyltrimethylammonium bromide (CTAB), TEOS, and aminopropyltriethoxysilane (APTES) in a facile room temperature (RT) method of making HSM having a mean diameter of 30.6 μm [27].

7.3.3.2.2 HSM Based on the W/O Emulsions Cores Route

In 1994, Wilcox et al. disclosed the first process for making HSM by a sol-gel technique involving emulsification, with the aid of a low HLB surfactant, Span® 80, of an aqueous colloidal silica sol in an organic phase like 2-ethyl-1-hexanol [28]. The W/O emulsion is subsequently dehydrated with *n*-butanol to form hollow gelled spheres, which are finally recovered by filtration, drying, and calcination. Later, other materials have been used to form and stabilize the starting W/O emulsion. Park et al. used hydroxypropyl cellulose (HPC) and added polyethylene glycol (PEG) or polyvinylpyrrolidone (PVP) in the water phase to stabilize the W/O emulsion and to favor the formation of a hollow structure [29]. Singh et al. found that decreasing the calcination temperature from 700° to 500° enhances the yield of hollow spheres [30]. The authors proposed that silica shells form via hydrolysis and polycondensation of TEOS at the aqueous–oil interface, and HSM forms upon removal of water and ethanol during calcination. Oh et al. synthetized HSM functionalized with magnetic particles (HSFM) without templates using a W/O emulsion containing magnetic nanoparticles which were synthesized by the redox reaction of iron salts [31]. Mesoporous HSM can be prepared by ultrasonication of an EtOH-ammonia solution/soybean oil emulsion in the presence of CTAB as a template [32]. The sol-gel reaction of TEOS was achieved only at the interface of the emulsion droplets, resulting in HSM after washing with acetone. Microfluidic processes can be used to obtain size-controllable microspheres from hollow, to partly hollow, to solid [33]. Raspberry-like HSM with mesoporous shells can be obtained by using triblock copolymer surfactant such as poly(ethylene oxide)-poly(propylene oxide)-poly(ethylene oxide) (EO-PO-EO) as a template and *n*-octane as core material [34]. Table 7.1 summarizes the key steps of the development of HSM based on W/O emulsions cores.

TABLE 7.1
Key Advances in the Development of HSM Based on W/O Emulsions Cores

Years	Reference	Surfactant	Silica Precursor	Additives in Water Phase	Oil Phase	Size (μm)
1994	[28]	Span 80	Colloidal silica sol	—	2-ethyl-1-hexanol	1–30
2003	[29]	HPC	TEOS	PEG or PVP	Parafin	1–10
2007	[30]	Span 80	TEOS	EtOH	Parafin	5–60
2009	[31]	—	TEOS	FeO$_x$		N/A
2009	[32]	CTAB	TEOS	EtOH/NH$_3$	Soybean	N/A
2010	[33]	CTAB	TEOS			N/A
2010	[34]	EO-PO-EO	TEOS	—	*n*-octane	0.8

Note: Chemical notation—HPC, hydroxypropyl cellulose; CTAB, cetyltrimethylammonium bromide; PEG, polyethyleneglycol; PVP, polyvinylpyrrolidone; EtOH, ethanol.

7.3.3.2.3 HSM Based on Solid Cores Route

In 1996, Seliger et al. [35] communicated the first sacrificial way to make HSM by burning off a polystyrene (PS) core material, which acted as a shell template. Since then, many authors used PS as a templating material [35–47]. To favor silica deposition onto PS particles, some authors used cationic polystyrene (CPS) [48–53]. Other synthetic latexes can be used like polydiallyldimethylammonium chloride (PDADMAC) [54], poly(methylmethacrylate) (PMMA) [55], styrene-2-(methacryloyl)ethyltrimethylammonium chloride copolymer (P(St-EA-MAA) [56,57], polyvinyl pyridine [58] and poly(N-isopropylacrylamide) (PNIPAM) [59]. Biotemplates have also been used such as pollen grains [60], *Saccharomyces cerevisiae* yeast cells [61], and *Staphylococcus aureus* bacteria cells [62]. An inorganic core of calcium carbonate, $CaCO_3$, has also been used [63]. The core materials are usually removed by burning around 600°C, but they can also be eliminated by solvents such as dimethylsulfoxide (DMSO) or tetrahydrofuran (THF) [54,52] or simply by water in the case of PNIPAM [59]. The positively charged polystyrene-containing template P(St-EA-MAA) is self-dissolved as the silica layer is formed [56,57].

While the ambient room temperature, atmospheric pressure, and water-based process is by far the most utilized, HSM can also be obtained using supercritical (SC) carbon dioxide to carry TEOS into cross-linked PS microspheres [39].

The shell precursors are most often TEOS. To functionalize the shell, TEOS can be used in combination with other alkoxysilanes to form new organic–inorganic hybrid microspheres. Popular combination are TEOS and APTES [48,52,57], TEOS and 3-mercaptopropyltriethoxysilane (MPTS)[49], TEOS with methacryloxypropyltrimethoxysilane (MPS) [51], or TEOS with hexadecyltrimethoxysilane ($C_{16}TMS$) [44]. Hybrid HSM can be further functionalized. An amino-functionalized ($-NH_2$) HSM (HSM-NH_2) has been polyethylene glycolated (PEGylated) with methoxy polyethylene glycol propionic acid (mPEG-COOH) [53].

Magnetic HSM can be prepared by seeding polymerization of TEOS and iron salts [35,65].

TEOS can be used in combination with substances that helps its adsorption onto the solid core like CTAB [38,40–42,47], dodecylamine [50], and the cationic comonomer 2-(methacryloxy)ethyltrimethyl ammonium chloride (DMC) [51]. In 2011, Wei et al. proposed a unified mechanism for the formation of hollow silica shells with and without a templating surfactant like CTAB [40]. CTAB is mandatory to template PS beads only when the reactive medium itself gives discrete silica microspheres. In contrast, in a medium which itself generates highly aggregated silica microspheres, HSM are successfully templated without any cationic surfactant. Liu et al. studied the influence of surfactant CTAB concentration on the surface morphology of HSMs templated by PS beads [38]. They showed that cetyltrimethylammonium cations (CTA^+) preferentially assemble with silica species to form silica-CTA^+ composite nanoparticles. Since the zeta potential of silica-CTA^+ composite nanoparticles is smaller than that of pure silica nanoparticles, these composite nanoparticles encounter less repulsion when they are deposited on the surface of PS beads and orient close to each other. As more CTAB is added, the silica-CTA^+ nanoparticles are less negatively charged, such that more compact and smooth HSMs are obtained.

In a subsequent investigation, the authors studied the influence of ammonia concentration on the formation of HSMs via PS beads templating [41]. HSMs could only be templated in water–ethanol–ammonia–TEOS media with low concentrations of ammonia. The authors showed that in template-free media, the aggregation of silica particles decreases with ammonia concentrations. The increase in ammonia promotes spontaneous nucleation and condensation of silica oligomers to form silica particles. Since the repulsive forces between the particles as well as those repulsive forces between particles and templates are relatively large, this hinders assembly of silica primary particles on the templates to form silica shells. Table 7.2 summarizes the key steps of the development of HSM based on solid cores.

7.3.3.2.4   HSM Based on Surfactant Micelles Cores Route

The use of surfactant micelles to template HSM is a more recent development [27,65]. The most common templating surfactants used are swollen CTAB [27], octylamine [65,66], the nonionic polymeric surfactant, Pluronic® F127 ($EO_{106}PO_{70}EO_{106}$) [67], 1-dodecyl-3-methylimidazolium bromide [68], and polyethylene glycol [69]. In the case of octylamine, surface depressions, cracks, and nonhollow microspheres have been observed that can be alleviated by decreasing the octylamine content and increasing the acid content. Liu et al. hypothesized that because acid conditions favor hydrolysis to condensation, an easier mobility of hydrolyzed silica species is created which in turn smoothens the shell surface [66]. At RT and neutral pH synthesis conditions, the unhydrolyzed and hydrophobic TEOS forms an O/W emulsion with the aid of Pluronic F127 surfactant [67]. The use of 1-dodecyl-3-methylimidazolium bromide leads to the synthesis of worm-like HSM [68].

CTAB templated HSM can be rapidly synthetized by salt bridging in a surfactant-aided aerosol process [70]. The hypothesis is that the salt "locks in" the surfactant (CTAB) through salt bridging, thus preventing the ability of CTAB to template silica by electrostatic attraction. Table 7.3 summarizes the key steps of the development of HSM based on surfactant micelles cores.

7.3.3.2.5   HSM Based on O/W Pickering Emulsion Cores Route

In 2007, Shen et al. synthetized magnetic HSM by using an O/W Pickering emulsion whereby the droplets were stabilized by modified Fe_3O_4 nanoparticles (MFN). The nanoparticles acted as templates and sol-gel technology was used with a low-temperature drying step [71]. A certain contact angle of MFN was required in order to form the O/W Pickering emulsion. The size of the oil droplets can be decreased by increasing the amount of MFN. Later, the authors loaded the magnetic HSM with ibuprofen and proved their efficacy as a drug carrier having a slow release effect [72].

Pu et al. modified HSM by first treating the silica spheres with APTES, then grafting epoxide-containing nitroxide radicals to the spheres, followed by radical polymerization [73]. The technique allows for controllable size, high mechanical strength polymer layers, and good compatibility with organic matrices. Table 7.4 summarizes the key steps of the development of HSM based on O/W Pickering emulsion cores.

TABLE 7.2
Key Advances in the Development of HSM Based on Solid Cores

Years	Reference	Core	Shell Precursors	Shell Material	Size (μm)
1996	[35]	PS	TEOS/Fe$^+$	SiO$_2$ and Fe$_3$O$_4$	0.035
1998	[54]	PDADMAC	Colloidal SiO$_2$	SiO$_2$	0.7–1
2003	[55]	PMMA	TEOS	SiO$_2$	N/A
2004	[36]	PS	TEOS	SiO$_2$	0.035
2006	[56]	P(St-EA-MAA)	TEOS	SiO$_2$	1.2
2008	[48]	CPS	APTES/TEOS	Hybrid SiO$_2$/NH$_2$	0.125
2008	[49]	CPS	TEOS/MPTS	Hybrid SiO$_2$/S-H	N/A
2009	[60]	Pollen	TEOS	SiO$_2$	150
2009	[61]	*S. cerevisiae*	TEOS	SiO$_2$	4
2009	[62]	CaCO$_3$	TEOS	SiO$_2$	2–5
2009	[50]	CPS	TEOS/C$_{12}$NH$_2$	Hybrid SiO$_2$/NH$_2$	N/A
2009	[37]	PS	TEOS	SiO$_2$	0.04
2009	[38]	PS	TEOS/CTAB	SiO$_2$	0.5–4
2010	[39]	PS	TEOS	SiO$_2$	SC CO$_2$
2011	[40]	PS	TEOS/CTAB	SiO$_2$	N/A
2011	[51]	CPS	TEOS/MPS	CPS/SiO$_2$	N/A
		CPS/SiO$_2$[a]	TEOS/DMC	CPS/SiO$_2$/CPS/SiO$_2$	
2011	[58]	Polyvinyl pyridine	TEOS	SiO$_2$	0.18
2011	[41]	PS	TEOS/CTAB	SiO$_2$	N/A
2011	[42]	PS	TEOS/CTAB	SiO$_2$	N/A
2011	[52]	CPS	APTES/TEOS	Hybrid SiO$_2$/NH$_2$	1.3
2011	[53]	CPS	mPEG-COOH	Hybrid SiO$_2$/ mPEG-COOH	N/A
2011	[43]	PS	TEOS	SiO$_2$	N/A
2012	[44]	PS	TEOS/C$_{16}$TMS	Hybrid SiO$_2$/C$_{16}$	N/A
2012	[59]	PNIPAM	TEOS	SiO$_2$	N/A
2012	[57]	P(St-EA-MAA)	APTES/TEOS	Hybrid SiO$_2$/NH$_2$	1–10
2012	[45]	PS	TEOS	SiO$_2$	5–60
2012	[63]	CaCO$_3$	TEOS	SiO$_2$	N/A
2012	[46]	PS	TEOS	SiO$_2$	N/A
2013	[64]	TEOS	TEOS/Fe$^+$	SiO$_2$ and Fe$_2$O$_3$	N/A
2014	[62]	*S. aureus*	TEOS	SiO$_2$	0.7
2015	[47]	PS	TEOS/CTAB	MCM 48	N/A

Note: Chemical notation—PS, polystyrene; PDADMAC, polydiallyldimethylammonium chloride; PMMA, poly(methylmethacrylate); P(St-EA-MAA), styrene-2-(methacryloxy)ethyltrimethyl-ammonium chloride copolymer; CPS, cationic polystyrene; PNIPAM, poly(*N*-isopropylacrylamide); APTES, aminopropyltriethoxysilane; MPTS, 3-mercaptopropyltriethoxysilane; CTAB, cetyltri-methylammonium bromide.

[a] Double layered HSM.

TABLE 7.3
Key Advances in the Development of HSM Based on Surfactant Micelles Cores

Years	Reference	Surfactant	Shell Precursors	Shell Material	Size (μm)
2001	[27]	CTAB	TEOS and APTES	Hybrid SiO_2/NH_2	30.6
2007	[65]	Octylamine	TEOS	SiO_2	0.035
2009	[66]	Octylamine	TEOS	SiO_2	0.7–1
2008	[67]	Pluronic F127	TEOS	SiO_2	N/A
2010	[68]	1-dodecyl-3-methylimidazolium bromide	TEOS	SiO_2	0.035
2012	[69]	PEG	TEOS	SiO_2	150
2013	[70]	CTAB	$SiO_2/FeCl_3$	SiO_2	N/A

Note: CTAB, cetyltrimethylammonium bromide; Pluronic® F127, $(EO_{106}PO_{70}EO_{106})$; PEG, polyethyleneglycol.

TABLE 7.4
Key Advances in the Development of HSM Based on O/W Pickering Emulsion Cores

Years	Reference	Shell Precursors	Shell Material	Size (μm)
2007	[71]	HSM/MFN	SiO_2/Fe_3O_4	0.35
2010	[72]	HSM/MFN	SiO_2/Fe_3O_4	0.35
2009	[73]	HSM/APTES/epoxy	SiO_2/epoxy-amine resin	N/A

Note: MFN, modified Fe_3O_4 nanoparticles; APTES, aminopropyltriethoxysilane.

7.3.4 CORE-SHELL MICROCAPSULES

The first description of the use of an O/W emulsion to template-forming a silica shell around a water insoluble liquid, mesitylene, was published by Stucky et al. [74]. Development of these silica-based, core-shell microcapsules for encapsulating lipophilic actives first occurred in industrial laboratories. Starting from about 2002, reports from the academia appeared (see Figure 7.8).

In February 1998, Yoshioka et al. encapsulated 2-ethylhexyl *p*-methoxycinnamate (EHMC), an organic sunscreen, by using a surface-active silane, a copolycondensate of the hydrolyzate of *N*-[2-hydroxy-3-(3′-trihydroyxsilyl)propoxy]propyl hydrolyzed collagen [75]. No surfactant templating was used; instead, the surface-active silane played both the role of emulsifier and shell precursor in combination with MTES and phenyltriethoxysilane (PhTES). This formed a microcapsule after the hydrolysis and condensation reactions occurred at the active/water interface.

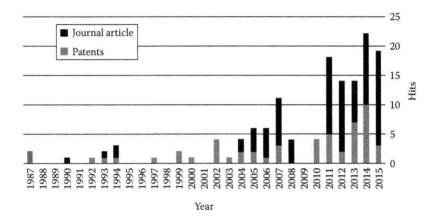

FIGURE 7.8 Number of publications found in SciFinder database using the keywords "core-shell silica microcapsules from O/W emulsions." Search conducted on March 2015.

7.3.4.1 The *In Situ* Route

Dauth et al. disclosed encapsulation of a nonionic, surfactant-stabilized emulsion using methyltrimethoxysilane (MTMS) and aminoethylaminopropyltrimethoxysilane (AEAPTMS). The reported microencapulation yield was poor, with a range from 38% to 87% [76].

The next key milestone occurred when Sol-Gel Technologies Ltd., a spin-off company of the Hebrew University of Jerusalem, patented a process for making sol-gel microcapsules [77]. The process consisted of blending a lipophilic active to be encapsulated (organic sunscreens, OS) with sol-gel precursors to arrive at *in situ* core-shell microcapsules (see Figure 7.9).

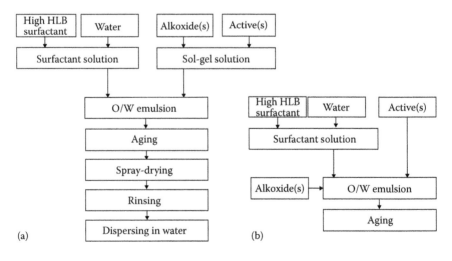

FIGURE 7.9 *In situ* (a) and *ex situ* (b) process for making core-shell microcapsules, drawn based on the descriptions of US6303149B1 and EP1471995B1, respectively.

This technique entraps water-insoluble liquid material called the dopant. The dopant is first mixed with the sol-gel precursor, generally TEOS, to form a sol-gel solution. The sol-gel solution is then emulsified with the aid of cetyltrimethyl-ammonium chloride (CTAC). The volume fraction of the droplets is roughly (or approximately) 10%. The hydrolysis and condensation of the sol-gel precursors present at the O/W interface build an amorphous silica shell from the interface toward the core of the microcapsule. The process occurs first in acidic pH and proceeds in basic pH. The acidic pH aids a higher rate of hydrolysis while the basic pH allows a more complete condensation. A film-forming polymer like polyvinyl-pyrrolidone (PVP) is added to the suspension to prevent aggregation of the micro-capsules. In a third step, the suspension is spray-dried or freeze-dried. Next, the microcapsule powder is washed and dispersed in water to obtain an alcohol-free aqueous suspension having a final volume fraction of about 30%. This process is a true sol-gel process. It goes through a gelation step followed by removal of the mother liquor. This process is called the *in situ* route due to the fact that the shell precursors are added into the dispersed phase before emulsification. Both hydrolysis and condensation of the alkoxysilanes occur at the O/W interface after diffusion of the precursors from the oil phase.

Some modifications of this technique have been disclosed like using a cationic additive to improve deposition of microcapsules from body wash compositions [78,79], or using precursors other than TEOS, such as ethylpolysilicate (EPOS), MTES, or MTMS [80]. Colloidal silica (Ludox® TM50) and nonionic surfactant like polyoxyethylene (20) sorbitan monooleate (Tween® 80) along with a sol-gel precursor have been reported to make thicker and therefore more impervious shell walls [81,82]. Actives other than organic sunscreens (OS) have been encapsulated by this process, such as fragrances, perfumes, or flavors. This was achieved by dissolving the actives in paraffin or polyvinylether waxes prior to encapsulation [83].

7.3.4.2 The *Ex Situ* Route

In 2002, Marteaux disclosed the *ex situ* route for encapsulation whereby the shell precursors are combined with the continuous phase [84]. The shells are generated in an O/W emulsion having a positive zeta potential on the droplet surface using a one-step process without gelation and without removal of the mother liquor (Figure 7.9b). The oil phase can be any water-insoluble liquid material. Optionally, the core-shell microcapsules can be harvested in a powdery form, for example, by spray-drying or freeze-drying (Figure 7.10).

Many variants of this process exploiting a positive zeta potential as the shell templating mechanism have been reported. One approach consisted of prehydrolyzing TEOS and using APTES as a gelling agent [85]. Another approach used cationic cellulose polymeric surfactant (CCPS) and a mixture of formic acid and acetic acid. The later mixture allows faster microencapsulation kinetics [86–88]. Still another approach used TEOS and an alkoxysilane having either amino or quaternary ammonium substituted alkyl group (QUATS) as the precursor. This resulted in improved deposition of the microcapsules onto negatively charged surfaces like textiles fabrics, hair fibers, and skin [89].

(a)

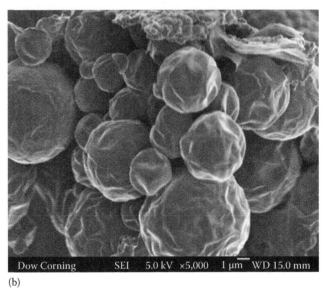

(b)

FIGURE 7.10   SEM of spray-dried (a) and freeze-dried (b) microcapsule suspensions. White scale bar represents 10 μm (a) and 1 μm (b).

CTAC has also been used to template fragrance-containing microcapsules obtained by the *in situ* and *ex situ* methods [90]. Using CTAB, researchers from Altachem claimed improvements for leach-proof benzoyl peroxide (BPO) microcapsules by the partial alkylation, from 2% to 25%, of the silica shell [91].

Other developments of the *ex situ* route have been reported such as thermosensitive microcapsules of volatile compounds [92] or crystallizable oils having melting temperature below 100°C [93]. Nonionic surfactants like nonylphenoxy-polyethoxyethanol (NP-9) have been used as templating agents in combination with trialkoxysilane (TAS) and aminopropylsilane (APS) as shell precursors [94]. Table 7.5 summarizes the key advances in the development of core-shell microcapsule technologies.

TABLE 7.5
Key Advances in the Development of Core-Shell Type Microcapsules

Years	Reference	Route	Core Material	Templating Agent	Shell Precursors	Shell Material
1996	[74]	*In situ*	Mesithylene	CTAB	TEOS	SiO_2
1998	[75]		OS	*N*-[2-hydroxy-3-(3′-trihydroyxsilyl)propoxy]propyl hydrolyzed collagen, MTES/PhTES		
1998	[76]	*Ex situ*	PDMS	Nonionic surfactants	MTMS/ AEAPTMS	$SiO_2/NH_2/Me$
1999	[77]	*In situ*	OS	CTAC	TEOS	SiO_2
2002	[84]	*Ex situ*	WILM	CTAC	TEOS/TAS	SiO_2/R
2003	[85]	*Ex situ*	OS	APTES	TEOS	SiO_2/NH_2
2005	[94]	*Ex situ*	Dyes	NP-9	TAS/APTES	$SiO_2/R/NH_2$
2007	[92]	*Ex situ*	WILM	CTAC	TEOS/TAS	SiO_2/R
2008	[86]	*Ex situ*	OS	CCPS	TEOS	SiO_2
2008	[89]	*Ex situ*	WILM	CTAC	TEOS/APTES TEOS/QUATS	SiO_2/NMe_4
2008	[90]	*In and ex situ*	Fragrances	CTAC	TEOS	SiO_2
2008	[78]	*In situ*	OS	CTAC	TEOS	$SiO_2/$ polyquaternium 4
2008	[80]	*In situ*	OS	Protective colloid	TEOS/EPOS/ MTES/MTMS	SiO_2/Me
2009	[91]	*Ex situ*	BPO	CTAB	TEOS/TAS	SiO_2/R
2009	[79]	*In situ*	OS	Protective colloid	TEOS/EPOS	$SiO_2/$ quaternium 80
2009	[81]	*In situ*	BPO/ Squalane	CTAC	TEOS/Ludox TM 50	SiO_2
2009	[93]	*Ex situ*	N/A	CTAC	Fumed SiO_2	SiO2
2010	[87]	*Ex situ*	OS	CCPS	TEOS	SiO_2
2011	[84]	*In situ*	Fragrances	CTAC	TEOS/MTES	SiO_2/Me
2011	[88]	*Ex situ*	Fragrances	CCPS/ APTES	TEOS/TAS	SiO_2/R
2012	[82]	*In situ*	Oils	Tween 80	TEOS/VTES	SiO_2/R

Note: OS, organic sunscreen; PDMS, polydimethylsiloxane; WILM, water insolubleliquid material; BPO, benzoyl peroxide; CCPS, cationic cellulose polymeric surfactant; APTES, aminopropyl-triethoxysilane; TAS, trialkoxysilane; MTMS, methyltrimethoxysilane; MTES, methyltri-ethoxysilane; VTES, vinyltriethoxysilane; QUATS, quaternary ammonium substituted alkyl group; EPOS, ethylpolysilicate.

7.3.5 POLYNUCLEAR MICROCAPSULES

In 2006, Nakahara et al. disclosed a method of making hollow silica particles via the water-glass route, starting from a water-in-oil-in-water (W/O/W) multiple emulsion [95]. The internal water phase was composed of water-soluble silicates and the actives to be encapsulated were emulsified into a continuous oil phase. The obtained W/O emulsion was further emulsified into a water external phase to arrive at the W/O/W emulsion. Giant biomolecules, cells, or viruses can be included and were claimed to be preserved for long periods of time in the hollow particle.

Another approach was to use sol-gel precursors like alkoxysilanes as the initial oil phase. The process started from a water-in-alkoxysilane-in-water multiple emulsion and ended with polynuclear microcapsule suspension having a water-in-silica-in-water or water-in-silsesquioxane-in-water structure (Figure 7.11) [96].

The inventors describe the use of this process to encapsulate biocatalysts. The internal water phase contains the biocatalyst, which is preferably a water-soluble enzyme and its cofactor, and the external phase contains the substrate. The polynuclear microcapsule is acting like a microbioreactor, wherein the substrate can diffuse into the internal water phase, becomes transformed by the biocatalyst into a product that can diffuse out of the capsule and into the external water phase. The internal water phase contains *Aspergillus niger* catalase, an oxidoreductase enzyme, catalyzing the transformation of hydrogen peroxide into water and oxygen. A significant improvement of enzyme half-life time from 2 weeks to 1 year at room temperature has been observed.

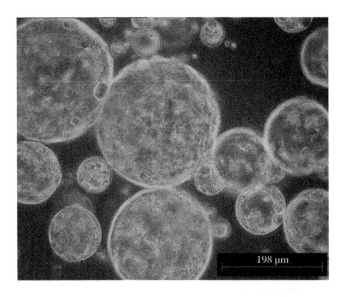

FIGURE 7.11 Optical micrograph of W/SiO$_2$/W polynuclear microcapsule suspension.

7.4   INDUSTRIAL EXAMPLES

7.4.1   Applications of Colloidal Silsesquioxanes

As previously mentioned, colloidal silsesquioxane resins were discovered more than 50 years ago, and their commercial utility has really not been developed to any appreciable extent. The reason for this is not known, but it may be due to the low efficiency of their preparation. The reason for this is not known; it may be due to the low efficiency relative to the cost of their preparation. The active silsesquioxane content achievable is typically 5%–10%. This is significantly lower than the oil content in most commercial O/W emulsions prepared with similar cost structure.

Some reported uses for colloidal silsesquioxanes include textile treatment to repel dirt and soiling [97–99]. Colloidal silsesquioxanes can also be used as reinforcing filler for silicone water-based elastomers and to modify organic latexes. [100,101]. Aqueous dispersions of hybrid particles and "nanopowders," both consisting of silicone resin cores and acrylic shells that were prepared from colloidal silicone resins and emulsion polymerization of acrylic monomers, were investigated as impact modifiers for coatings and plastics [102]. Dispersions of colloidal silsesquioxanes in organic solvents have been prepared by removing water from aqueous dispersions while simultaneously adding organic solvent [103].

7.4.2   Hollow Spheres

7.4.2.1   Hollow Glass Microspheres

Hollow glass microspheres are used as void fillers for lightweight sealants, adhesives, edge sealers, ceramics, and composites. Hollow glass microspheres used in composites can be surface treated with organofunctional silanes such as N-phenylaminosilanes, epoxysilanes, alkylchlorosilanes and alkylaminosilanes to improve their dispersion and interfacial adhesion.

In aerospace applications, hollow glass microspheres have been used as part of ablative coatings and heat shielding for rockets, space vehicles, and missiles. Their radio-frequency transparent properties are exploited into stealth specialty paints and coatings and also in constructing radomes. They are also used in abradable seals in jet engines for improved efficiency.

7.4.2.2   Hollow Silica Microspheres

HSM are useful as lightweight fillers in composites, reflective and luminescent coatings, thermal insulation, rechargeable batteries, and heterogeneous catalysts. There is interest in the encapsulation of nuclear fusion materials. Low dielectric constant materials are of high interest in electronics, and HSM provides a route to low dielectric constant materials by introducing controlled porosity.

In the medical field, hollow magnetic particles are used for magnetic field triggered drug delivery. Diagnostics based on hollow magnetic particles are promising for ultrasound biomedical imaging. Preliminary studies demonstrated the potential use of the immunomagnetic microspheres for two purposes in male infertility. First potential use is to aid the specific removal of antisperm antibodies and sperm cells containing antisperm antibodies from semen of infertile males *in vitro*. Second potential utility is for the separation of sperm cells from epithelial cells [35].

In environmental applications, HSM have been reported to be of interest as adsorbents for rapid removal and detoxification of dyes, heavy metal ions, and organic pollutants in both water systems and waterways. Highly sensitive microsensors were grafted onto hollow spheres to obtain hollow-chemosensors that were able to detect Hg^{2+} and separate the latter from the water stream with high efficiency [104].

In the area of biotechnology, enhanced enzymatic activity of *Candida rugosa* lipase has been observed after it was immobilized on pore-enlarged HSM and further cross-linked [105].

HSM can be employed as the stationary phase in thin-layer chromatography [106]. Thin layers of HSMs coated on glass slides can separate model mixtures of methyl red and dimethyl yellow. The preparation of homogeneous thin-layer chromatography plates is easier with HSM than with common silica gels.

7.4.3 CORE-SHELL MICROCAPSULES

7.4.3.1 Sun Protection

The first commercial application for silica-based microcapsules was to encapsulate organic sunscreens (OS). OS have an unpleasant greasy feel and potentially skin irritant. Therefore, there is a need to prevent direct contact between OS and skin. As OS can (or "can potentially" or "are suspected to") penetrate human skin, entrapping OS into a larger object significantly reduces the risk of percutaneous permeation [77]. More specifically there is a need to physically separate EHMC, the most widely used UV-B sunscreen, and benzyldimethoxybenzoylmethane (BMBM), the most widely used UV-A sunscreen as their combined mixture is photo-unstable. These technical challenges can be overcome by a silica-based, core-shell microcapsule. Indeed OS are polar oils that are good solvents for organic-based polymer and shell material in general. Their dosage level in a sun protection product is in the range of 5%–15% by weight. As a consequence, only high payload microcapsules like core-shell types can be envisaged. The shell must be thin, to allow UV light to transmit and become absorbed by the OS, and the shell needs to be impervious to the OS at the same time.

In 2002, Seiwa Kasei Co. Ltd. launched Silasoma®, a product range of core-shell microcapsules containing coencapsulated EHMC and BMBM and made from silsesquioxane-peptide (hydrolyzed silk) shells. The microcapsules are typically 2 μm in size, and they are useful for protecting both skin and hair [75]. Besides protecting skin from harmful rays from the sun, these products are nonirritating and leave the skin with a pleasant sensory feel. In mid-2002, Merck launched Eusolex® UV-Pearls™ under license from Sol-Gel Technologies Ltd. [77].

7.4.3.2 Construction Chemicals

TEOS can encapsulate phase change material (PCM) such as *n*-octadecane with high encapsulation efficiency and good thermal stability, good phase-change performance and antiosmosis properties [107]. When silica-encapsulated phase change materials are undergoing cooling, the silica shell contributes to a widening of the phase transition temperature range.

Shi et al. developed new self-healing materials, called passive smart microcapsules that hold promise for "crack-free" concrete or other cementitious composites [108]. Crack propagation triggers microcapsule breakage and releases the healing agent and a catalyst into the microcracks.

Lecomte et al. disclosed a patent covering processes for increasing hydrophobicity of porous materials like clay, bricks, gypsum, lime, or wood-based substrates [109,110]. Treatment of construction substrates with *ex situ* prepared core-shell microcapsules provides water repellency performance that is superior to treatment using emulsions of organopolysiloxane or alkoxysilane.

De Schijver et al. disclosed a method of making hollow mesoporous silica spheres containing benzoyl peroxide (BPO), a catalyst used to formulate one component polyurethane (PU) foam [91]. BPO is released by bursting the leach-proof organically-modified silica microcapsule upon being dispensed from a pressurized can.

7.4.3.3   Textiles

Dow Corning Corporation launched the DS 9000 Multifunctional Additive, a polydimethylsiloxane containing microcapsule suspension for textile treatment obtained by the *ex situ* process. It provides the treated fabric with superior hydrophobicity, quick drying, softness, and luxurious hand. The cold treatment reduces the harshness of flame-retardant-treated fabrics without negatively impacting fire resistance (Figure 7.12).

7.4.3.4   Pharmaceuticals

Sol-gel Technologies Ltd. used their patented *in situ* process to microencapsulate benzoyl peroxide (BPO) crystals [111]. BPO is effective to treat acne (acne vulgarius) and rosacea but it is a skin irritant. Therefore, the porosity of the silica shell of the microcapsule has been designed to limit direct contact of BPO while controlling its crystals dissolution by skin lipids. A first product, Cool Pearls™ BPO anti-acne kit, was commercialized in 2009 in the United States.

7.5   CONCLUDING REMARKS

The hydrolysis and condensation of alkoxysilanes are usually run at RT and do not generate toxic by-products. This mild and environmental friendly chemistry leads to many silicon-based material compositions and functionalities. In a large excess of water, physico-chemical parameters like pH, ionic strength, and the solubility of alkoxysilanes upon hydrolysis and condensation are creating a diversity of colloidal suspension of silica or silsesquioxane microspheres. The use of surfactants allows the design of these particles into hollow silica, silsesquioxane spheres, core-shell, and polynuclear microcapsules. The easy combination with ferromagnetic materials improves target delivery of actives and fast recuperation under induced magnetic field. Additional morphologies can be obtained by using advanced physical processes such as supercritical fluid extraction.

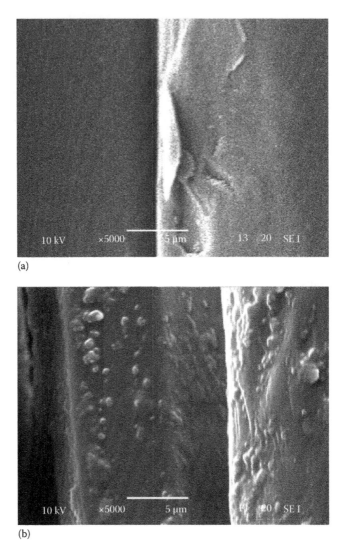

FIGURE 7.12 Untreated Trevira® CS polyester fibers (a) and the same fibers treated with 40 g/L of DS 9000 Multifunctional Additive (b). White scale bars represent 5 μm.

As a consequence of their versatile morphologies, silica, and silsesquioxane dispersions are used in many applications to meet numerous technical needs or to provide new benefits.

Colloidal silsesquioxanes have been used in textile treatment to prevent dirt and soiling and as reinforcing filler for silicone water-based elastomers. HSM are useful in several areas like material science, medical devices, diagnostics, environmental remediation of heavy metals, and separation technologies. Inorganic core-shell microcapsules are used in advanced UV protection for skin, in textile treatments, and for hydrophobing cement. Clinical studies of active pharmaceutical actives encapsulated with silica are ongoing and show promise in the topical treatment of acne, psoriasis, and rosacea. Polynuclear microcapsules can act as a protective and reusable microreactor in fermentation and cell culture.

From basic silica particles to active containing objects of different morphologies, silica and silsesquioxane suspensions are versatile by providing multiple benefits from a low energy consuming and sustainable chemistry.

ACKNOWLEDGMENTS

The authors thank Debbie Bailey for providing the SEM images in Figure 7.5 and Dr. Steven Swier for providing the AFM image in Figure 7.6. The authors thank the Dow Corning Corporation for support during the preparation of this manuscript.

REFERENCES

1. Brinker, C., *Journal of Non-Crystalline Solids*, 1998, *100*, 31–50.
2. Stöber, W., Fink, A., Bohn, E., *Journal of Colloid and Interface Science*, 1968, *26*, 62–69.
3. Perry, R. J., Adams, M. E., *Silicones and Silicone-Modified Materials*, ACS Symposium Series, 2009, *729*, 533–543.
4. Cekada, J., Weyenberg, D. R., US3433780, 1969.
5. Schmidt, H., *Journal of Non-Crystalline Solids*, 1985, *73*, 681.
6. Avnir, D., Levy, D., *Journal of Physical Chemistry*, 1984, *88*, 5956.
7. Carturan, G., *Journal of Molecular Catalysis*, 1989, *57*, L13.
8. Livage, J., *Journal of Physics: Condensed Matter*, 2001, *13*(33), R673–R691.
9. Beck, J., WO9111390, 1991.
10. Schacht, S., Huo, Q., Voigt-Martin, I. G. et al., *Science*, 1996, *273*, 768–771.
11. Iler, R., *The Chemistry of Silica*, Wiley, New York, 1979.
12. Brinker, C., Scherrer, G., *Sol-Gel Science: The Physics and Chemistry of Sol-Gel Processing*, Academic Press, Boston, MA, 1990, p. 153.
13. Pierre, A., *Introduction to Sol-Gel Processing*, Springer Verlag, Berlin, Germany, 1998, p. 4.
14. Ciriminna, R., Fidalgo, A., Pagliaro, M., *Chemical Review*, 2013, *113*, 6592.
15. Leung, K., Criscenti, L., *Journal of the American Chemical Society*, 2010, *131*, 18358.
16. Handy, B., Baiker, A., *Studies in Surface Science and Catalysis*, 1991, *63*, 239–246.
17. Bey, A. E., US4424297, 1984.
18. Liles, D. T., US8715828, 2014.
19. Kolbe, G., *Das Komplexchemische Verhalten der Kieselsäure*, PhD dissertation, Friedrich-Schiller Universität, Jena, Germany, 1956.
20. Giesche, H., *Fine Particles: Synthesis, Characterization and Mechanisms of Growth*, Surfactant Science Series, Vol. 92, Sugimoto, T. (ed.), Marcel Dekker, New York, 2000, pp. 126–146.
21. Esquena, J., Pons, R., Azemar, N. et al., *Colloids and Surfaces A: Physicochemical and Engineering Aspects*, 1997, *123–124*, 575–586.
22. Kaiser, C., Hanson, M., Giesche, H. et al., *Fine Particle Science and Technology: From Micro to Nanoparticles*, Pelizzetti, E. (ed.), Kluwer, Dordrecht, the Netherlands, 1995, pp. 71–84.
23. Kimura, H., US4528390, 1985.
24. Adams, M. E., US5801262, 1989.
25. Veatch, F., US2978339, 1957.

26. Ravindra, S., *Journal of the American Ceramic Society*, 1991, *74*(8), 1987–1992.
27. Fowler, C., Khushalani, D., Mann, S., *Journal of Materials Chemistry*, 2001, *11*(8), 1968–1971.
28. Wilcox, D., US5492870, 1994.
29. Park, J., Oh, C., Shin, S., Moon, S., Oh, S., *Journal of Colloid and Interface Science*, 2003, *266*(1), 107–114.
30. Singh, R., Garg, A., Bandyopadhyaya, R., Mishra, B., *Colloids and Surfaces, A: Physicochemical and Engineering Aspects*, 2007, *310*(1–3), 39–45.
31. Oh, C., Lee, Y., Jon, C. et al., *Colloids and Surfaces, A: Physicochemical and Engineering Aspects*, 2009, *337*(1–3), 208–212.
32. Yoon, H., Hong, J., Park, C., Park, D., Shim S., *Materials Letters*, 2009, *63*(23), 2047–2050.
33. Li, D., Guan, Z., Zhang, W., Zhou, X., Zhang, W., Zhuang, Z., Wang, X., Yang, C., *ACS Applied Materials and Interfaces*, 2010, *2*(10), 2711–2714.
34. Wang, X., *Guangzhou Huagong*, 2010, *38*(7), 94–96.
35. Margel, S., Gura, S., Bamnolker, H., Nitzan, B., Tennenbaum, T., Bar-Toov, B., Hinz, M., Seliger, H., *Proceedings of the First International Conference on Scientific and Clinical Applications of Magnetic Carriers*, Rostock, Germany, 1997, pp. 37–51.
36. Shin, H., Park, Y., Woo, J., Chang, K., Park, Y., *Kongop Hwahak*, 2004, *15*(1), 65–69.
37. Wang, J., Dong, Q., *Cailiao Kaifa Yu Yingyong*, 2009, *24*(4), 42–47.
38. Liu, S., Wei, M., CN200910258303, 2009.
39. Chen, Z., Li, S., Xue, F., Sun, G., Luo, C., Chen, J., Xu, Q., *Colloids and Surfaces, A: Physicochemical and Engineering Aspects*, 2010, *355*(1–3), 45–52.
40. Wei, M., Van Oers, C., Hao, X., Qiu, Q., Cool, P., Liu, S., *Microporous and Mesoporous Materials*, 2011, *138*(1–3), 17–21.
41. Liu, S., Wei, M., Cool, P., Van Oers, C., Rao, J., *Microscopy and Microanalysis*, 2011, *17*(5), 766–771.
42. Qiu, Q., Diao, Z., Van Oers, C., Kwong, F., Cool, P., Liu, S., *International Journal of Materials Research*, 2011, *102*(12), 1488–1493.
43. Liu, S., Wei, M., Rao, J., Wang, H., Zhao, H., *Materials Letters*, 2011, *65*(13), 2083–2085.
44. Wang, M., Chen, Y., Ge, X., Ge, X., *Chinese Journal of Chemical Physics*, 2012, *25*(1), 120–124.
45. Liu, S., Qiu, Q., Wang, H., *Faming Zhuanli Shenqing*, CN102380345, 2012.
46. Cheng, X., Li, J., Li, X., Zhang, D., Zhang, H., Zhang, A., Huang, H., Lian, J., *Journal of Materials Chemistry*, 2012, *22*(45), 24102–24108.
47. Hao, Z., Qiao, N., Zhang, X., Cheng, J., *Faming Zhuanli Shenqing*, CN104477925, 2015.
48. Yang, Z., CN200810035588, 2008.
49. Yuan, J., Wan, D., Yang, Z., *Journal of Physical Chemistry C*, 2008, *112*(44), 17156–17160.
50. Feng, X., Jin, W., Yang, C., *Wujiyan Gongye*, 2009, *41*(9), 18–20.
51. Cao, S., Hao Z., Jin, X., Sheng, W., Li, S., Ge, Y., Dong, M., Wu, W., Fang, L., *Journal of Materials Chemistry*, 2011, *21*(47), 19124–19131.
52. Hu, H., Zhou, H., Liang J. et al., *Journal of Colloid and Interface Science*, 2011, *358*(2), 392–398.
53. Hu, H., Zhou, H., Du, J. et al., *Journal of Materials Chemistry*, 2011, *21*(18), 6576–6583.
54. Caruso, R., Möhwald, H., *Science*, 1998, *282*(5391), 1111–1114.
55. Wang, D., *Chinese Chemical Letters*, 2003, *14*(12), 1306–1308.
56. Chen, M., Wu, L., Zhou, S., You, B., *Advanced Materials*, 2006, *18*(6), 801–806.
57. Gu, Y., Niu, K., Ke, L., Duan, F., Chen, M., *Gongneng Cailiao*, 2012, *43*(5), 669–672.
58. Su, Y., Yan, R., Dan, M., Xu, J., Wang, D., Zhang, W., Liu, S., *Langmuir*, 2011, *7*(14), 8983–8989.

59. Wang, M., Chen, Y., Ge, X., Ge, X., *Chinese Journal of Chemical Physics*, 2012, *25*(1), 120–124.
60. Cao, F., Li, D., *Biomedical Materials*, 2009, *4*(2), 025009/1–025009/6.
61. Weinzierl, D., Lind, A., Kunz, W., *Crystal Growth and Design*, 2009, *9*(5), 2318–2323.
62. Ebrahiminezhad, A., Najafipour, S., Kouhpayeh, A. et al., *Colloids and Surfaces B: Biointerfaces*, 2014, *118*, 249–253.
63. Mao, Z., Hu, L., Zhao, Q., Gao, C., *Chemical Research in Chinese Universities*, 2012, *28*(3), 546–549.
64. Kovacik, P., Singh, M., Stepanek, F., *Chemical Engineering Journal*, 2013, *232*, 591–598.
65. Xu, J., Sui, X., *Guisuanyan Tongbao*, 2007, *26*, 1193–1196.
66. Liu, S., Wei, M., Sui, X., Cheng, X., Cool, P., Tendeloo, G., *Journal of Sol-Gel Science and Technology*, 2009, *49*(3), 373–379.
67. Liu, J., Fan, F., Feng, Z., Zhang, L., Bai, S., Yang, Q., Li, C., *Journal of Physical Chemistry C*, 2008, *112*(42), 16445–16451.
68. Zhao, M., Gao, Y., Zheng, L., Kang, W., Bai, X., Dong, B., *European Journal of Inorganic Chemistry*, 2010, *6*, 975–982.
69. Wang, L., Du, F., *Qingdao Keji Daxue Xuebao, Ziran*, 2012, *33*(3), 229–232.
70. Olasehinde, G., He, J., Drenski, M., McPherson, G., John, V., Zhang, Y., *Abstracts of Papers, 246th ACS National Meeting and Exposition*, Indianapolis, IN, 2013, IEC-29.
71. Shen, S., Wu, W., Guo, K., Meng, H., Chen, J., *Colloids and Surfaces A: Physicochemical and Engineering Aspects*, 2007, *311*(1–3), 99–105.
72. Wang, Z., Wu, W., Zhang, K., Chen, J., Zhang, P., *Beijing Huagong Daxue Xuebao, Ziran Kexueban*, 2010, *37*(3), 110–114.
73. Pu, H., Qin, Y., Yuan, J., Wan, D., CN101543756, 2009.
74. Schacht, S., Huo, Q., Voigt-Martin, I. G., Stucky, G. D., Schüth F. D., *Science*, 1996, *273*, 768–771.
75. Yoshioka, M., JP4106398, 1998.
76. Dauth, J., Daubzer, B., EP0941761B1, 1998.
77. Magdassi, S., US6303149B1, 1999.
78. Traynor, D., WO2008144734, 2008.
79. Habar, G., FR2937248, 2008.
80. Habar, G., Bernoud, T., WO2012/004461, 2009.
81. Toledano, O., Sertchook, H., Loboda, N. et al., US2011/0177951, 2009.
82. Gosselin, M., US2014/0341958, 2014.
83. Dreher, J, WO2011/124706, 2011.
84. Marteaux, L., EP1471995, 2002.
85. Seok, S., US6855335, 2003.
86. Bone, S., EP2080552A1, 2008.
87. Viaud-Massuard, M.-C., FR2965190, 2010.
88. Bone, S., WO2013/083760, 2011.
89. Marteaux, L., Roidl, J., Severance, M. et al., WO2010045446, 2008.
90. Popplewell, L., EP2196257A2, 2008.
91. De Schrijver, A., US20120199671, 2009.
92. Marteaux, L., Zimmerman B., US9089830, 2009.
93. Schmitt, V., Mathieu, D., Rénal, B. et al., WO2011012813, 2009.
94. Barbe, C., Finnie, K., Kong, L. et al., WO2006/133519, 2005.
95. Nakahara, Y., *Journal of Biomedical Materials Research, Part A*, 2007, *81A*(1), 103–112.
96. Marteaux, L., Zimmerman, B., US8435560, 2007.
97. Mohrlok, S. R., Yates, G. P., US3493424, 1970.
98. Johnson, M. T., Patel, V., US7320956, 2008.

99. Heller, A., Hashemzadeh, A., US7981961, 2011.

100. Cekada, J., US3355406, 1967.

101. Cekada, J., Weyenberg, D. R., US3445415, 1970.

102. Kozakiewicz, J., Ofat, I, Trzaskowska, J., *Proceedings of the 39th Annual Water-Borne, Higher-Solids, and Powder Coatings Symposium*, Rawlins, J. W., Storey, R. F. (eds.), University of Southern Mississippi, Hattiesburg, MS, 2012, 386–398.

103. Ma, C., Yamamoto, M., Oi, F., US8088863, 2012.

104. Cheng, X., Li, J., Li, X., Zhang, D., Zhang, H., Zhang, A., Huang, H., Lian, J., *Journal of Materials Chemistry*, 2012, *22*(45), 24102–24108.

105. Ren, L., Jia, H., Yu, M., Shen, W., Zhou, H., Wei, P., *Biotechnology and Bioprocess Engineering*, 2013, *18*(5), 888–896.

106. Qin, L., Wang, H., Liu, S., *Arabian Journal of Chemistry*, Ahead of Print.

107. Zhang, H., Wang, X., Wu, D., *Journal of Colloid and Interface Science*, 2010, *343*(1), 246–255.

108. Shi, X., *Issue Nanotechnology in Cement and Concrete*, 2013, *2*, 9–17.

109. Lecomte, J.-P., WO2013164381, 2013.

110. Lecomte, J.-P., WO2013166280, 2013.

111. Toledano, O., Sertchook, H., Loboda, N., Bar-Simantov, H., Shapiro, L., Abu-Reziq, R., WO2008093346, 2008.

8 Dispersion Processes of Silicone Antifoams

Steven P. Christiano

CONTENTS

8.1 INTRODUCTION

Foam formation is a widespread problem in industrial processes. Agitation, turbulent flows, distillation, and pressure differences occur frequently in chemical processing and bring about the entrainment of gas into process liquids and the formation of foam [1]. Foam reduces unit productivity by occupying volume. It can also reduce the quality of production. For example, during the jet dyeing of fabric, foam can attach to parts of the cloth, ruining the uniformity of dye absorption and increasing reject rates. A foam head can also act as a barrier against mass transport or heat transfer. For example, a thick foam head reduces the rate of carbon dioxide volatilization from ethanol fermentation broth, reducing the efficiency of the biological processing of sugars.

Foam is a gas dispersed in a liquid. Foam can be formed through the mechanical mixing of gas, typically air, into a liquid, such as occurs during pouring, pumping, or mixing of a process liquid. Gas can also be generated inside the liquid through chemical or biological action, such as the formation of carbon dioxide bubbles upon the release of pressure over a carbonated beverage. In some cases, gas is intentionally introduced into a solution, such as the bubbling of oxygen into a fermenting solution to promote aerobic fermentation.

Surface-active materials need to be present to stabilize foam. One commonly experienced type is low molecular weight synthetic surfactant such as utilized in textile washing, shampoos, and personal care products. Other types of surface-active materials are also encountered. Fine hydrophobic particulates act to stabilize foam in ore slurries or in phosphoric acid manufacture. Biological molecules, including macromolecules such as starches and proteins, can stabilize foams in potato processing [2,3], sugar processing, wastewater treatment, food processing, fermentation, paper pulp processing, and paper making [4], to name a few examples.

Foam is problematic in a variety of aqueous solutions but is also experienced in nonaqueous liquids such as in petroleum during gas-oil separation, in hydraulic and transmission fluids, in diesel fuels, and in petrochemical processing and distillation [5]. Foam stability in nonaqueous foams is attributed to surface dilatational viscosity and elasticity derived from the presence of surface-active species [6–9]. Application of antifoams to a variety of aqueous and nonaqueous media was reviewed by McGee [10].

A variety of solutions are employed to ameliorate foam problems. One approach is to minimize gas entrainment by improved process design. In some cases, mechanical systems are used to separate the foam, or to rupture foam using ultrasonic devices, water jets, heating bars, and even moving mechanical arms.

Chemical approaches to foam destruction are widely used and can be separated into two general types. *Defoamers* are sprayed onto a pre-existing foam head to destabilize it. Oftentimes, surfactants, alcohols, or oils are used. In contrast, *antifoams* are added into the foaming liquid where they disperse and then destabilize the bubbles before a foam head can fully form.

There is a widening interest in understanding antifoam mechanism. Several excellent reviews elucidate the mechanistic steps and relate them to physical–chemical fundamentals [11–14]. Recent reviews also highlight kinetics effects and aqueous films [15,16]. In holding with their commercial significance, there is also a very large body of patent literature that describes many of the practical aspects of preparation, formulation, and delivery of antifoams.

Unfortunately, this body of literature does not establish the relationship between antifoam preparation and mechanistic fundamentals. This chapter will attempt to forge that link between the practical aspects of antifoam preparation and delivery with the mechanistic fundamentals. In order to best accomplish this goal, it is necessary to focus on one type of antifoam technology, specifically silicone/silica-based antifoams and their use in aqueous-surfactant-stabilized foam.

The details of the foaming process, including the nature of the foaming medium and rate of foam generation, play a significant role in setting the conditions under which foam control must occur. Therefore it is necessary to delineate factors that contribute to the foam conditions. Each process will be different, requiring proper design of the antifoam compound for each situation. Design includes antifoam compound formulation but more significantly efficient performance is strongly dependent on the dispersion of the antifoam compound, and dispersion of the hydrophobic solids within the compound. Finally, a novel microscopic technique is introduced that provides an enhanced understanding of antifoam–foam dynamic spreading which contributes significantly to antifoam dispersion.

8.1.1 MAXIMUM TIME FOR FOAM DRAINAGE

Typically, there is a maximum volume of foam, V_{max}, that can be tolerated in any given industrial process, oftentimes limited by the headspace volume or containment volume that can accommodate the foam prior to overflow of the equipment. Alternatively, there is a maximum foam volume above which the process loses efficiency. In consumer products, the maximum foam volume can be based on the desired level of foam dictated by user preference. In any case, it is highly desirable to achieve a level of *controlled foaming* whereby the foam volume is held below V_{max}. Note that the limitation is almost always on the volume of foam, not on the number of bubbles. As such, it is the foam volume that must be controlled.

We can describe the foam volume, V, as a function of time over which foam generation has occurred, t, and the average volumetric rate of foam formation $\Delta V/\Delta t$. Thus, $V = (\Delta V/\Delta t)t$. Similarly, for the maximum foam volume,

$$V_{max} = (\Delta V / \Delta t)\, t_{max} \tag{8.1}$$

On this basis, the maximum time of foam generation, t_{max}, is defined, assuming $V = 0$ at time $t = 0$.

In many systems $\Delta V/\Delta t$ is not constant. The volumetric rate of foam formation will be variable over time, either due to changes in the process or changes in foamability of the solution over time. Therefore, it is beneficial to generalize Equation 8.1 and define an instantaneous volumetric rate of foam formation, $v(t)$, to which t_{max} relates.

$$V_{max} = \int_{0}^{t=t_{max}} v \, dt \tag{8.2}$$

Conceptually, t_{max} is the length of time allowed for an antifoam to diminish the volumetric foam growth in order to keep it under control. For a great many surfactant-based industrial processes the foam is very stable, with foam heads lasting hours or days with little reduction in volume. Therefore, t_{max} provides a convenient timescale on which to consider foam formation, drainage process in foam films, and foam rupture processes contributed by antifoaming agents.

The volumetric rate of foam formation is dictated by the particulars of the foaming process. Each process will have an inherent volumetric rate of gas/liquid mixing. Secondly, the foaming solution conditions, such as the surfactant type and concentration, pH, temperature, and electrolyte concentrations, will dictate how efficiently gas is captured and stabilized as foam. These factors are defined by the functional aspects of the process. Additionally, the size distribution of bubbles formed depends on the type of mechanical mixing and surfactant stabilization of newly forming bubble surfaces.

Foam volume reduction requires breakage of bubbles to release gas to the air over the foaming system. Ruptures that simply create larger bubbles are not effective at diminishing the volume of foam. Volume reduction requires rupture of a significant fraction of bubble films in the foam head to allow release of gas to the air. Therefore, there is a relationship between the process and its bubble size distribution and the number of bubble films that must be ruptured to achieve volumetric reduction.

The use of antifoams allows greatly enhanced volumetric rates of foam rupture. Through application of a sufficient dose of antifoam it is possible to enhance the rate of foam rupture to offset the rate of foam formation, allowing the attainment of controlled foaming. Necessarily, the antifoam must be effective at the bubble film thicknesses prevailing in the foam head at drainage times less than or equal to t_{max}. Many factors defined by the foaming solution, including type and amount of surfactant, dictate the rate of bubble film thinning and consequently the bubble film thickness at which rupture must occur.

In order to understand this relationship better, we must understand the role of surfactants in stabilizing foam structures and controlling bubble film thickness over time.

8.2 STRUCTURE AND STABILIZATION OF FOAM

Foam is thermodynamically unstable relative to bulk phases of solution and gas, because the energy contribution from the increased surface area is significant relative to the total energy of the system. The free energy for foam includes surface area as an extensive variable in the system:

$$\Delta F = \sigma \Delta A \qquad (8.3)$$

where ΔF is the increase in free energy of the system at constant temperature upon formation of new surface area, ΔA, at surface tension σ. There is a small amount of pressure–volume work resulting from compression of gas into bubbles [17,18]. However, this contribution is rather small.

Based on Equation 8.3 a reduction in area upon breaking of bubbles reduces the energy of the system and will be thermodynamically favored. Pure liquids do not support significant foam formation. Any bubbles formed exist temporarily and as liquid drains away they rupture to the air. However, the presence of surface-adsorbed species *kinetically* stabilizes foam.

8.2.1 FOAM FORMATION

The generation of a stable layer of foam, generally referred to as a foam head, can be split into two different regimes: (1) bubble formation and new surface stabilization and (2) foam head formation and drainage.

In bubble formation, as illustrated in Figure 8.1, a gas is dispersed into the aqueous liquid through mechanical or chemical means. As the nascent gas/liquid surface is formed, surface-active materials present in solution quickly adsorb and pack together (inset in Figure 8.1). The adsorbed surfactant layer imbues the surface with physical properties (surface viscosity, surface elasticity, surface charge, etc.) which significantly contributes to stability of the resulting foam.

Simple sparging of nitrogen through a surfactant solution can allow measurement of foam formation rates, foam stability, and maximum foam height [19]. Foam production is dependent on surfactant concentration, generally allowing formation of a measurable foam head at about 5% of the surfactant's critical micelle concentration (CMC), with increasing surfactant concentration increasing foamability. For example, Karakashev et al. [20] report nearly complete capture of the air volume sparged into a solution at about 15% of the CMC in testing performed with sodium octylsulfate as the surfactant.

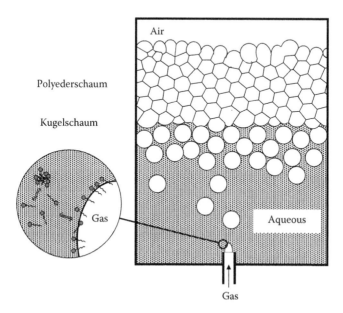

FIGURE 8.1 Evolution from bubble formation to a well-drained foam head. Inset illustrates the equilibration of surfactant molecules from micelles to individual molecules in solution and subsequent adsorption on the expanding bubble surface.

8.2.2 FOAM HEAD FORMATION

The second regime, foam head formation and drainage, describes the concentration of bubbles into a foam head. Driven by buoyancy, the dispersed gas bubbles rise upward and collect at the top of the liquid, establishing a layer of foam as illustrated in Figure 8.1. The foam is initially composed of roughly spherical bubbles, sometimes referred to as kugelschaum [21]. As more liquid drains from the foam, the volume fraction of gas increases and the bubbles become distorted, forming a nearly close-packed array, with fine liquid films between them. Well-drained foam is often referred to as polyederschaum reflecting the bubbles' polyhedral shape.

The fine liquid films formed in draining foam are diagrammed in Figure 8.2 where the dark regions are the aqueous liquid. Where neighboring bubbles flatten against each other, they form a thin film referred to as a bubble lamella (not readily visible in Figure 8.2). Where three bubbles come together, they form aqueous channels known as Plateau's border (see inset in Figure 8.2) that are roughly triangular in cross section but with inward curving sides. The bubbles adjust shape and position to establish a 120° orientation of their lamellae to balance surface forces. Where four Plateau borders connect together, they form vertices, sometimes referred to as nodes [22,23]. Rearrangement of adjacent bubbles orients the Plateau borders at the tetrahedral angle (109.47°), where they join the vertex. The system of bubble films (lamellae, Plateau borders, and vertices) forms a continuous network allowing liquid to drain out of the foam head to the underlying solution.

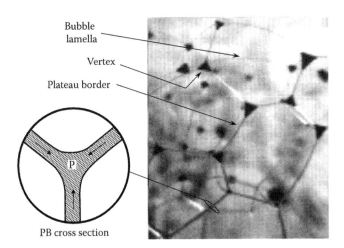

FIGURE 8.2 Polyhedral well-drained foam showing film structures of vertices and Plateau borders. The aqueous phase in this image is black. Bubble lamellae are present but not readily visible in this image. Inset illustrates the cross-sectional structure of the Plateau border with arrows indicating aqueous solution flow from lamellae to the Plateau border.

Liquid drains out of the foam driven by a combination of gravitational and capillary forces. Gravitational drainage is driven by the pressure gradient due to varying height of the liquid in the foam head above the underlying bulk liquid. At hydrostatic equilibrium the liquid pressure is given by

$$\frac{dP_d}{dz} = \Delta\rho g \qquad (8.4)$$

where
   P_d is the pressure of the liquid as a function of vertical position, z
   z is taken from top to bottom in the foam head
   $\Delta\rho$ is the density difference between the gas and the liquid
   g is the gravitational constant [24]

Capillary forces arise within the Plateau border due to curvature of the liquid surfaces. The resulting capillary pressure is calculated using the Laplace equation:

$$\Delta P_c = \sigma\left(\frac{1}{R_1} + \frac{1}{R_2}\right) \qquad (8.5)$$

where
   σ is the surface tension
   R_1 and R_2 are the principle radii of curvature

As illustrated in Figure 8.2 (inset), the capillary pressure within the Plateau border is considered from the vantage point P. From P, the radius of curvature within the plane of the page, R_1, is bending sharply away and making R_1 strongly negative, resulting in a negative contribution to ΔP_c. The second radius, R_2, measured orthogonal to the page, is positive and large, making the second term vanishingly small. Therefore, the negative curvature of R_1 results in a capillary suction arising within the Plateau borders.

Drainage results in the movement of aqueous solution from the foam head and consequent thinning of bubble lamellae, and a reduction of the cross-sectional size of Plateau borders and vertices over time. In many foaming situations, lamellar film drainage process are quite long and can result in the formation of foam heads that are stable for hours or days. In order to avoid these delays, it is necessary to circumvent the foam stabilizing mechanisms through the use of antifoam additives.

8.3 ANTIFOAMING MECHANISM

Hydrophobic oils simply dispersed into the foaming medium may be used as antifoams, but they are inefficient or may actually stabilize foam by collecting in Plateau borders of vertices and diminishing the rate of drainage. In contrast, *antifoam compounds* are formulated systems of hydrophobic oil, sometimes referred to as the carrier oil, containing dispersed hydrophobic solids. Modern theories of antifoaming tend to be focused on this type of compound due to its high efficiency.

Consideration of antifoam function generally starts from the point of dispersed small droplets of antifoam compound present in the aqueous foaming medium as it approaches the surface of a bubble film. The key mechanistic steps of antifoam action can be summarized as utilizing an entering–bridging–rupture cycle [11,12], as illustrated in Figure 8.3.

Entering is illustrated in Figure 8.3a and b. The antifoam droplet emerges from the aqueous phase to displace some of the air/water surface, thus forming a new

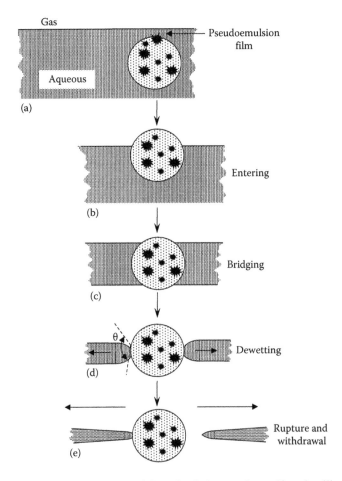

FIGURE 8.3 Schematic of the essential mechanistic steps for antifoaming illustrating the entering (a, b), bridging (c), dewetting and thinning (d), rupture, and withdrawal (e) mechanistic steps.

surface that is the antifoam compound/air. The entering coefficient, E, as utilized by Ross [28] quantifies the free energy change for the entering process:

$$E = (\sigma_{w/a} + \sigma_{w/o}) - \sigma_{o/a} \qquad (8.6)$$

where

$\sigma_{w/a}$ and $\sigma_{o/a}$ are the surface tensions of the aqueous phase and the antifoam compound, respectively

$\sigma_{w/o}$ is the interfacial tension between the aqueous phase and the antifoam compound

The terms within brackets quantify the surfaces present in the initial condition and $\sigma_{o/a}$ the surface present after entering has occurred. The entering coefficient must be positive, signifying a reduction in total surface and interfacial energy, for the entering process to be thermodynamically spontaneous. A larger positive value for E does not confer any advantage, it is sufficient for E to be positive.

Aveyard et al. [25] point out that the actual values of E depend on the state of the system in relation to equilibrium. Under dynamic conditions, surfactant based surface and interfacial tension reduction is less effective than when true equilibrium is attained. Therefore, it is useful to define an initial entering coefficient, E_I, to describe the maximum dynamic limit which is too fast for surfactant adsorption to occur. E_I is defined in terms of the surface and interfacial tensions of water and bulk antifoam in the absence of surfactant to simulate the dynamic limit of no surfactant adsorption. Then the actual entering coefficient experienced under dynamic conditions will lie between this limiting value and the entering coefficient, E as calculated in Equation 8.6 using equilibrium values.

A positive entering coefficient is necessary in order for entering to be thermodynamically spontaneous, but this does not imply anything regarding the rate at which the entering process occurs. Entering is generally thought to be the rate determining step of the overall antifoaming process [26]. Therefore, it is the *rate* of entering that determines the overall rate of bubble rupture, and ultimately, the volumetric rate of foam rupture. The rate at which entering occurs is thought to depend on the stability of the *pseudoemulsion film* (arrow in Figure 8.3a), which will be discussed here.

Spreading is illustrated in Figure 8.4. In our case, the oil is the antifoam compound. Spreading is the extension of the oil over the aqueous foaming solution to form a macroscopically thick film. The thermodynamics of this process are quantified using the spreading coefficient, S, first defined by Harkins [27] and later applied to antifoaming by Ross [28]:

$$S = \sigma_{w/a} - (\sigma_{w/o} + \sigma_{o/a}) \qquad (8.7)$$

where $\sigma_{w/a}$, $\sigma_{w/o}$, and $\sigma_{o/a}$ are as defined above. The spreading coefficient must be positive for the spreading process to be energetically favorable. For a negative value of S, the oil drop exists in a partial wetting condition and forms a lens on the aqueous surface that adopts a shape balancing gravity, surface, and interfacial tensions.

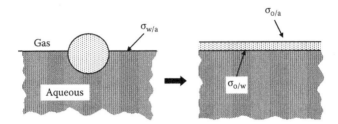

FIGURE 8.4 Schematic of spreading of antifoam oil droplet to form a macroscopically thick (duplex) film. The spreading coefficient, S, is dependent on the surface tensions of the oil, $\sigma_{o/a}$, the aqueous phase, $\sigma_{w/a}$, and the interfacial tension between the oil and aqueous phase, $\sigma_{o/w}$.

When S is positive, the oil drop spontaneously spreads to form a macroscopically thick oil layer. Further spreading on a given volume of oil will result in thinning of the oil layer. The extent of spreading depends on the interaction potential, $V(h)$, between the oil surface and the oil–solution interface, where h is the thickness of the oil layer [29]. The interaction potential may limit the spreading at a given thickness associated with a repulsive interaction potential.

A positive spreading coefficient does not define the rate of spreading. Xi et al. [30] measured the spreading rate of polydimethylsiloxane (PDMS) oils on sodium dodecylsulfate solutions. In the case of complete wetting, the spreading oil front extends outward in a roughly circular shape as a function of time. The spreading rate for positive S is often modeled as a power law function $R = kt^n$, where R is the radius of the perimeter of the spreading layer, and t is time. Experimentally, n ranges from ¼ to ¾.

Bergeron and Langevin [31] proposed a model based solely on Marangoni driven surface tension gradients that was able to accurately predict the power-law behavior observed for expanding monolayers of PDMS.

Binks and Dong [32] have investigated spreading of PDMS oils on nonionic surfactant solutions and demonstrated a reduction in surface tension upon spreading and an initial positive spreading coefficient. Aveyard et al. [33] elucidated a relationship between alkane chain length and entry and spreading of alkane fluids over aqueous surfactant solutions and related this to foam stability.

As discussed by Garrett [34], no correlation has been found between the magnitude of the spreading coefficient and the ability to rupture foam. This could be interpreted to imply that spreading is not critical to rupturing aqueous foam. Robinson and Woods [35] used a similar, thermodynamic approach comparing E and S to defoaming action and established a better correlation between foam rupture and the entering coefficient than with the spreading coefficient.

Koczo et al. [36] showed that spreading coefficients do not correlate with the antifoaming effectiveness of oils, especially at higher surfactant concentrations. Garrett [11] found no correlation between S and antifoam action and suggested that oil spreading does not significantly contribute to the kinetics of foam rupture. Denkov [16] in a broad survey of the literature concluded that there is no correlation between spreading coefficient for a given oil and antifoaming activity. He also observed that very fast antifoams can operate without spreading-fluid entrainment [37] and that oil spreading is not a necessary condition for antifoam activity. Denkov notes that the observed correlation between the oil spreading ability and antifoaming efficiency for many systems suggests that spreading could facilitate the foam destruction process. Pelton and Goddard [38] argue that spreading is a necessary but not sufficient requirement for antifoam effectiveness.

It should also be emphasized that both E and S are exceedingly difficult to measure correctly, especially regarding the state of equilibration of the system. This is especially true for viscous polymeric antifoam compounds.

Bridging is illustrated in Figure 8.3b and c. In Bridging, the antifoam droplet which is entered on one side of the bubble film, spans the thickness of the bubble film and enters the opposite aqueous/air surface. Most authors suggest that the oil drops containing solid particles rupture aqueous foams by bridging [13] and adopting a gas/droplet/gas position spanning the bubble film.

Inherently, the dimensions of the antifoam droplet normal to the plane of the bubble film must be equal to or greater than the bubble film thickness. Only when the film has drained sufficiently, will the antifoam droplet be of sufficient size to enter the air interface on the second side of the film essentially allowing the droplet to contact air-to-air across the film.

The thermodynamics of successful bridging, as first defined by Garrett [39], elucidate the relative stability of an antifoam lens bridging the bubble lamella to form a gas/antifoam/gas configuration:

$$B = \sigma_{w/a}^{2} + \sigma_{w/o}^{2} - \sigma_{o/a}^{2} \qquad (8.8)$$

where B is the bridging coefficient, and the surface and interfacial tensions are as defined. The derivation of Equation 8.8 assumes that the antifoam droplet adopts an equilibrium lens configuration bridging the bubble film (Figure 8.3c does not show an equilibrium lens shape). The lens lacks mechanical stability if the air/water/antifoam contact angle formed by the tangents to the air water and oil water surfaces is >90°. This condition is satisfied for B > 0, indicating an unstable bridging configuration. High curvature at the antifoam/water surface induces an unbalanced capillary pressure in the form of a local Laplace pressure that induces flow of the aqueous film away from the antifoam lens into the bubble film.

Garrett's bridging coefficient argument is based on the ability of the antifoam to adopt an equilibrium lens shape. In many cases, this will not occur. Frye and Berg [63] suggested that the presence of solids within the antifoam particle increases the immersion depth of the oil lens due to its nondeformability. This effect might allow bridging to occur in a thicker bubble film and at a lesser extent of foam drainage allowing faster rupture.

Dewetting is envisioned in Figure 8.3d. As described by Frey and Berg [40] based on its low surface tension, the antifoam droplet will be poorly wet by the aqueous solution of the bubble film. Consequently the aqueous film will adopt a high contact angle where it meets the antifoam droplet. The aqueous phase/droplet/air contact angle, θ, measured through the aqueous phase is defined in Figure 8.3d.

Dippenaar [41] directly observed bridging of water lamellae by hydrophobic particles using cinematography. For solid particles spanning the bubble film, the solid particles can either stabilize or destabilize the film. If the particle is sufficiently hydrophobic, the aqueous phase/solid /air contact angle is high, imparting convex curvature to the aqueous film, driving localized thinning of the bubble film. If, however, the particle is hydrophilic, the film has a concave meniscus, and the capillary pressure has the effect of stabilizing the film. Thus, the particles must exceed a critical (receding) contact angle to act as foam breakers.

Garrett [39] and Frye and Berg [42] suggested an analogous mechanism with oil drops in the bridging position. It is argued [42] that the high contact angle produces a local convex curvature of the aqueous film near the droplet. That curvature induces a positive Laplace pressure local to the point of bubble film/antifoam droplet contact. The local pressure increase is relieved by flow of aqueous fluid from the film surrounding the antifoam droplet. This results in a local thinning of the film adjacent to the droplet. As the local film thickness decreases, the contact lines sweep longitudinally along the surface of the droplet ultimately achieving a single line of contact around the droplet's perimeter.

Rupture as illustrated in Figure 8.3d. It is then presumed that small mechanical forces imposed on the film by pressure differences between the bubbles on either side of the film or by vibrations of the film will impart sufficient force to cause the contact line between the aqueous film and the droplet's perimeter to disengage from the surface of the antifoam droplet (see arrow in Figure 8.3d) rupturing the bubble film. Alternatively, rupture has been observed to occur upon breakage of the antifoam droplet under elongation [12].

Withdrawal of the aqueous film is initiated as soon as a break in the bubble film occurs since surface tension is no longer balanced in the perforated bubble lamella. The break originates with mechanical disconnection from the antifoam droplet. Withdrawal results in a very rapid and complete retraction of the bubble lamella into the surrounding Plateau borders (Figure 8.3e). As a result, the thin film is removed from between the bubbles, causing two bubbles to coalesce into one.

The ultimate disposition of the antifoam compound droplet in contact with the bubble film upon rupture impacts efficiency of the process. Denkov [12] proposes that the compound can be withdrawn with the film, presumably allowing it to participate in future bubble ruptures.

8.4 SILICONE ANTIFOAM FORMULATION AND DELIVERY

Antifoam compounds consist of a nonpolar oil containing dispersed hydrophobic solids. The patent literature is replete with compositions containing a wide variety of oils and solids, complex reaction, and dispersion schemes. This range of formulation is beyond the scope of this chapter but a brief summary will allow the reader to appreciate the breadth of chemistries applied:

Nonpolar oils: Nonpolar oils, sometimes referred to as carrier fluids, are selected to be insoluble in the foaming medium. Commonly, these are vegetable oils (castor oil, soy bean oil, tall oil), paraffinic oils such as low cost petroleum oils, fatty esters, or silicone oils. In some cases polyglycols or polyglycol surfactants are used. These are selected to be above their cloud point under application conditions. Surfactants such as blocky ethylene oxide/propylene oxide (EO/PO) polyglycols, alkyl polyglycols, silicone polyethers [43], or mixtures are utilized [44]. Branched synthetic paraffins like polyisobutylene have also found some minor applications.

Hydrophobic solid particles: The hydrophobic solids are milled into or formed within the nonpolar oils to form an antifoam compound. The hydrophobic particles are selected to be insoluble in the oil and in the foaming medium. They are processed to generate particles that contain sharp needle-like protrusions or asperities. Examples of solids used are semicrystalline paraffin waxes, hydrophobic silica, precipitated hydrophobic polymers, or precipitated calcium salts of fatty acids or of fatty phosphate esters [45,46]. Rather elaborate milling/temperature drop processes are used to create particles of ethylene bis-stearamide with the crystal morphology bearing sharp protrusions and proper particle size distribution [47,48]. The hydrophobic solids content in an antifoam compound is typically 1–10 wt.%.

The patent literature describes a wide variety of ancillary chemistries designed to augment antifoam performance. In some cases, these are components within the antifoam compound itself or dosed in the emulsion delivery system for the antifoam. In other cases, they are dosed separately from the antifoam. These augmenting agents promote the antifoaming process through a number of mechanisms. In some cases, the agents act as foam modifiers designed to reduce surface viscosity to speed drainage. In other cases, the components are waxy materials that provide delayed antifoam release for increased longevity of performance. Discussion of this range of formulation is beyond the scope of this chapter.

8.4.1 SILICONE FLUIDS

Silicone fluids are commonly used nonpolar oils. PDMS contain repeating dimethyl siloxy monomers, and are capped with trimethylsiloxy groups:

$$(CH_3)_3 SiO - \left[(CH_3)_2 SiO - \right]_n Si(CH_3)_3 \tag{8.9}$$

PDMS are fluids up to very high molecular weights, attributed to low intermolecular interactions and the high flexibility of the siloxane ($-Si-O-Si-$) polymer backbone. PDMS fluids have very low surface tension, about 21 mN/m, which contributes to positive entering and spreading coefficients against a wide variety of surfactant systems, making them useful in a broad range of antifoaming applications [49].

Diorganosiloxane fluids containing other than methyl are also discussed in the antifoam patent and technical literature. These fluids contain $[-R^1R^2SiO-]$ monomers where R^1 and R^2 represent alkyl, phenyl, or other organic groups. The diorganosiloxy moieties may be used in homopolymers or in combination with dimethylsiloxy groups to form random or blocky copolymers. Diorganosiloxane fluids of this type are often reserved for specialty applications due to their higher cost.

PDMS used in antifoam compounds are typically linear in structure and of intermediate viscosity, ranging in kinematic viscosity from about 100 to 60,000 mm^2/s, with higher molecular weight providing higher viscosity. The fluids can be composed of a single molecular weight distribution or can be blends of fluids of different viscosities, giving a multimodal molecular weight distribution.

In some cases, viscoelastic PDMS fluids are useful. Branched fluids contain side chains of significant molecular weight, often large enough to contribute to entanglement. Cross-linked silicone fluids form a network structure through interconnection of polymer chains. The degree of interconnection will be quite low and often these networks are swollen by simple linear fluids. In either case, these types of fluids exhibit viscoelasticity. Used alone or blended with intermediate viscosity linear silicone fluids, antifoams based on viscoelastic silicone fluids are claimed to provide enhanced performance.

Several chemical techniques are used to create viscoelastic fluids. One method is to use platinum-catalyzed addition chemistry. As typically practiced, modest molecular weight linear copolymers containing diorgano $[-R_2SiO-]$ and organohydrido $[-R,HSiO-]$ siloxy monomers are used as cross-linkers. The hydride groups react with vinyl groups which are terminal or pendant on higher molecular weight diorgano siloxanes by a process of platinum catalyzed addition [50]:

$$\equiv Si-H + CH_2=CH-Si\equiv \xrightarrow{Pt} \equiv Si-CH_2CH_2-Si\equiv \qquad (8.10)$$

where $\equiv$ represents organo and oxo groups bonded to the silicon but not involved in the reaction.

Another approach is to promote condensation reactions between silanol fluids. This chemistry is typically practiced with polydiorganosiloxane fluids terminated with silanol groups $[\equiv SiOH]$ in combination with a cross-linker, bearing pendant silanol or alkoxy groups. In one example [51], the cross-linker used is an alkoxy methyl siloxane copolymer, of the type:

$$\left(RO\right)_2\left(CH_3\right)SiO-\left[\left(RO\right)\left(CH_3\right)SiO-\right]_n Si\left(CH_3\right)\left(OR\right)_2 \qquad (8.11)$$

with n typically less than 10, and R being methyl. Hydrolysis forms methanol and a silanol group terminal or pendant on the cross linker. Silanol condensation reactions, typically catalyzed by metal species such as tin, titanium, or zinc complexes [49], results in formation of a siloxane bond and liberation of one water:

$$\equiv Si-OH + HO-Si \equiv \xrightarrow{M} \equiv Si-O-Si\equiv + H_2O \qquad (8.12)$$

Silanol condensation reactions can also utilize silanol groups present on silicone resins. Silicone resins are condensed organopolysiloxanes that, on average, contain fewer organo groups and more siloxane linkages on each silicon. For example, $RSiO_{3/2}$ units and SiO_2 units in a ratio of from 0.5:1 to 1.2:1, where R denotes a monovalent hydrocarbon group of 1–6 carbon atoms [52]. Typically silicone resins of this type contain silanol groups accessible for reaction allowing the resin to serve as a cross-linking site in the branched or a network structure.

Viscoelastic silicone fluids are used under conditions where degradation of compounds based on linear silicones would occur rapidly. Antifoam compounds based on viscoelastic silicone fluids are claimed to be resistant to emulsification and degradation in surfactant-rich foaming situations [53,54]. They are often used under high shear conditions experienced in centrifugal pumps, spray heads, and the like. They can also offer advantage under chemically demanding conditions that degrade the silicone such as solution pH of either acidic or basic extremes and at elevated temperatures. For example, paper pulp manufacture operates at pH 11–12.5 and temperatures ranging to 90°C, requiring the use of robust antifoam compounds [55].

8.4.2 HYDROPHOBIC SOLIDS

The presence of hydrophobic particles within the antifoam compound strongly enhances the rate of foam rupture. To perform this function, the particles need to be present at the right concentration, be strongly hydrophobic, be of the right size distribution for the foaming medium and bear sharp surface irregularities.

High surface-area silicas of either a precipitated or fumed origin are used in formulating silicone antifoam compounds. Precipitated silicas are formed through a two-stage aqueous process of formation of silica sols and then, aggregation and intergrowth driven by changes in solution electrolyte concentration and/or pH. Adjustment of solution conditions then fosters light aggregation of the silica particles forming an open-branched structure. Continued deposition of silica from solution covalently links the primary particles together into permanent aggregates. These silicas have Brunauer–Emmett–Teller (B.E.T.) surface areas of 35–190 m^2/g and average aggregate size of 3–50 μm [56]. Silicas of this type are sold under the Sipernat, ZeoSil, or Ultrasil brand names.

Fumed silicas, sometimes referred to as pyrogenic silica, are formed using flame hydrolysis of silicon tetrachloride, or from quartz sand vaporized in a 3000°C in an electric arc furnace [57], or more recently, fumed silicas are produced commercially by burning organosilanes using a hydrogen/oxygen flame [58]. Primary silica particles form through the condensation of SiO_2 from the vapor phase. Typically the primary particles are in the size range of 10–21 nm [59]. Through collisions the primary particles fuse together forming aggregates with an open "grape-cluster" structure, as illustrated in Figure 8.5 [60]. The aggregate is the finest particle to which fumed silica can be dispersed. A typical mean aggregate size for fumed silica is 0.2–0.3 μm [61], or could be as wide as 0.1–1 μm [62]. Fumed silicas are typically formed with nitrogen B.E.T. surface areas in 70–380 m^2/g. Major global producers are Evonik (who sells it under the name Aerosil®), Cabot Corporation (CAB-O-SIL®), Wacker Chemie (HDK®), and OCI (Konasil®).

Extensive hydrophobic treatment of the silica is required for antifoaming performance. The foam rupturing effect of surface hydrophobicity and irregular shape have been extensively verified for small solid particles such as powdered Teflon [41]. Frye and Berg [40,42] have tested angular, irregularly shaped glass particles that were ground and then hydrophobically treated. Those with contact angles less than 90° were ineffective.

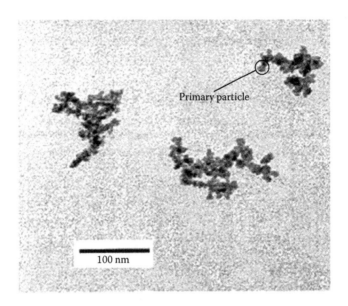

Primary particle

100 nm

FIGURE 8.5 Transmission electron microscopy (TEM) image of dispersed silica aggregates highlighting a primary particle. (Modified from Davis, S. et al., WO/2010/028261, 2010 [Cabot].)

Garrett [11] explained this on the basis of the contact angles, claiming that an oil drop with particles can rupture the pseudoemulsion film if, for spherical particles:

$$\theta_{AW} > 180° - \theta_{OW} \tag{8.13}$$

where
θ_{AW} is the gas/solid/aqueous contact angle
θ_{OW} is the oil/solid/aqueous contact angle

Thus if $\theta_{OW} > 90°$ and $\theta_{AW} < 90°$, the contact angles satisfy this condition. Frye and Berg [63] and Aveyard et al. [64] found that θ_{OW} is much higher than 90° in mixed antifoam–surfactant solution systems.

As formed, both precipitated and fumed silica particles are hydrophilic due to the presence of siloxane and silanol groups exposed at the particle surface. Hydrophobic treatment is accomplished using several chemical techniques aimed at covalently linking hydrophobic groups to the silica surface.

Silanes, such as dimethyldichlorosilane, or octyltrimethoxysilane, or silazanes are used in commercial scale treatment of fumed silicas [61]. Reaction using hexamethyldisilazane, $(Me_3Si-N-SiMe_3)$ utilizes surface-adsorbed waters to hydrolyze the Si–N bond, liberating ammonia, and capping surface silanols with trimethylsiloxy groups [65]. Treatment is often accomplished in a fluidized bed at elevated temperature [66].

Hydrophobic treatment is also commercially accomplished using polymeric siloxanes. The silica is coated with PDMS [67] or dispersing it into bulk silicone

oil [68,69] at 100°C–150°C for several hours to complete the hydrophobic modification. It is theorized that silanols on the silica surface react to degrade the adsorbed
silicone fluid, forming siloxane bonds with fragments of the fluid, covalently linking
loops of polydimethylsiloxane to the surface. This reaction eliminates some surface
silanols and covers others, rendering the silica hydrophobic. Proper treatment results
in the attachment of a thick layer of tightly associated polydimethylsiloxane onto the
silica surface [70].

Methyl silsesquioxane resins [71] have been shown to bind to the hydrophilic
surface of fumed silicas, and impart a high level of hydrophobicity to the silica [72].
Silsesquioxane resins typically contain a few silanol groups which, presumably, condense with silanols present on the silica surface or simply hydrogen bond with the
silica surface effectively covering the hydrophilic silica surface, rendering the silica
hydrophobic.

8.4.3 SILICONE–SILICA INTERACTION

The interaction between PDMS and high surface area silica, even after extensive
hydrophobic treatment of the silica is significant. DeGroot and Macosko [73] examined the interaction of polydimethylsiloxane with silicas of varying hydrophobicity
after simply mixing the silica into the PDMS at room temperature. They demonstrated that a layer of PDMS formed on the fumed silica that was strongly bound and
was not removed with carbon tetrachloride washes. The bound layer was demonstrated to be approximately 7–10 wt.% of the silica and was observed with a variety
of PDMS molecular weight fluids. Hydrophobic treatment reduced the amount of
bound polymer but did not eliminate it.

Nuclear magnetic resonance (NMR) relaxation and diffusion studies on hydrophobic silicas in PDMS demonstrated that portions of the PDMS have very different
relaxation behaviors. It was concluded that segments of the PDMS tightly adsorb
onto the silica resulting in lower motion and significantly changed relaxation properties. The structure of the adsorbed layer is modeled as a three-state system comprising bound polymer segments, polymer entangled or restricted by the adsorbed layer,
and free bulk polymer [74].

On the basis of these studies, there are strong interactions between silicone fluid
and the surface of silica which are not readily reversible. Therefore, any discussion
of the role of hydrophobic silica particles in an antifoam must consider the presence
of bound silicone fluid on its surface.

8.4.4 ANTIFOAM COMPOUND PREPARATION

Commercial antifoam compounds typically contain 3 to about 10 wt.% hydrophobic
silica. For example Garrett et al. [11,75] varied the content of a hydrophobically
treated precipitated silica in a mineral oil to prepare a series of antifoam compounds
varying in silica content. These compounds were tested for efficacy in a cylinder-
shaking test using a commercial 0.5 g/dm^3 of C_{10-14} alkyl benzene sulfonate (approximately 1.4 mM) as the foaming solution. The results are illustrated in Figure 8.6,
where F is the fraction of foam at cessation of shaking in the sample containing

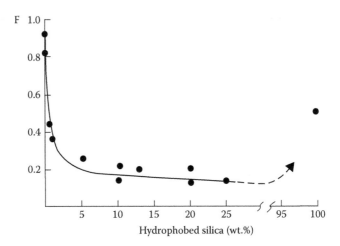

FIGURE 8.6 Antifoam effectiveness, F, with varying concentration of hydrophobic silica (Sipernate D17) in mineral oil in a cylinder shake test with a 1.4 mM sodium alkylbenzene sulfonate solution. (Reproduced from Garrett, P.R., The mode of action of antifoams, in: *Defoaming: Theory and Industrial Applications*, Garrett, P.R., Ed., Surfactant Science, Series, Vol. 45, Marcel Dekker, New York, 1993, p. 1. With permission.)

antifoam relative to the foaming solution alone. As the weight percent of hydrophobic silica is increased, there is a sharp decline in F signifying an increased antifoaming efficacy. The decline begins at only a few percent of silica and plateaus at about 5–10 wt.% silica with little additional gain above that level.

Fumed or precipitated silicas have a significant void volume because of their open structure. Simply mixing these materials into the carrier fluid entrains a significant volume of air. Deaeration of the system is often accomplished by applying vacuum as the silica powder is mixed into the fluid, during the wet-out stage [76]. Once the silica is properly wet out, shear is applied to reduce silica particle size.

Shearing forces reduce the particle size of the silica agglomerates predominantly by separating the aggregates. The rate of size reduction is dependent on the fluid molecular weight, and filler concentration [77]. Hydrophobically modified silica is typically easier to disperse into silicone fluid because the hydrogen bonding holding agglomerates together is diminished. Therefore hydrophobic modification is often done prior to or during application of shear to decrease the time and energy input needed for size reduction. Davis et al. [60] systematically varied silica structure and surface effects and related these changes to size distribution as achieved by high-shear mixing into mineral oil. Figure 8.7 illustrates a well-sheared system with a number-average distribution that is fairly wide with a maximum at just over 200 nm.

High shear is often applied on an industrial scale using rotor-stator type shearing devices. Units of this type feature a rotating shaft fitted with modestly angled blades (rotor) that are precisely machined to move within a stationary, cylindrical, perforated screen (stator). The small gap between the rotor and stator and high blade speed generates very high shear rates. The angle of the rotor blades pumps fluid through the shearing zone. Devices of this type are available from Ross, Greerco, Ika, or Silverson.

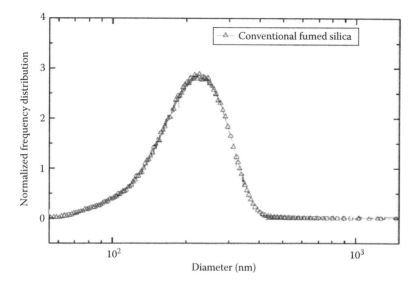

FIGURE 8.7 Distribution of fumed silica agglomerates after high shear treatment in mineral oil showing the maximum frequency size of about 210 nm. (Reproduced from Davis, S. et al., WO/2010/028261, 2010 [Cabot].)

Shearing is ceased when the compound passes a test for reduction in particle size. In some cases, this can be as simple as passing a test using a Hegman grind gauge. More sophisticated measurements can be based on laser light scattering particle size analysis. For this analysis, the antifoam compound is dissolved to a target concentration in a solvent like toluene prior to measurement. In some cases, the shearing is done on a concentrate which is then diluted with more silicone fluid and mixed to uniformity to complete preparation of the compound.

Shear and mixing will provide for dispersion of the hydrophobic particles with a volume-average particle size with a value of about 2–10 μm as being quite typical for antifoam compounds. It should also be noted that too broad a distribution or a distribution that tails to very large particle size can result in solids sedimentation in the compound, which can be problematic.

Currently there is little systematic information linking hydrophobic solids particle size or size distribution to antifoaming efficacy, but this is changing in that it is possible to demonstrate the role of hydrophobic solids in destabilizing the pseudo-emulsion film (see here).

8.5 ANTIFOAM DISPERSION AND DELIVERY

Very rarely is an antifoam compound dosed into the foaming system as the neat compound. Typically the compound is formulated into a delivery system such as liquid-in-liquid dispersions and liquid-in-solid dispersions. These offer the advantage of providing the antifoam compound in a predispersed form designed for optimum performance in the foaming process.

Antifoam delivery systems perform several essential functions:

- The antifoam compound must be delivered to the foaming system in a particle size distribution that meets the needs of the foaming process. The delivery system should stabilize the antifoam compound in this predispersed form so that the compound and the particle size distribution are not degraded during storage and handling.
- The delivery system should enhance ease of use. Neat antifoam compounds are difficult to work with due to their viscous, greasy consistency. In comparison, an emulsified compound can be easily measured and dosed or can be pumped to deliver to remote injection points where foaming occurs. Automated dosing systems can be set to provide foam control at a set dosing rate or in response to foam sensors.
- In some cases the delivery system provides additional function such as slow release benefits, effectively metering in the antifoam compound with time. For example, it is possible to encapsulate emulsion droplets in nonaqueous emulsions through surface crystallization of waxy stearate esters [78]. In liquid-in-solid delivery forms moderately soluble polymers or waxy materials can slow release [79].

8.5.1 EMULSION DELIVERY SYSTEMS

Liquid-in-liquid dispersions are commonly used because they provide a convenient means to deliver antifoam compound. The antifoam compound is dispersed into a second liquid that serves as the continuous phase. That liquid is chosen to be inexpensive and readily soluble/dispersible into the foaming medium. One common approach is to emulsify a silicone antifoam compound into water to provide an emulsion containing a nominal-weight-percent antifoam compound such as 10% or 20%.

A variety of nonaqueous antifoam emulsions are also known where a liquid other than water is used as the continuous phase in the emulsion. A wide variety of oils, such as paraffin or vegetable oils are used. Similarly a variety of glycols, such as polyoxypropylene or polyoxyethylene glycols, surfactants, such as alcohol ethoxylates are used. Glycol continuous emulsions are useful delivery systems for low water content liquid heavy-duty detergent concentrates [80,81]. In some cases, the antifoam compound may be predispersed directly into a detergent's surfactant concentrate [82,83]. Solvent-based systems can be used to create aqueous emulsions upon water dilution [84].

The selection of surfactants used in forming and stabilizing silicone emulsions is discussed elsewhere in this book. However, stabilization of antifoam compound emulsions can sometime be rather challenging due to the presence of dispersed hydrophobic solids within the oil phase. These solids can significantly diminish the coalescence stability of antifoam emulsions. For example, van Boekel and Walstra [85] demonstrated that the coalescence stability of paraffin oil-in-water emulsions was decreased by the presence of crystals in the oil phase, sometimes by as much as six orders of magnitude. They observed that instability correlated with the presence of crystals within the oil phase near to the oil–water interface. They reasoned that some of the crystals protruded into the aqueous phase as to pierce the aqueous film between two approaching droplets promoting droplet coalescence. They hypothesized that van der Waals interactions between the crystals and the aqueous phase favored particle concentration near the oil–water interface.

To overcome this limitation, the type and concentration of surfactants selected to stabilize antifoam-compound emulsions will be somewhat different than typically used in silicone oil emulsions. There are a wide variety of surfactant types and combinations in preparing silicone antifoam emulsion as exemplified in the patent literature. A rather typical approach is to use a blend of ethoxylated nonionic surfactants. One surfactant will be selected to have a lower degree of ethoxylation and the second with higher ethoxylation, with the latter providing improved coalescence stability. This approach is often supplemented by the inclusion of other surfacants such as fatty acid esters or fatty ethers such as isopropyl myristate, or di-n-octyl ether [86]. As another rather typical approach, Azechi and Itagaki [87] describe use of a combination of sorbitan and polyethylene glycol esters of C_{12-18} fatty acids. It is often the case that surfactants of this type based on unsaturated fatty oils are preferred. Oftentimes rather high levels of surfactant are required to attain stability. In some cases, liquid-crystalline surfactant systems may be involved in stabilizing silicone antifoam emulsions, yet there is little systematic information in the literature.

It should be noted that Azechi and Itagaki [87] utilized a thickener of carboxyalkyl cellulose and biocides to stabilize their emulsion. Thickeners are used to enhance storage life of antifoam emulsions by diminishing the rate of emulsion droplet movement under gravity. This practice is necessary in antifoam emulsions because the particle size distribution must be set for optimal antifoaming performance, and not necessarily for long-term emulsion stability. Sedimentation leads to formation of a concentration gradient in the container, sometimes referred to as "creaming" when the dispersed oil phase concentrates at the top of the emulsion and results in inconsistent dosage of antifoam or ultimately in the formation of a completely phase-separated top layer of antifoam compound. In the simplest case, thickeners simply increase the viscosity of the emulsion-continuous phase and slow the rate of emulsion droplet settling. In preferred systems, the thickener establishes a yield point under very low shear rates that is sufficient to eliminate particle sedimentation/creaming. Shear thinning allows easy pumping and handling of these types of systems. Thickeners can be cellulose derivatives such as carboxy methylcellulose (Dow Chemical), or can be cross-linked polyacrylates such as Carbopol® (Noveon).

Storage may last up to a year. A successful antifoam emulsion will provide stability against creaming, coalescence, and biological growth meeting the time/temperature conditions to be faced during storage. Freeze-thaw instability leading to coalescence or aggregation can also be challenging in some systems [88].

Stability during dilution is also a factor to consider. Dilution can occur during an initial "letdown" prior to use, such as in a day tank, or dilution can be at the point of injection of the emulsion into the foaming medium. In poorly designed antifoams, the droplets will aggregate together or coalesce ruining the distribution desired for efficient foam destruction. In other cases dilution, diminishes surfactant stabilization allowing formation of an oily separated layer upon dilution.

Antifoam compound-in-liquid dispersions are often prepared using relatively simple turbine blade mixing to impart enough shear to allow emulsification without degrading the antifoam compound by separating the hydrophobic particles from the silicone carrier oil. There are a variety of processing approaches used to aid in this kind of emulsification. In some cases, the emulsification is done in a highly concentrated system and then diluted down with more continuous phase. This approach is sometimes beneficial because it imparts higher shear to more efficiently reduce particle size of the silicone antifoam compound.

Higher shear devices are sometimes useful. High pressure homogenizers allow for smaller mean particle size and can often greatly narrow the particle size distribution and reducing tailing toward larger particle size. This can be useful in applications such as in antifoams for paints where large particle size silicone droplets can negatively impact surface appearance of the paint.

8.5.2 Solid Delivery Systems

Antifoam delivery can also be based on solid granulated forms. A granulated antifoam can be easily dosed into batch processes by hand as workers identify a foaming problem. Granulated antifoams also find significant utility when blended with other powders, such as powdered detergents, so they are dosed at a fixed ratio relative to the powder.

Granulated antifoams must meet several basic criteria. They must be free-flowing powder and nonpacking or noncaking for easy use. The antifoam is released into the foaming medium upon the dissolution of the granulate requiring sufficient solubility in the foaming medium. When dosed into detergent powders the size and density of the antifoam granule must be set to avoid segregation of the antifoam which would cause it to concentrate at the top or bottom of the powder due to vibration-driven sifting during transport and handling.

Granulated antifoams are prepared by applying the antifoam compound to a solid carrier, generally selected to be inert and readily water soluble or water dispersible. For example, cellulose ether and water-soluble inorganic salts such as sodium, potassium or magnesium sulfates, carbonates, polyphosphates, silicates, and aluminosilicates [89].

The antifoam loading within the granulate needs to below the liquid-holding capacity of the inert carrier to avoid the formation of sticky particles that might form lumps. Therefore, selection of the inert carrier with high porosity provides for higher loading of antifoam while still providing a free-flowing powder. The inert carrier will contribute significantly to the physical properties of the powder and will establish the resistance of the particle to moisture pickup, and will impart crush resistance to the granule.

In more sophisticated systems, the solid antifoam compound/inert carrier granule is treated with a coating that encloses the silicone antifoam within the particle. The added coating can be a waxy material [90] or water-soluble ingredient such as a polymer which coats the entire particle, effectively encapsulating the antifoam [91]. Encapsulation is beneficial because it eliminates spreading or bleeding of the silicone from the granulate particles. Thus, the antifoam compound composition and particle size distribution within the solid is maintained, avoiding loss of activity during storage [92].

Encapsulated silicone antifoam granulates can be prepared using a number of different mechanical systems [93] such as the vertical high shear agglomerator of the Schugi type or horizontal high shear Rubberg-Mischtechnik type. It is also possible to utilize fluidized bed systems such as offered by Glatt, or ploughshear or paddle type mixers such as offered by Lodige. These systems provide means for delivering the antifoam compound and subsequent encapsulant onto moving particulates and offer a controlled temperature profile for solvent removal, in the case of polymer encapsulation, or cooling of waxy materials to finalize the form of the encapsulant.

8.5.3 ANTIFOAM DROPLET SIZE DISTRIBUTION

Regardless of the delivery type, it is necessary for that system to establish the anti-foam droplet size distribution during manufacture, stabilize that distribution during storage, and reproducibly deliver the correct antifoam particle size distribution to the foaming solution for optimum performance.

In liquid-in-liquid delivery systems the antifoam droplet size distribution is derived from mechanical emulsification. Such processes typically produce a droplet size distribution well described as a log-normal distribution [94,95]:

$$P = \frac{1}{\sigma\sqrt{2\pi}} \exp\left[-\frac{(\ln x - \ln x_m)^2}{2\sigma^2}\right] \tag{8.14}$$

where

P is the probability of finding an emulsion droplet of diameter x

x_m is the number average mean diameter

σ is the logarithmic standard deviation which describes the breadth of particle size distribution

Qualitatively, a log-normal distribution differs from a Gaussian distribution in that it has a greater probability at the extremes of the distribution. Using σ and x_m as adjust-able variables, the log-normal distribution fits aqueous silicone–silica antifoam com-pound emulsions quite well [96].

8.6 FOAM FILM THINNING

The combination of gravitational drainage and capillary suction pulls liquid from the bubble lamellae resulting in a decrease in thickness over time. Lamella thinning can be split into three stages based on the physical–chemical processes dominat-ing in different thickness regimes: (1) Hydrodynamic drainage of thick films, based on bulk liquid flow (2) Thin film drainage of bubble lamella under gravitational and capillary suction, and (3) As drainage progresses in a stable bubble lamella, the film ultimately reaches a critical film thickness, h_{Cr} [97]. According to Vrij [98,99], the wave-like surface corrugations attributed to thermal fluctuations spontaneously grow leading to film rupture.

The vast majority of the time is spent in regimes 1 and 2 so an improved understanding of the factors that impact these regimes is important to the antifoaming design for a particular foaming system.

8.6.1 HYDRODYNAMIC DRAINAGE OF BUBBLE LAMELLAE

As two equally sized bubbles approach each, generally under a significant and long-standing force such as buoyancy, they flatten against each other to form a bubble lamella. Key events associated with this interaction are schematically illustrated in Figure 8.8a through c. As the drainage continues, the bubbles move closer and the area of the lamella grows larger.

During the initial stage of bubble lamella formation, the rate of thinning is governed by the limited flow rate of bulk solution from between the bubbles. Hydrodynamic drainage of this type can be viewed with respect to a limiting case where immobile surfaces provide zero velocity at the surface. Liquid flow attains a parabolic velocity profile as the parallel surfaces approach one another. Reynolds [100] was the first to formulate the rate of film drainage (V_{Re}) between two flat and

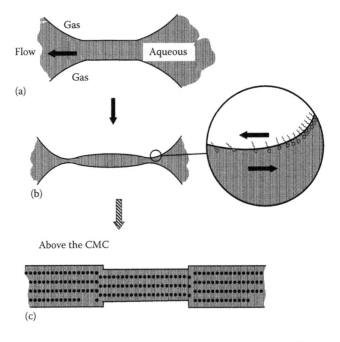

FIGURE 8.8 Cross-sectional view of the formation and drainage of a bubble lamella formed as bubbles approach and undergo hydrodynamic drainage leading to flattening of the bubbles (a). This illustrates deviation of the liquid film from planarity with formation of a dimple (b). The inset (greatly magnified) illustrates the proposed flow-induced surfactant concentration gradient at the bubble periphery. The cross-hatched arrow indicates a change in magnification of the view in moving to (c) which illustrates micelles as black dots layering within the bubble film.

rigid surfaces. As modified by Sheludko [101,102] to consider thermodynamic contributions to capillary pressure:

$$V_{Re} = -\frac{dh}{dt} = \frac{8h^3}{3\mu R_f^2} \Delta P_c \qquad (8.15)$$

where
   h is half of the film thickness
   t is time
   μ is the viscosity of the film liquid
   R_f is film radius, and the net capillary pressure causing drainage is ΔP_c

The net capillary pressure $\Delta P_c = P_c - \Pi(h)$, where P_c is the capillary pressure and Π is the disjoining pressure which sums several thermodynamic interactions between opposing sides of the thinning film (vide infra). Gross deviations from the assumed planar geometry occur for $R_f > 0.01$ cm and film thickness, h, of less than about 200 nm.

Surface immobility presents a resistance to flow and therefore represents the low limit for flow rate [103–105]. Both theoretical and experimental measurements demonstrate that drainage between foam film surfaces is generally much more rapid than would be calculated using Reynold's model.

In hydrodynamically draining bubble lamellae, the planar surfaces are mobile due to surface flows of the adsorbed surfactant layer. Bikerman [106] reported that the stability of foams depends on surface viscosity, which is also strongly correlated with the drainage of foams. Sharma and Ruckenstein [107] suggest that $R_f^{0.8}$ is a closer fit to experimental results than R_f^2 used in Equation 8.15. Manev [108] demonstrated that V/V_{Re} is dependent on surfactant concentration and typically V/V_{Re} about equals 2–5.

Observation shows that with time, drainage of the bubble film deviates from planarity, with faster thinning at the perimeter of the film resulting in the formation of a thicker region near the center, typically referred to as a "dimple" (Figure 8.8b). Liquid flow imposes stresses on the surfactant layer, in the direction of liquid flow (arrow to the right in Figure 8.8b, inset) and imparting surface shearing flow and surface-dilating deformations to the surface film. Surface dilatation creates a surfactant concentration gradient and a corresponding surface-tension gradient which opposes further surfactant-movement resisting flow (arrow to the left in Figure 8.8b) by lowering surface mobility. Restricted flow at the film perimeter leads to a build-up of liquid in the center of the film and formation of a dimple area [109]. The time required to reach this condition in typical aqueous foams is about 2 s with the thickness at the perimeter of the film of about 1 μm [137].

The surface-tension gradient in the thinning film, can be characterized by the dimensionless elasticity number, E_s which is defined for one surface-active component [109] by

$$E_s = E_o \frac{R_f}{\mu D} \qquad (8.16)$$

where

    μ is bulk viscosity of the liquid
    D is diffusivity of the surfactant
    E_o is the Gibbs elasticity

$$E_o = -\left(\frac{d\sigma}{d\ln\rho^s}\right)_o \tag{8.17}$$

where

    ρ^s is surface density of the surfactant
    σ is the surface tension

The surface-adsorbed layer can create less mobile and more stable film surfaces if E_s is high enough.

In this regard, the type and concentration of surfactant, solution pH, electrolyte, and temperature conditions, can play a significant role in establishing the surfactant adsorption and packing behavior and therefore the attainment of surface properties relevant to hydrodynamic drainage of foam. For example Shah et al. [110] demonstrated a strong correlation between surface area/molecule, surface shear viscosity, and foam stability in mixed surfactant systems of sodium dodecyl sulfate/dodecanol.

Several authors have analyzed the drainage of flat, thin films and have taken into account the surface viscosity, bulk and interfacial diffusion [111], shear and dilatational viscosities [112], and bulk and interfacial mass transfer [113]. For example, Malhotra and Wasan [114] developed a generalized model that considers the kinetics of adsorption–desorption of surfactants, surface and bulk diffusion, surface rheological properties, and flow in both film and bulk phases. The generalized model predicted results that were in fair agreement with the experimental data. Jain and Ruckenstein [115] and Gumerman and Homsy [116] reported that surface rheological properties may also considerably stabilize a draining bubble film by imparting rigidity to liquid-film surfaces.

Similarly, Huang et al. [117] demonstrated that the dilatational modulus (or the dilatational elasticity) of a series of α-olefin sulfonate (AOS) solutions depended on the length of the alkyl chain: $C_{16} > C_{14} > C_{12}$, where C_{12}AOS has a negligible elasticity. Because the dilatational modulus measures the ability of the surface to develop a surface-tension gradient, this one property well characterizes a surfactant solution relative to film and foam stability.

Liquid drainage through the Plateau border is similarly impacted by the properties of the liquid surface. Leonard and Lemlich [118] introduced a model for liquid drainage through a Plateau border including surface viscosity effects coupling bulk and surface flows. They developed a dimensionless mobility parameter $M = a\mu/\mu_s$, where a is the radius of curvature, μ_s is the surface shear viscosity and μ is the shear viscosity of the bulk, determines the mobility of the channel surface. Koehler et al. [119] used confocal microscopy to measure the velocity and trajectory of latex particles through a Plateau border in sodium dodecyl sulfate (SDS) and protein (BSA) solutions. The drainage behavior is determined by both the surface properties and by the geometry of the system.

8.6.2 THIN FILM STABILITY

As the plane-parallel bubble film continues to drain, the thermodynamic properties of the bubble lamellae begin to control the rate and extent of further thinning. Those properties start to become relevant as the thickness of the film reaches approximately 100 nm [13,109].

The thermodynamic interactions operating within thin films provide an excess stability that resists thinning. Derjaguin and Kusakov [120] introduced the term "disjoining pressure" to describe this stabilizing effect, which is manifested as an excess pressure. Disjoining pressure, Π, is the surface force per unit area of the film surfaces. Following the approach used in classical DLVO theory [121,122], contributions from individual forces can be summed together to provide the net disjoining pressure. The individual forces differ in magnitude and sign and differ in the length scale over which they operate. Thus, the net sum can provide a disjoining pressure that is a function of film thickness, h, with the sign and magnitude of $\Pi(h)$ changing as a function of film thickness, as schematically illustrated in Figure 8.9. In this illustration, as the film thins, decreasing h, Π increases, stabilizing the film against further thinning, and at lower thickness (arrow), Π decreases with h, indicating thickness instability of the film.

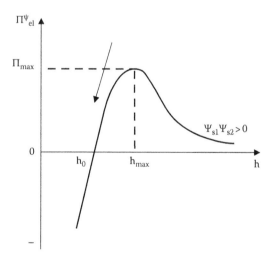

FIGURE 8.9 Electrostatic disjoining pressure, Π, for films of dissimilar surfaces at fixed surface potential as a function of film thickness, h. $\Pi(h)$ can have thicknesses at which disjoining pressure increases with decreasing thickness (positive $d\Pi/dh$) providing thickness stability, and negative $d\Pi/dh$ (arrow) where film thickness is unstable and spontaneous thinning will occur. (Reprinted with permission from Bergeron, V., Fagan, M.E., Radke, C.J., *Langmuir*, 9, 1704. Copyright 1993 American Chemical Society.)

A more detailed delineation of disjoining pressure is beyond the stated scope of this chapter, and is well-described elsewhere [123]. However, in summary, significant contributions to disjoining pressure are attributed to

- *Electrostatic interactions* arise from overlap of the ionic double layers formed through adsorption of ionic surfactants on bubble lamella surfaces. This interaction contributes a repulsive disjoining term, Π_{El}, increasing in magnitude as film thickness decreases.
- *Steric interactions* are contributed by polymeric surfactants adsorbed on the bubble lamella surfaces. Polymer strands extend into solution from the lamella surface. Lamella thinning increases polymer concentration within the lamella and overlap which is unfavorable, contributes a positive term to disjoining pressure, Π_{Steric}, stabilizing the film against further thinning.
- *Oscillatory structural forces* are contributed by surfactant micelles or other fine particles entrapped within the bubble lamella [124,125]. Mutual repulsion between the particles organizes them into layers. Breakup of the organized layers limits the rate of film thinning and contributes a positive disjoining term, Π_{Osc} term to the overall disjoining pressure. Nikolov [126–129] demonstrated layer lifetimes of up to minutes. Micellar layering as observed in typical aqueous foam films will result in multiple thickness transitions and may require 2–4 min [139].

We can consider the total disjoining pressure as a sum of factors [123]:

$$\Pi_{Tot} = \Pi_{vdw} + \Pi_{El} + \Pi_{Steric} + \Pi_{Osc} + \cdots \tag{8.18}$$

In summary, there are many factors that contribute to bubble lamella drainage rates and thickness as a function of time. With many foams, the time of foam drainage to spontaneous rupture is far too long for the limitations to tolerate foam. Antifoams are used to circumvent foam stabilization mechanisms to greatly increase the volumetric foam rupture rate. Antifoam droplets must intervene at film thicknesses prevailing at drainage times $t \leq t_{max}$. To meet this need with efficiency, the antifoam droplet size distribution must be optimized for the task.

8.6.3 ANTIFOAM DROPLET SIZE DISTRIBUTION AND PERFORMANCE

Dippenaar [41] proposed that film rupture occurred when the antifoam particle meets the requirements of size, shape, and hydrophobicity were met. Shearer and Akers experimentally demonstrated a very strong relationship between antifoam droplet size and foam area reduction [130]. Prins [131] demonstrated that the foam half-life of a sodium caseinate solution depended on the size of soybean oil droplets. In this system, there was a maximum destabilizing effect at mean drop diameter of 3–4 μm.

Dippenaar [41] argued that at a given frequency at which a single particle ruptures a film, f_r, then the rate of film destruction is a function of the total number of particles.

Similarly, Garret [11] discussed antifoam particle size distribution and foam film thickness as related to foam rupture kinetics. Thus, it would seem a simple matter of generating the maximum number of antifoam droplets from a given dosage of antifoam compound that possess a diameter equal to or greater than the prevailing bubble lamella thickness at $t \leq t_{max}$.

However, Denkov [12,132] recently segregated antifoams into fast-acting and slow-acting compositions. The fast-acting antifoams were observed to operate within bubble lamellae. These are thought to form unstable bridges across the lamellae that rupture the film within seconds of formation, resulting in rapid destruction of the foam head in tens of seconds. The fast action of these antifoams is attributed to a rapid coincidence of lamella thickness with antifoam-droplet diameter occurring in the thinning lamellae.

The slow-acting antifoams were unable to rupture the lamellae and were observed to be excluded from the thinning lamella. Wasan et al. [133] observed movement of hydrophobic particles or oil droplets out of thin, supported surfactant-stabilized films during drainage. The particles quickly moved into the adjoining menisci (Plateau borders). Similarly, the exclusion of antifoam, semicrystalline wax particles from thinning bubble film to the perimeter of the bubble was observed by Aronson [134]. Apparently the slow acting type of antifoam composition operates in the Plateau borders or vertices instead of within the lamellae.

Koczo et al. [135] proposed that the bridging takes place not in the foam films but in the Plateau borders. This is proposed to occur at a given point in the foam drainage, the drops get trapped in the thinning Plateau borders. Continued application of capillary pressure increases the force pushing the antifoam droplet against the bubble surface, destabilizes the pseudoemulsion film and initiates entering on one side of the Plateau border. Droplets smaller than the cross section of the Plateau border are simply swept through the draining Plateau borders to the underlying solution.

Therefore there is an optimum mean droplet size that provides the correct balance between size and number of antifoam droplets for each location of action. There is a very large difference in the size of a particle needed to span a lamella at an estimated 0.1–2 μm, compared to the 5–50 μm required to span a Plateau border. Thus, at a given mass dosage of antifoam, the number of antifoam droplets of a spanning diameter for action in a bubble lamella is 2–3 orders of magnitude larger than for droplets sized to act in a Plateau border. Based on a simple dosage calculation, the impact of location of action is expected to have a profound effect on mass efficiency.

In summary, for maximum efficiency, the antifoam droplet dispersion must deliver a distribution of droplet sizes suitable for lamella rupture appropriate to $t \leq t_{max}$ for the foaming process, the rates of drainage afforded by the surfactant system, and the location of action of the antifoam.

Of significance, the slow-acting antifoams were shown to possess higher barriers to entry [136,137]. Therefore, it is critical to control the barrier to entry in the system to allow the antifoam to operate within the lamellae. Dispersion of hydrophobic solids is critical to overcoming that barrier.

8.7   HYDROPHOBIC SOLIDS DISPERSION

In the previous sections, we learned that precipitated or fumed silica solids are hydrophobically modified and sheared into silicone fluids to prepare antifoam compounds. To prepare the most efficient antifoam compound, it is necessary to disperse solids that are sufficiently hydrophobic into a carrier fluid so as to provide the maximum number of solid particles that are above an effective size. That effective size is thought to depend on the thickness and properties of the pseudoemulsion film.

8.7.1   KINETICS AND PSEUDOEMULSION FILM STABILITY

For entering to occur, it is necessary to rupture the pseudoemulsion film, which is the thin aqueous film formed as the antifoam droplet approaches the aqueous/gas surface from the aqueous side, as illustrated in Figure 8.10. The pseudoemulsion film is asymmetrical in that it has different phases adjacent to it, namely antifoam compound on one side, gas on the other. Thus it must be distinguished from the gas/aqueous/gas arrangement in bubble films or oil/aqueous/oil films present in emulsions. The asymmetrical nature of the film is expected to have a significant impact on thick film drainage and thin film thermodynamic properties of the film.

Racz et al. [138] demonstrated that rupture of the pseudoemulsion film is rate controlling in the foam rupture process.

8.7.2   ASYMMETRICAL FILM STABILITY

The stability of the pseudoemulsion film is expected to be influenced by many of the same factors that are important to drainage and stabilization of bubble films. However, the bulk phases adjacent to the pseudoemulsion film have very different physical and chemical properties, imposing asymmetry onto the film. For example, the details of surfactant adsorption and packing density will depend on the surface.

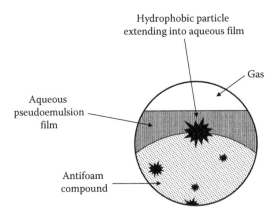

FIGURE 8.10   Illustration of the pseudoemulsion aqueous thin film formed between an oil/hydrophobic solids antifoam compound and gas, schematically illustrating the proposed position of hydrophobic particles and their extension into the aqueous pseudoemulsion film.

The oil–aqueous interface will exhibit a significantly different surfactant-adsorption isotherm than the aqueous/air surface. Since both surfaces are at equilibrium with the same aqueous solution, at some surfactant concentrations, we can expect very different levels of surfactant adsorption onto these two surfaces. This difference will be reflected in the magnitude of charge formation, surface viscosity, and surface dilatational properties, to name just a few manifestations of this relatively simple difference.

It is reasonable to expect significant differences in the thick film hydrodynamic drainage regime. In this regime, the velocity of fluid motion will depend on surface mobility. The viscosity of the antifoam compound is far higher than that of air and will contribute to significantly reduced mobility to that side of the pseudoemulsion film.

In the thin film regime, the pseudoemulsion film will be dominated by disjoining pressure terms as discussed above for thin bubble films. Most notably,

Electrostatic contributions to disjoining pressure (Π_{El}) are to a large degree determined by surfactant adsorption. Asymmetry of surfactant adsorption is expected because the air/water bubble surface is different than the oil–water interface. Even with that asymmetry, as demonstrated by Kulkarni et al. [26], electrical double-layer repulsion can represent a significant potential energy barrier against antifoam droplet entering. Recent model calculations by McCormack et al. [139] point out some of the assumptions that must be applied to predict repulsions between dissimilarly charged surfaces.

Oscillatory contribution to disjoining forces (Π_{Osc}) arises from the presence of fine particles or micelles in the thinning pseudoemulsion aqueous film. Lobo and Wasan [140] observed the drainage and stability of pseudoemulsion films prepared from nonionic surfactant (Enordet AE1215-30 ethoxylated alcohol) solutions with *n*-octane without hydrophobic solids as the oil. At 4 wt.% surfactant, far above the surfactant's CMC, they observed that the pseudoemulsion film thinned in a stepwise fashion, exhibiting thickness transitions, similar to those observed in bubble films. They attributed that thinning to micellar stratification comparable to the behavior for this surfactant in foam films. Similarly, Bergeron et al. [141] observed stratified thinning in pseudoemulsion films. Bergeron et al. [142] provided the experimental measurement of pseudoemulsion film disjoining pressure isotherms for several combinations of oils and surfactant solutions. They provide a generalized model for pseudoemulsion film rupture (entering) based on pseudoemulsion stability as related to $\Pi(h)$ as it withstands imposed capillary pressures.

As shown by Bergeron et al. [142] the pseudoemulsion film will have a thickness dependent stability similar to that depicted in Figure 8.8, regardless of the thermodynamics as indicated by the entering coefficient.

The film trapping technique (FTT) was developed by Hadjiiski et al. [143] as a tool providing measure of the energy barrier for pseudoemulsion film rupture and oil droplet entry. In the FTT technique, micron sized oil or antifoam compound droplets are trapped between a glass surface and an aqueous surface inside a capillary tube. The air pressure within the capillary is increased to force the aqueous surface against the droplets. The pressure at which the oil or antifoam-compound droplet enters the surface is the critical capillary pressure, P_C^{CR}, and is indicative of the barrier for entry. FTT allows systematic variation of the aqueous, oil, and antifoam compound conditions and provides quantitative comparison of these effects.

Significantly, for a given antifoam compound the P_C^{CR} can depend very strongly on the surfactant type and concentration in the aqueous solution. For example, Denkov et al. [144] measured P_C^{CR} for a silicone oil drop to be 28 Pa against a solution of AOT but was >200 Pa for a solution of Triton X-100. FTT measurements using dodecane droplets against a sodium dodecyl sulfonate in 12 mM NaCl showed significant systematic variation with surfactant concentration. At 0.16 mM (0.32 CMC) P_C^{CR} was 10 Pa. Between 0.5 and 9 mM (1–18 times CMC), P_C^{CR} steadily increased from 40 to 150 Pa. Above 9 mM P_C^{CR} steeply increased to more than 400 Pa.

Much more needs to be understood regarding the contributions to pseudoemulsion film disjoining pressure. However, one might reason that factors associated with the foaming system (surfactant type, concentration, electrolyte concentration, etc.) and with the physical chemical properties of the antifoam compound will impact the pseudoemulsion film thickness and magnitude of Π at maximum disjoining pressure.

8.7.3 Hydrophobic Silica Dispersion and Pseudoemulsion Film Stability

Garrett et al. [145] observed in testing a model antifoam compound containing paraffinic carrier oil and trimethylsilane-treated silica that the presence of particles adhering to the oil droplet surface facilitated entering into the surface of a sodium alkyl benzene sulphonate solution. Garrett's interpretation was that the particles assisted by rupturing the asymmetrical oil/water/gas pseudoemulsion films.

Wasan et al. [133] established that for an oil drop to enter a bubble surface the pseudoemulsion film must rupture. Koczo et al. [137] studied foams containing emulsified oil where the pseudoemulsion film was stable. The presence of dispersed oil droplets actually increased foam stability by occupying space in the Plateau borders and vertices and diminishing the rate of foam drainage. This is consistent with the common observation that liquid oils alone are relatively ineffective foam control agents. Under these conditions, entering and spreading do not take place even though E and S may both be positive because the droplets are unable to overcome the repulsive energy barrier and penetrate to the air–water interface.

Racz et al. [140] demonstrated that the presence of hydrophobic solid particles in the antifoam compound had a profound destabilizing effect on the pseudoemulsion film. Pseudoemulsion film stability was measured experimentally in the form of capillary pressure needed for rupture. They formed an aqueous surfactant (0.06 M sodium dodecyl sulfate) film on the tip of a capillary tube filled with silicone oil. Then, the oil was slowly pushed out of the capillary forming a model silicone oil/aqueous/air pseudoemulsion film as the oil exuded from the tip of the capillary. Capillary pressure was measured indirectly based on the shape of the oil at the top of the tube [133].

The pseudoemulsion film between pure silicone oil (without particles) and air was stable and did not rupture before maximum capillary pressure was attained (hemispherical shape). However, when the silicone oil contained hydrophobic silica particles, the film was much less stable, rupturing at very low capillary pressures. The authors explored silicone oil viscosity, hydrophobic silica type, and varied silica concentration. They found that there was a concentration of hydrophobic silica at which the pseudoemulsion film stability decreased rather sharply, rupturing at intermediate pressures. The authors termed the hydrophobic solids concentration at which this steep decrease in film stability occurs as the "critical solids concentration." The critical solids concentration was shown to lie between 0.1 and 0.01 wt.% of hydrophobic solids in silicone fluid, for various formulations, as shown in Table 8.1 [138].

The critical solids concentration identified in the capillary experiment also was reflected in a measure of foam breaking efficiency of the silicone oil/hydrophobic silica compound. The foam half-life, as illustrated in Figure 8.11, was shown to increase sharply as solids concentration was decreased below a level (arrows in Figure 8.11), indicating that the antifoam compound foam-breaking efficiency diminished as the solids concentration was adjusted below the critical solids concentration.

One possible reason for the decrease in pseudoemulsion film stability is that hydrophobic particles extend from the surface of the antifoam droplet into the aqueous phase. The exact details of pseudoemulsion film rupture are not known, but it is assumed that the extension of the particles must be significant relative to the thickness of the pseudoemulsion film in order to strongly affect film stability [13] (see Figure 8.10).

One likely factor influencing particle extension is the size of the hydrophobic particles. Smaller particles will be less likely to extend as far from the droplet surface.

Denkov [16] captured an optical microscopic image of a silicone/silica antifoam compound droplet suspended in an aqueous surfactant solution. This image, reproduced in Figure 8.12, shows dark irregularly-shaped silica particles present within

TABLE 8.1

Critical Hydrophobic Solids Concentration in Silicone Oil Required to Rupture the Pseudoemulsion Film at Intermediate Capillary Pressure

PDMS Oil Viscosity (mm²/s)	Solid	Capillary ID (mm)	Critical Solids Conc. (wt.%)
5	T-500	0.22	0.1
5	T-250	0.32	0.03
200	T-250	0.22	0.01

Source: Reprinted with permission from Rácz, G., Koczo, K., Wasan, D.T., *J. Colloid Interface Sci.*, 181, 124–135. Copyright 1996 American Chemical Society.

Note: PDMS is polydimethylsiloxane of the kinematic viscosity shown. T-500 and T-250 are Tullanox hydrophobic silicas (Tulco Co.) with 0.2 and 2 μm particle size, respectively.

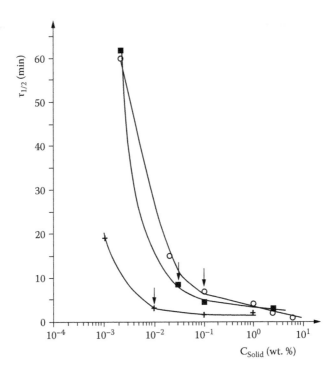

FIGURE 8.11 Effect of hydrophobic solids concentration in mixed antifoams on the foam half-life ($\tau_{1/2}$). Key: **+**, 200 mm²/s polydimethylsiloxane with T-250 (2 µm) solids; ■, 5 mm²/s polydimethylsiloxane with T-500 (0.2 µm) solids; o, 5 mm²/s polydimethylsiloxane with T-500 solids. Antifoam concentration was 200 ppm. Arrows indicate the respective critical solids concentration needed to break the pseudoemulsion film at the tip of the capillary. (Reprinted from *J. Colloid Interface Sci.*, 181, Rácz, G., Koczo, K., Wasan, D.T., 124–135. Copyright 1996, with permission from Elsevier.)

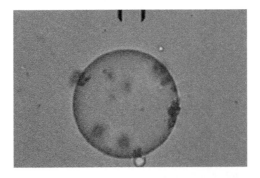

FIGURE 8.12 Optical microscopic image of silicone/silica antifoam droplet suspended in surfactant solution. Scale bar at top indicates 32 µm. (Reprinted with permission from Denkov, N.D., *Langmuir*, 20, 9463. Copyright 2004 American Chemical Society.)

the silicone oil but around the perimeter of the droplet [16]. This places the silica in the right location to protrude with its adsorbed layer of silicon fluid away from the droplet surface. If the protrusion of the particle from the antifoam droplet surface can significantly span the thickness of the pseudoemulsion film, then it might be sufficient to rupture the pseudoemulsion film.

The hydrodynamic drainage properties and thin film energetics of the pseudoemulsion thin film will determine its thickness and therefore the level of particle extension required for efficacy. Currently there is little systematic information describing hydrophobic particle extension. One can presume that factors such as particle concentration and mean particle size and particle size distribution would impact the probability of having a particle with sufficient extension being on the antifoam droplet surface pointed toward the pseudoemulsion film. On this basis, the dispersion of the hydrophobic particles is expected to play a significant role in establishing antifoam compound efficiency.

8.8 DYNAMIC FOAM–ANTIFOAM INTERACTIONS

The antifoam mechanism as described in Section 8.3 provides a framework for examining entering and bridging thermodynamics and pseudoemulsion film rupture and entering kinetics. However, other aspects of the mechanism are still not well understood. Specifically, the role of spreading in the overall mechanism has not been well-defined. Previously unpublished evidence [97] based on detailed observations of dynamic antifoam–foam interactions, demonstrates that spreading can play a significant role in dispersing of antifoam across bubble surfaces. Newly visualized modes and aspects of spreading offer a possible correlation to bubble rupture.

Please see the URL address at the end of the chapter for the location of supplemental information. This site contains a large number of image files that can be viewed serially and will strongly enhance intuitive understanding of antifoam compound spreading.

8.8.1 Experimental Approach

A new experimental device, a shallow foam flow well slide, schematically illustrated in Figure 8.13, fixes the location of the antifoam–foam interaction and utilizes video microscopy to capture the key events. The foam was pumped beneath a sessile droplet of silicone antifoam compound fixed in position by wetting attachment to the cover glass. The well slide was mounted on a video microscope, which captured images at 30 frames per second (fps) and 50× magnification. Differential interference contrast (DIC) optics allowed enhanced resolution of details of antifoam spreading that had not previously been observed. DIC highlights differences in optical density and allows resolution of structural details not easily observed in other modes. This effect highlights real differences in the sample, but some degree of caution is needed when interpreting the actual physical dimensions of the sample. Secondly, Köhler illumination, which focuses the collimated light from the condenser lens onto the same focal plane as the magnification lensing, was utilized to further enhance visualization in this technique.

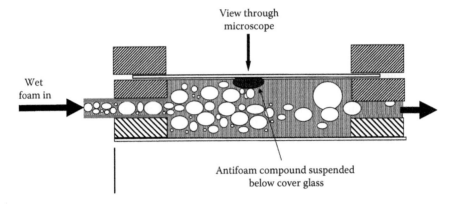

FIGURE 8.13 Schematic cross-sectional view of the foam flow well slide showing the position of the antifoam droplet pendant from the coverslip. Premade foam is pumped through the cell using a syringe pump. Video microscopic view is from above at 50× magnification.

A tiny drop (~2 nL) of antifoam compound was deposited on the underside of the coverslip using a needle point. Aqueous foam was generated using a 1.0 wt.% solution of C_{12-15} alkyl ether sulfate containing an average of three ethylene oxides (Witcolate LES 60). Wet foam was generated just prior to use. A syringe pump pushed the foam through the well slide at a constant rate of 24 mL/min. An image of a NIST traceable graticule was used to establish the size bar on the image.

Simple-model antifoam compounds were prepared for this study and are summarized in Table 8.2. Linear PDMS fluids of 1,000 and 12,500 mm²/s viscosity (Dow Corning) were used as received. One model antifoam compound was prepared using a branched silicone fluid. The branched silicone fluid was prepared following Aizawa's Example 3 [146] using 12,500 mm²/s silanol terminated fluid. All three model compounds incorporated 5.0 wt.% of Cabosil TS-530 (Cabot) hydrophobically modified fumed silica, used as received. The silica was sifted into the fluid and mixed by hand and then subjected to high shear using a Greerco homomixer (model 1 L), rotor–stator high-shear mixer until a uniform distribution of silica was attained.

TABLE 8.2

Model Antifoam Compounds

Antifoam Compound	Hydrophobic Solids[a] Content (wt.%)	Polydimethylsiloxane Fluid		
		Type	Kinematic Viscosity (mm²/s)	Fluid Content (wt.%)
E1000	5	Linear	1,000	95
E12500	5	Linear	12,500	95
E Br	5	Branched	>30,000	95

[a] Cabot TS-530.

8.8.2 Antifoam Spreading

The view through the microscope is illustrated in Figure 8.14. The region between the coverslip and the bubble is filled with aqueous surfactant solution flowing left to right across the image. The bubbles have their top surfaces partly flattened against the underside of the cover glass. The dark/light/dark banding in the middle of the image is associated with the bubble lamella, which is curving slightly top to bottom in the image. It is roughly orthogonal to the coverslip and disappears downward out of the focal plane. Therefore, the ultimate lamella thickness is not resolved in this image. The dark banding is due to light reflection off the bubble surfaces and away from view. The roughly triangular shapes seen on either end of the lamella are cross sections of the Plateau borders.

The antifoam droplet (compound E1000) is on the right side of the image in Figure 8.14. The main body of the droplet is about 200 μm in diameter. Small circular structures inside the droplet are tiny air bubbles entrained in the antifoam compound during placement on the coverslip. The antifoam droplet is sessile in that the droplet's base maintains an area wetting the coverslip, which holds it in a fixed location. The area wetting the coverslip is generally not discernible in these images, yet the liquid flow does not move the antifoam droplet.

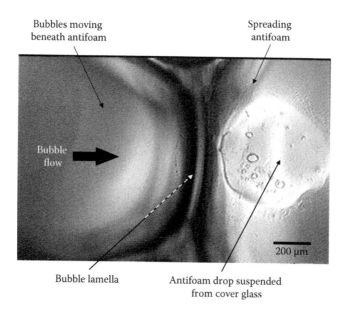

FIGURE 8.14 View in the foam flow well slide experiment at 50× magnification. Bubbles and solution flow left to right. A partly drained bubble lamella lies top to bottom in the image. Bubble surfaces curve down toward the lamella, but the ultimate thinnest dimension lies below the focal plane of the image. The antifoam droplet, which accidentally contained small air bubbles, is suspended below the coverslip, traversing the foaming solution. The antifoam has entered the bubble surface below it. Spreading antifoam (arrow) is observed on the surface of the bubble and descending into the lamella region.

At the beginning of the experiments, as the well slide filled with aqueous surfactant solution, the antifoam droplet was observed to change shape, partially dewetting the cover glass to adopt a fixed three-phase contact angle against the coverslip. In some circumstances, the change in drop shape left residue in a circular pattern on the coverslip as the bulk of the droplet retracted to an equilibrium shape.

Closer examination of Figure 8.14 shows that around the perimeter of the main antifoam droplet, there is an irregularly-shaped spreading film with a somewhat granular texture. This is attributed to the droplet having entered into the bubble surface from the aqueous side, implying a process very similar to the generally envisioned antifoam mechanism (see Figure 8.3b). The irregular film is antifoam compound emanating from the perimeter of the droplet and spreading on the surface of the bubble. Because the film was readily visualized using optical microscopy, the spread layer was macroscopic in thickness. Figure 8.14 also shows that the thick spreading layer extends from the top left of the main droplet, extending across the bubble surface and descending downward into the bubble lamellar region.

The foam flow well slide experiment provides an excellent view of spreading of silicone antifoam compound over bubble surfaces. Spreading, by necessity requires that entering has occurred. In some cases, changes in antifoam droplet shape suggest that entering is occurring, the method at 30 fps, unfortunately, does not capture the details of the entering process. In addition, it should be recognized that dynamic processes of the type observed in this experiment are nonequilibrium events, meaning that entering and spreading coefficients are not well-defined and are likely not comparable from one observed entering/spreading event to the next. However, observation of the occurrence of both entering and spreading conclusively demonstrates that these are spontaneous events with positive coefficients.

Figure 8.15 shows compound E12500 entered into a bubble surface with macroscopically thick layers of fluid spreading at the droplet perimeter. These flows have a worm-like or vermicular appearance. Formation these vermiculi suggests that there is a degree of complexity to the fluid flow as the fluid transitions from the body of the droplet to spread across the bubble surface. In comparison, Figure 8.16 shows a droplet of 12,500 mm²/s PDMS fluid having entered into a bubble surface and spreading. In Figure 8.16, there are rather subtle changes in curvature at the right and left sides of the droplet indicating that entering had occurred and that the fluid was spreading across the bubble surface. The spreading film is smooth and lacks the complex flow observed with the antifoam compound. This suggests that the vermicular flow patterns observed with E12500 compound arise due to the presence of dispersed silica in the antifoam compound.

8.8.3 Formation of Derivative Particles

Figure 8.17 contains images of compound E1000 spreading across a bubble surface. In the top frame, the droplet had entered the bubble surface and was undergoing macroscopic flows around its perimeter. In some regions, the flows manifested the vermicular appearance. Lines emanating from the droplet suggest spreading of antifoam compound across the bubble surface. Along the top right, flow lines are

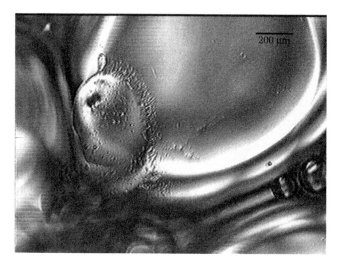

FIGURE 8.15 Foam flow well slide experimental image showing initial spreading of anti-foam compound E12500 on a bubble surface. Spreading antifoam undergoes a complex flow pattern at its perimeter related to spreading on the bubble surface.

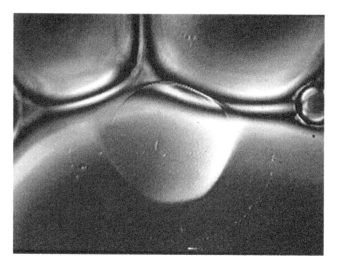

FIGURE 8.16 Foam flow well slide experimental image showing spreading of polydimethyl-siloxane fluid (12,500 mm²/s) that does not contain silica across a bubble surface. A change in shape at the perimeter of the droplet on its left and right sides indicates that the droplet has entered into the bubble surface and is undergoing spreading.

observed descending down the bubble surface. Interestingly, in Frame 2 there are flow lines at the upper right of the droplet showing flow of compound away from the droplet with the flow lines curving as they descend into the lamellar region between bubbles. This observation illustrates that spreading flow can carry antifoam fluid some distance from the point of entering into a region between bubbles.

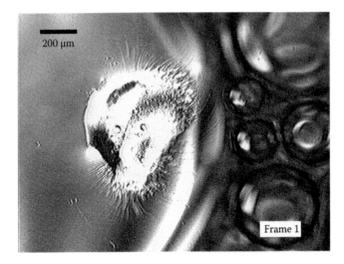

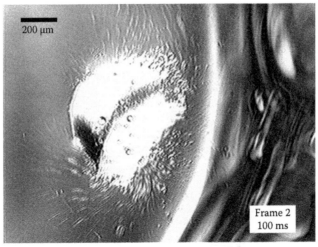

FIGURE 8.17 Foam flow well slide experimental images showing spreading of antifoam compound E1000 across a bubble surface. Frame 1 is an early stage in the interaction complex flow patterns at the droplet–bubble contact line and features lines emanating from the antifoam droplet's perimeter consistent with spreading of antifoam compound. Frame 2, captured 100 ms later, and indicates that there is a change in the appearance of the spreading antifoam layer. Point-like entities are observed as blurred lines indicative of rapid movement. See text for further discussion.

Frame 2 captures the same droplet on the same bubble but 100 ms later. There is continued outward flow emanating from the perimeter of the antifoam drop-let. Point-like features are observed spreading rapidly across the bubble surface. The point-like features are approximately microns to tens of microns in diameter and must be sufficiently optically different from the surrounding spread layer to be readily observable. The point-like features were only observed in silicone–silica compounds and not in silicone fluid alone. The term *derivative particles* is proposed to describe the point-like features.

It is likely that the derivative particles are particles of silica entrained in the spreading antifoam film. Possibly these are larger particles so they are simply more visible microscopically. However, the particles were not initially visible. One possible explanation is that thinning of the antifoam layer as it spreads radially away from the main droplet allows visualization of the particles. There is no direct information on the nature of the spread film at this point. However, one possible interpretation is that larger silica particles would locally deform the spreading film as they begin to exceed thickness of the spreading film. In this circumstance, they would protrude through the film imposing local curvature on the spread film making them visible under DIC microscopy. The particles are not visible in Frame 1 because the spreading film is too thick to be deformed by the particles.

As observed in Frame 2, fluid spreading resulted in the rapid dissemination of derivative particulates across the bubble surface. The derivative particles were blurred because they were moving too rapidly for the camera to clearly capture their image. Measurements on the movement of derivative particles in this series of frames shows that they traveled a distance of 100 μm in 198 ms, demonstrating that the derivative particles were traveling at a velocity in the mm/s range.

In summary, these observations demonstrate that antifoam entry can result in antifoam spreading, driving rapid movement of derivative particles across the bubble surface and resulting in movement hundreds of microns from the original point of entering. This form of antifoam dispersion can play a significant role in the overall antifoaming process.

It remains to be demonstrated if more realistically sized antifoam compound droplets, of 5–10 μm diameter would behave qualitatively the same as well. Potentially, smaller antifoam droplets would not exhibit the volume of spreading flow and dissemination of such a large numbers of derivative particles.

8.8.4 SPREADING AND RUPTURE

In many cases, spreading is observed to disperse antifoam compound toward bubble lamellae or Plateau borders. Figures 8.14 and 8.17 Frame 1 clearly show this type of movement.

Figure 8.18 contains multiple images labeled by frame number and the elapsed time. This sequence was captured late in the experiment so the antifoam droplet was diminished in size and, unfortunately, is slightly out of focus. The frames are not sequential images but are marked in time elapsed from Frame 1, and were selected to illustrate key events. Frame 1 shows the antifoam droplet after it entered into the bubble surface. Surrounding the droplet is a visible halo of a macroscopically thick spread film extending tens of microns from the antifoam perimeter. There is a graininess to the spread film attesting to the irregularity of thickness. Frame 3 (133 ms) shows that the movement of the bubbles has taken the antifoam droplet, which is fixed in location, close to the lamella between bubbles labeled A and C. By Frame 4 (167 ms), the bubble labeled B has disappeared from view, with a consequent change in the shape of bubble A. In Frame 5, the bubbles have repositioned themselves.

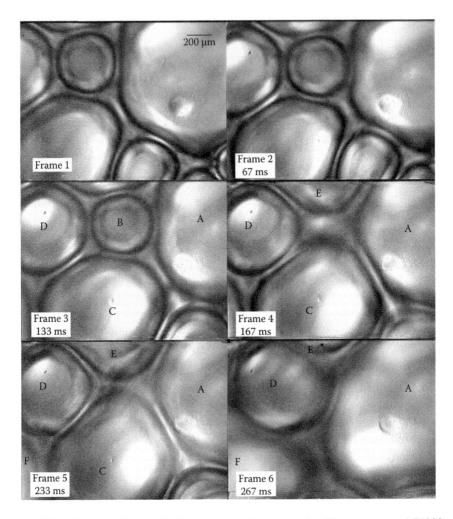

FIGURE 8.18 Foam flow well slide experimental images of antifoam compound E1000. Frame 1 shows the antifoam droplet entered into bubble A and is surrounded by a macroscopically thick spread film. Rupture of the lamella between bubbles A and C are detailed in Frames 3–6. Extension of the spread film into the lamellar region correlated with bubble film rupture.

Sequentially, in Frame 6, bubble C has ruptured with a consequent change in the shape/size of bubble A. The shape change and the nonequilibrium positioning of the surrounding bubbles suggests that bubble C had coalesced into bubble A. The position of the antifoam droplet provides a position of reference to help the viewer judge the rapid extension of bubble A.

Figure 8.18 clearly illustrates that antifoam entering can take place in a location, hundreds of microns from the location of rupture. It also shows that hundreds of milliseconds can separate the entering step from bubble rupture.

Figure 8.18 also illustrates one example where there is a correlation between the proximity of a spread film containing derivative particles and a clearly observed bubble rupture.

Speculatively, this correlation is suggestive of bridging in the lamella region. However, there is no direct image of derivative particles in or near the lamella or of them bridging.

8.8.5 BUBBLE SURFACE DEFORMATION

Figure 8.19 contains an image of antifoam compound E1000 interacting with a bubble. As the bubble moved beneath the antifoam there was a change in reflectivity of the bubble surface, creating a white region of high light transmittance (arrow). This change in reflection indicates that the bubble surface was deformed as it interacted with the antifoam droplet.

Deformation of the bubble surface was quite commonly observed. Interaction with an antifoam droplet is expected to be governed by pseudoemulsion thick film hydrodynamic interactions and to be dominated by bulk solution flow. The large antifoam droplet used in these experiments, presents a very large R_f (see Equation 8.15), which strongly impedes drainage. Thus, this experiment over-emphasizes the timescale of drainage expected for typically sized antifoam droplets.

The deformation observed in Figure 8.19 clearly illustrates that bubble surfaces and compound droplets are fluid and are readily deformed under the influence of the relatively modest forces acting on them during the interaction. Thus, while

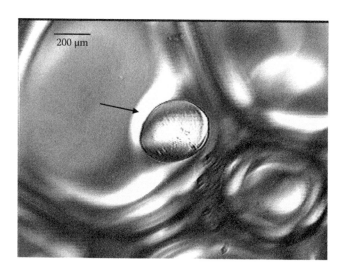

FIGURE 8.19 Foam flow well slide experiment using antifoam compound E12500 interacting with a bubble surface. The bright region to the left of the antifoam droplet (arrow) resulted from deformation of the bubble surface as the bubble moved left to right and interacted with the antifoam droplet. Localized changes of the bubble shape decreased the light reflection and increased light transmission from below.

conceptually easier to model these interactions using planar surfaces and spherical droplets, this simple geometry will not be realized in actual foam.

For this particular interaction, the deformed state was fairly long-lived, lasting more than 400 m. This interaction never resulted in antifoam entering and the bubble simply moved past the antifoam. As clearly illustrated in this encounter, not every antifoam–bubble interaction results in antifoam entering.

8.8.6 Branched Silicone Compound

The observations detailed above were made using antifoam compounds based on linear fluids. Figure 8.20 illustrates that compound EBr, prepared using viscoelastic branched silicone fluid (see Table 8.2), exhibits a markedly different behavior during interactions with foam. The EBr compound was repeatedly observed to form strands that stretched hundreds of microns long while interacting with bubbles.

In Figure 8.20, flow is from left to right as in all the previous images, yet a kind of swirling flow occurred that temporarily reversed the orientation of flow. That strange flow is possibly caused by bubble ruptures off camera that precipitated a reorganization of the bubble head. The EBr compound has entered into the bubble surface (arrow 1). It has remained attached to the coverslip as shown by location 2. Movement of bubbles imposed stresses on the antifoam compound. Rather than simply breaking, as compounds based on linear fluids tended to, EBr tolerated a great deal of strain without rupturing. The compound underwent a process of elongation to form a roughly cylindrical strand hundreds of microns long.

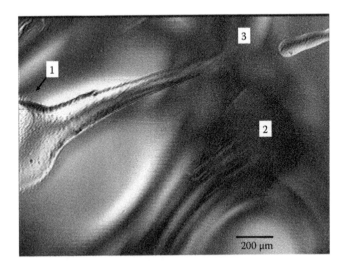

FIGURE 8.20 Foam flow well slide experimental image showing antifoam compound EBr interacting in swirling foam. EBr was drawn into strands through stresses imposed by the moving bubbles. EBr has entered the bubble surfaces at arrow at 1 and point 3. Movement of the bubbles imposed strain onto the compound drawing it into a cylinder. Original attachment to the coverslip at was at position 2.

Closer examination of the elongation process is needed, but preliminary evaluation suggests that enormous strain rates occur in these systems. In one observation (not shown), a branched silicone antifoam compound formed a strand, which was extended into a cylinder 525 μm long in 166 ms, corresponding to a uniaxial elongational flow at the quite high strain rate of 500–1000%/s.

Typically, a cylinder of fluid would spontaneously break up into droplets upon elongation. A fluid cylinder will spontaneously rupture when the length of the cylinder is greater than its circumference [147]. Beyond that length the fluid cylinder undergoes a process known as Rayleigh disproportionation, where localized thinning creates rupture points along the cylinder breaking it into separate droplets. Apparently, the viscoelasticity of the branched silicone fluid significantly stabilizes the antifoam cylinder against thinning and rupture, and promotes formation of elongated strands.

Thus, the nature of the silicone fluid can play a critical role in preventing or delaying the breakup of the antifoam compound and is expected to contribute significantly to the maintenance of the antifoam dispersion in the foaming medium.

8.8.7 SUMMARY AND PROPOSED MECHANISM

A novel foam flow well slide video microscopic technique provides a unique method to study dynamic silicone antifoam–foam interactions. Although this technique is new, it provides an excellent means of capturing images of dynamic spreading events. Using this new technique, it was possible to clearly observe that antifoam spreading is a significant event with a rich array of phenomenology associated with the details of the spreading process that must be understood in relation to antifoaming function.

Spreading was observed to play a significant role in dispersing antifoam compound across bubble surfaces. Spreading carried the antifoam and its derivative particles significant distances across bubble surfaces and was observed to carry antifoam compound toward Plateau borders and bubble lamellae.

Observations suggest that entering can occur at a location widely separated in time and position from the point of bubble rupture. That separation appears to operate on a scale of hundreds of microns and hundreds of milliseconds. This time/distance separation has important implications regarding antifoam mechanism. The best way to explain the separation is to interpose spreading between entering and the mechanistic steps resulting in bubble rupture.

One way to envision a mechanism including spreading is summarized in the proposed antifoaming mechanism illustrated in Figure 8.21. As drainage decreases, the dimensions of the Plateau border, it entraps antifoam droplets of a sufficient diameter so they can no longer move out of the foam head (Figure 8.21a). Sustained pressure driven by continued drainage is expected to press the antifoam droplet against the bubble wall and deform the wall, following pseudoemulsion film hydrodynamic drainage, and then thin film drainage occurs. As depicted in Figure 8.21g, the pseudoemulsion film is ruptured based on the extension of the hydrophobic silica particle's extension into the aqueous film. The exact details of pseudoemulsion film rupture are not currently known. Entering occurs at Figure 8.21b. The cross-hatched arrow signifies an increased magnification at Figure 8.21c. Figure 8.21c illustrates

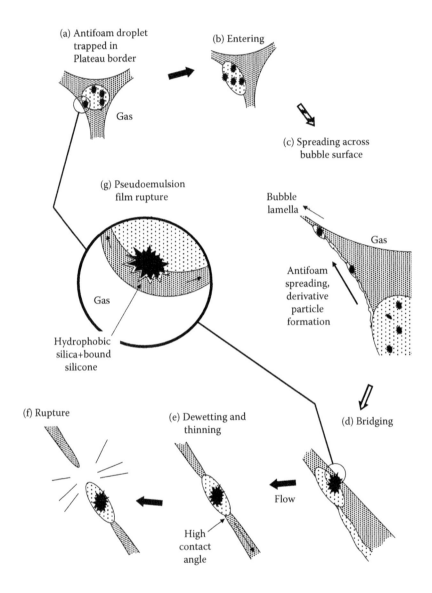

FIGURE 8.21 Modified antifoaming mechanism based on observations of spreading of derivative particles into the lamellae in foam. Inset (g) indicates the central role of pseudoemulsion film rupture to entering and bridging and features a deformed and nonplanar film. Cross-hatched arrows indicate an increase in magnification scale of the subsequent drawing.

spreading and dissemination of derivative particles across the bubble surface until they reach the bubble lamella. A second cross-hatched arrow signifies a second magnification of the view in Figure 8.21d. In Figure 8.21d, the size of the derivative particle with its hydrophobic silica particle exceeds the thickness of the bubble lamella, and it is pressed against the second bubble surface until rupture of a second pseudoemulsion film results in bridging. In Figure 8.21e, high contact angle between the aqueous bubble lamella and the antifoam bridge causes a high degree of curvature at the juncture between the film and the antifoam droplet, resulting in locally high Laplace pressure. This results in flow of liquid from the bubble film (arrow) and results in localized thinning. In Figure 8.21f, the area of wetting between the bubble lamella and antifoam bridge becomes very small. As a result, application of a small mechanical force, arising from a pressure differential between bubbles or thermal fluctuation the film, is sufficient to disengage the film from the antifoam particle. Rupture and withdrawal of the bubble film carries the remnants of the antifoam into the Plateau border and the neighboring bubbles coalesce.

The geometry of the well slide experiment provides a model of antifoam spreading from a position of entrapment within the Plateau border as described by Racz [140] and further discussed by Denkov [12]. The mechanism described earlier is similar to that proposed by Racz and is consistent with observations developed by Wasan [148] using a very different video method.

Further, the mechanism proposed above is not inconsistent with mechanisms featuring a single particle entrapped within a thinning bubble lamella which undergoes entering, bridging, and rupture. Instead, the "fast" antifoam particles described by Denkov [37] may operate as suggested but either in parallel to or alternatively to the mechanism proposed above. Conceptually, either of these mechanisms can yield a very similar end result, placing a hydrophobic particle in position resulting in bridging and rupture within the bubble lamella. It may be that this common end point is truly at the crux of what we mean by antifoam mechanism.

8.9 SUMMARY OF DISPERSION PROCESSES OF SILICONE ANTIFOAMS

In many industrial processes, greater efficiencies can be attained by achieving a condition of controlled foaming. This is accomplished using antifoam compounds to drastically increase the volumetric rate of foam rupture. The function of silicone/hydrophobic silica antifoam compounds and their delivery systems are strongly dependent on at least three different dispersion processes. The state of dispersion must be tuned to the end use for maximum efficiency of antifoaming:

- *Dispersion of the antifoam compound* is established and must be maintained in liquid-in-liquid or liquid-in-solid delivery systems. The droplet size distribution must be selected to provide the maximum number of appropriately sized droplets as dictated by the foaming process with its particular t_{max} and the surfactant properties which control foaming and bubble film thicknesses. Factors such as of antifoaming performance longevity and location of action of the antifoam can have a significant impact on selection of the dispersion particle size distribution. Often this optimum is established empirically.

- *Dispersion of hydrophobic solids* present in the antifoam compound is critically important to destabilization of the pseudoemulsion film. The hydrophobic solids provide a measureable reduction in capillary pressure needed for pseudoemulsion film rupture and foam breaking efficiency. The surfactant type/concentration strongly affects the energetic barrier to entering. It is reasoned that the size distribution and concentration of hydrophobic particles is important to pseudoemulsion film rupture. Observation that hydrophobic particles can be positioned at the antifoam droplet surface further supports the thinking that protrusion of the hydrophobic particles into the aqueous phase is likely the cause of disruption of the pseudoemulsion film.
- *Dynamic spreading processes* potentially play a significant role in antifoam dispersion. Video microscopic observations suggest that spreading drives the formation and rapid dissemination of derivative particles across the bubble surface. Preliminary observations suggest a possible correlation between spreading of antifoam into bubble lamellae and bubble rupture. Observations suggest that there can be a significant time/distance separation between the point of entering and the bubble rupture event.

For additional information please see the Taylor & Francis Group website https://www.crcpress.com/Silicone-Dispersions/Liu/p/book/9781498715553. The supplementary information provides details on the foam flow well slide construction and experimental methods. Numerous additional files containing time-marked images can be viewed. These files are organized according to the figure numbers used in this chapter. Please view the images in PowerPoint in rapid succession to gain a far greater understanding of the dynamic spreading and bubble movement over time.

ACKNOWLEDGMENTS

The author thanks Dow Corning Corporation for permission to publish this work and SiVance LLC, a subsidiary of Milliken & Company, for its support in preparing this chapter.

REFERENCES

1. Ross, S., *Chem. Eng. Prog.*, 1967, *63*, 41.
2. Christiano, S.P.; Fey, K.C., *J. Ind. Microbiol. Biotechnol.*, 2003, *30*, 13–21.
3. Pelton, R., *J. Ind. Microbiol. Biotechnol.*, 2002, *29*, 149–154.
4. Gerhartz, W., Ed., *Ullman's Encyclopedia of Industrial Chemistry*, 5th edn., VCH Publishers, New York, 1988, pp. 466–490.
5. Pape, P.G., *J. Pet. Technol.*, 1983, *June*, 1197–1204.
6. Criddle, D.W.; Meader, A.L., *J. Appl. Phys.*, 1955, *26*, 838.
7. McKendrick, C.B.; Smith, S.J.; Stevenson, P.A., *Colloids Surf.*, 1991, *52*, 47.
8. Callaghan, I.C.; Gould, C.M.; Hamilton, R.J.; Neustadter, E.L., *Colloids Surf.*, 1983, *8*, 17.

9. Callaghan, I.C., Antifoams for nonaqeous systems in the oil industry, Chapter 2 in *Defoaming: Theory and Industrial Applications*, Garrett, P.R. Ed., Surfactant Science Series, Marcel Dekker, New York, 1993, Vol. 45, p. 119.

10. McGee, J.B., *Chem. Eng.*, 1989, *96*, 131.

11. Garrett, P.R., The mode of action of antifoams, Chapter 1 in *Defoaming: Theory and Industrial Applications*, Garrett, P.R., Ed., Surfactant Science Series, Marcel Dekker, New York, 1993, Vol. 45, p. 1.

12. Denkov, N.D.; Marinova, K.G., Antifoam effects of solid particles, oil drops and oil-solid compounds in aqueous foams in *Colloidal Particles at Liquid Interfaces*, Binks, B.P.; Horozov, T.S., Eds., Cambridge University Press, Cambridge, U.K., 2006, pp. 383–444.

13. Wasan, D.T.; Christiano, S.P., Foams and antifoams: A thin film approach, Chapter 6 in *Handbook of Surface and Colloid Chemistry*, Birdi, K.S., Ed., CRC Press, New York, 1996.

14. Höfer, R.; Jost, F.; Schwuger, M.J.; Scharf, R.; Geke, J.; Kresse, J.F.; Lingmann, H.; Veitenhans, R.; Rewied, W., Foams and foam control, in *Ullmann's Encyclopedia of Industrial Chemistry*, Wiley VCH Verlag GmbH & Co. KGaA, Weinheim, Germany, 1988.

15. Hadjiiski, A.; Denkov, N.D.; Tcholakova, S.; Ivanov, I.B., Role of entry barriers in foam destruction by oil drops, Chapter 23 in *Adsorption and Aggregation of Surfactants in Solution*, Mittal, K.L.; Shah, D.O., Eds., Marcel Dekker, New York, 2002, pp. 465–500.

16. Denkov, N.D., Feature article in *Langmuir*, 2004, *20*, 9463–9505.

17. van Hess, H.C., *Ind. Eng. Chem. Fundam.*, 8, 464, 1969.

18. Ross, S., Foams, in *Kirk-Othmer Encyclopedia of Chemical Technology*, Pearce, E.M., Ed., 3rd edn., John Wiley & Sons, New York, 1980, pp. 127–145.

19. Bikerman, J.J., *Trans Faraday Soc.*, 1938, *34*, 634–638.

20. Karakashev, S.I.; Georgiev, P.; Balashev, K., *J. Colloid Interface Sci.*, 2012, *379*, 144–147.

21. Manegold, E., *Schaum. Straßenbau, Chemie und Technik Verlagsgesellschaft*, Heidelberg, Germany, 1953.

22. Matzke, E.B., *Am. J. Bot.*, 1946, *33*, 58.

23. Narsimhan, G.; Ruckenstein, E., Structure drainage and coalescence of foams and concentrated emulsions, Chapter 2 in *Foams, Theory, Measurements, and Applications*, Prud'homme, R.K.; Khan, S.A., Eds., *Surfactant Science Series*, Vol. 57, Marcel Dekker, New York, 1996, pp. 99–187.

24. Narsimhan, G.J., *Food Eng.*, 1991, *14*, 139.

25. Aveyard, R.; Binks, B.P.; Fletcher, P.D.I.; Peck, T.G.; Rutherford, C.E., *Adv. Colloid Interface Sci.*, 1994, *48*, 93.

26. Kulkarni, R.D.; Goddard, E.D.; Kanner, B., *J. Colloid Interface Sci.*, 1977, *59*, 468.

27. Harkins, W.D., *J. Chem. Phys.*, 1941, *9*, 552.

28. Ross, S., *J. Phys. Colloid Chem.* 1950, *54*, 429.

29. Brochard-Wyart, F.; di Meglio, J.M.; Quere, D.; de Gennes, P.G., *Langmuir*, 7, 335, 1991.

30. Xi, Y.; Cheng, Y.; Huang, X-D.; Ma, H-R., *Chin. Phys. Lett.*, 2007, *24*, 2345–2348.

31. Bergeron, V.; Langevin, D., *Phys. Rev. Lett.*, 1996, *76*, 3152–3155.

32. Binks, B.P.; Dong, J., *J. Chem. Soc., Faraday Trans.*, 1998, *94*, 401–410.

33. Aveyard, R.; Binks, B.P.; Fletcher, P.D.I.; Peck T.; Garrett, P.R., *J. Chem. Soc., Faraday Trans.*, 1993, *89*, 4313–4321.

34. Garrett, P.R., *Colloids Surf.*, 1994, *A85*, 159.

35. Robinson, J.V.; Woods, W.W., *J. Soc. Chem. Ind.*, 1948, *67*, 361.

36. Koczo, K.; Llyod, L.; Wasan, D.T., *J. Colloid Interface Sci.*, 1992, *150*, 492.

37. Denkov, N.D.; Cooper, P.; Martin, J.Y., *Langmuir*, 1999, *15*, 8514.

38. Pelton, R.H.; Goddard, E.G., *Colloids Surf.*, 1986, *21*, 167–178.

39. Garrett, P.R., *J. Colloid Interface Sci.*, 1980, *76*, 587.
40. Frey, G.C.; Berg, J.C., *J. Colloid Interface Sci.*, 1989, *127*, 222–238.
41. Dippenaar, A., *Int. J. Mineral Process*, 1982, *9*, 1.
42. Frey, G.C.; Berg, J.C., *J. Colloid Interface Sci.*, 1989, *130*, 54–59.
43. Kekevi, B.; Berber, H.; Yildirim, H., *J. Surf. Deterg.*, 2012, *15*, 73–81.
44. Németha, Z.; Ráczb, G.; Koczoc, K., *J. Colloid Interface Sci.*, 1998, *207*, 386–394.
45. Michaski, R.J.; Youngs, R.W., U.S. Patent 3,923,683, 1974 (Nalco Chemical).
46. McGee, J., *24th EUCEPA Conference*, Stockholm, Sweden, May 10, 1990.
47. Wachala, R.J.; Svetic, R.E., U.S. Patent 3,990,905, 1976 (Nalco Chemical).
48. Rosen, M.; Franklin, L.C., U.S. Patent 4,248,792, 1981 (Glyco Chemicals).
49. Noll, W., *Chemistry and Technology of Silicones*, Elsevier Science and Technology Books, Amsterdam, the Netherlands, 1968.
50. Kondou, K.; Okada, F.; Terae, N., U.S. Patent 4,741,861, 1988 (Shin-Etsu).
51. Aizawa, K.; Nakahara, H.; Sewa, S., U.S. Patent 4,639,489, 1987 (Dow Corning).
52. John, V., European Patent 0217501B1, 1991 (Dow Corning).
53. Becker, R.; Burger, W.; Rautschek, H., U.S. Patent 8,530,401, 2010 (Wacker Chemie).
54. Tonge, L.; Kidera, H.; Okada, R.; Noro, T.; Harkness, B., U.S. Patent 20030013808, 2003 (Dow Corning).
55. Andriot, M.; Chao, S.H., *Silicones in industrial applications*, Chapter 2 in *Inorganic Polymers*, De Jaeger, R.; Gleria, M., Eds., Nova Science, Hauppauge, NY, 2007.
56. *Handling of Synthetic Silica and Silicate*, Technical Bulletin Fine Particles 28, Evonik Brochure.
57. Brinkmann, U.; Ettlinger, M.; Kerner, D.; Schmoll, R., Synthetic amorphous silicas, in *Colloidal Silica Fundamentals and Applications*, Bergna, H.E.; Roberts, W.O., Eds., *Surfactant Science Series*, Vol. 131, pp. 575–588, Marcel Dekker, New York, 2006.
58. Rohr, D.F.; Buddle, S.T.; Wilson, P.R.; Zumbrum, M.A., U.S. Patent 5,340,560, 1993 (General Electric).
59. *Fumed Silica and Fumed Alumina in Coatings Applications*, Cabot Corporation, 2008.
60. Davis, S.; Sanchez Garcia, A.M.; Matheu, D.; Kutsovsky, Y.E., WO/2010/028261, 2010 (Cabot).
61. CAB-O-SIL®, *Fumed Silica for Silicone Elastomers*, Brochure Cabot Corporation, 2013.
62. Thomas, A., Synthetic silica, in *Surface Coatings: Volume 1 Raw Materials and Their Usage*, Springer Science and Business Media, December 6, 2012. See Table 30-1 p. 532.
63. Frye, G.C.; Berg, J.C., *J. Colloid Interface Sci.*, 1989, *130*, 54.
64. Aveyard, R.; Cooper, P.; Fletcher, P.D.I.; Rutherford, C.E., *Langmuir*, 1993, *9*, 604.
65. Slavov, S.V.; Sanger, A.R.; Chuang, K.T., *J. Phys. Chem. B*, 2000, *104*, 1021.
66. Ferch, H.; Leonhardt, W, Foam control in detergent products, Chapter 6 in *Defoaming Theory and Industrial Applications*, Garrett, P.R., Ed., Surfactant Science Series, Marcel Dekker, New York, 1993, Vol. 45.
67. Kroshwitz, J.I.; Howe-Grant, M. (Eds.), *Kirk-Othmer Encyclopedia of Chemical Technology*, Vol. 7, pp. 430–447, Wiley-Interscience, New York, 1993.
68. O'Hara, M.J.; Rink, D.R., U.S. Patent 3,560,401, 1967 (Union Carbide).
69. Rosen, M.R., U.S. Patent 4,076,648, 1978 (Union Carbide).
70. Ross, S., Nishioka, G., *J. Colloid Interface Sci.*, 1978, *65*, 216–224.
71. Clarson, S.J.; Semlyen, J.A., *Siloxane Polymers*, Prentice Hall, Upper Saddle River, NJ, 1993.
72. Willing, D.N., European Patent EP 163398, 1988 (Dow Corning).
73. DeGroot, J.V.; Macosko, C.W., *J. Colloid Interface. Sci.*, 1999, *217*, 86–93.
74. Cosgrove, T.; Turner, M.J.; Thomas, D.R., *Polymer*, 1997, *38*, 3885–3892.
75. Garrett, P.R.; Davis, J.; Rendall, H.M., unpublished work.
76. Raghavan, S.; Walls, H.J.; Khan, S.A., *Langmuir*, 2000, *16*, 7920–7930.
77. Aranguren, M.I.; Mora, E.; DeGroot, J.V.; Macosko, C.W., *J. Rheol.*, 1992, *36*, 1165.

9 Applications of Silicones in Cosmetic and Personal Care Products

Michael S. Starch

CONTENTS

9.1   INTRODUCTION

9.1.1   A Brief History of Silicone Applications in Personal Care Products

Personal care products are consumer products that are used for cleaning, grooming, care, and decoration of hair, skin, and other parts of the body. These products include shampoos and hair conditioners, hand and body lotions, antiperspirants and deodorants, sunscreens, body cleansers, and a wide variety of cosmetics. Today, silicones are used in nearly all categories of personal care products, and the global market for silicones in these applications is worth several hundred million dollars.

The use of silicones in personal care products, began in the 1950s when Avon and Revlon incorporated polydimethylsiloxane (PDMS) in their skin and hair care products. Silicones had been developed commercially a few years earlier, and formulators, then as now, are interested in new ingredients. The use of PDMS in skin care products provided novel esthetic properties compared to hydrocarbons such as mineral oils. PDMS oils are very hydrophobic, and this property allowed formulators to make formulations that provide a water-repellent effect, which helps protect the skin from waterborne irritants. PDMS became a key ingredient in antirash products for babies and hand-protection products for the hands. The Avon Company still sells a product called "silicone glove protective hand cream" [1]. Other properties of PDMS such as their tendency to spread spontaneously into thin films [2] and their lubrication properties led to applications in hair conditioning and grooming products.

As is often the case in the personal care industry, initial success with a new class of materials led to greater interest from the industry and applications in a wider range of personal care products. Silicone manufacturers responded by developing new silicone products or selling existing silicone products developed for other industries into new personal care applications. Silicones with phenyl groups incorporated into their structure became popular in the mid-1960s [3]. The relatively high refractive index of phenyl silicones can provide improved gloss and luster in hair care applications. Silicone Polyethers (SPEs) that were developed for use in making polyurethane foam in the late 1950s [4] were used in personal care formulations beginning in the 1970s. SPEs were of interest, partly because they possess very different solubility properties compared to PDMS, and their solubility can be adjusted by changing their composition. They are graft copolymers where a polyoxyalkylene such as polyethylene oxide (PEO) or polypropylene oxide (PPO) are attached to a PDMS backbone. Such copolymers are very surface active and their properties can be easily tailored by adjusting the ratio between the PDMS segment (strong hydrophobe) and the PEO/PPO segments (hydrophile). The first SPEs that were of interest to the personal care industry were SPEs with sufficient PEO content to render them water soluble. The combination of water solubility and surface activity of SPEs made them suitable for use in foaming products (e.g., shampoos) where they did not produce the defoaming effect of dimethyl silicones like PDMS. Water-soluble SPEs could also be included in clear, aqueous personal care formulations with no need for solubilizers or emulsifiers to stabilize the silicone.

Beginning in the late 1970s with the introduction of volatile silicones (i.e., cyclic dimethyl siloxanes), applications of silicone in personal care products increased rapidly and that led to the proliferation of new silicones, including SPEs that were designed solely for use in personal care products. These products included emulsifiers for water-in-oil (W/O [inverse]) emulsion formulations, as discussed in Section 9.3.3. These new silicone emulsifiers enabled novel types of personal care formulations and this led to further growth of silicone applications in the personal care industry. The 1980s and 1990s were a period of rapid proliferation of new silicone applications in the personal care industry. Many of the "new" silicones adopted by the industry during this period were already produced by silicone manufacturers for other uses, so only minor changes were needed to make them suitable for use in personal care formulations. Examples of such materials are silicone resins and silicone gums. In other cases such as silicone elastomers, the chemistry was well-known, but considerable technology development was required to produce ingredients that were useful for personal care applications.

9.1.2 Nomenclature for Silicones Used in Personal Care Products

In the United States and most other developed countries (e.g., Canada, Mexico, the European Union countries, Japan, China), personal care product labels include a list of ingredients contained in the product. One of the purposes for providing the ingredient disclosure is to allow consumers to determine if the product contains any ingredient that they wish to avoid (e.g., due to allergies). But these ingredient lists provide useful information for other interested parties as well. Formulators often study these ingredient lists to learn what sorts of ingredients are used in competitors' products and to begin the process of developing their own formulations. Ingredient suppliers study these lists to determine which personal care products use ingredients similar to the ingredient that they wish to promote for the same applications. So a brief discussion of the nomenclature for silicones seems warranted.

The practice of ingredient labeling for personal care products began in the United States in the late 1970s. After consultations with the U.S. Food and Drug Administration, the decision was made by an industry trade organization to develop a special nomenclature system for ingredient names. The organization was the Cosmetics, Toiletries, and Fragrance Association (CTFA), and their goal was to provide ingredient names that were succinct yet contain sufficient information to distinguish one ingredient from another. This nomenclature system eventually became known as the International Nomenclature for Cosmetic Ingredients (INCI). To give an example, a common quaternary ammonium compound used in hair conditioners is *N,N,N*-trimethyl-1-hexadecanaminium chloride (as it is called by the rules of formal chemical nomenclature). Its INCI name is *Cetrimonium Chloride*. In this chapter, INCI names will be capitalized and italicized. INCI names are assigned for each new ingredient that is offered for use in personal care products and the assignment is requested by the supplier who wishes to introduce the material. The INCI system is now administered by the Personal Care Products Council (PCPC), a successor organization to the CTFA.

As silicones were adopted for use in personal care formulations, they were assigned their own INCI names. The most common form of silicone, PDMS, has two different INCI names depending on terminal groups. Trimethylsiloxy ((CH_3)$_3$SiO–) end-blocked PDMS is called *Dimethicone*, while silanol (HO(CH_3)$_2$SiO–) end-blocked PDMS is called *Dimethiconol*. In the INCI nomenclature system, the *Dimethicone* and *Dimethiconol* names apply regardless of the molecular weight of the siloxane polymer. The exceptions to this are for very short chain oligomers, for example, *Trisiloxane*, which refers to octamethyltrisiloxane. A very important class of cyclic dimethyl siloxane oligomers that contain between four and seven siloxane units was assigned the INCI name *Cyclomethicone*. These materials were introduced to the personal care market in the mid-1970s and applications have grown to the point that they account for the largest volume of use among all silicones in this market. They are often designated by a shorthand notation: D_x, where x represents the number of dimethyl siloxanes units. As the number of applications for *Cyclomethicone* grew, the industry called for new INCI names to be assigned to the individual cyclic siloxanes: *Cyclotetrasiloxane* (for D_4), *Cyclopentasiloxane* (for D_5), and *Cyclohexasiloxane* (for D_6). These different *Cyclomethicones* differ in volatility as a function of molecular weight, so D_4 is the most volatile while D_6 is the least volatile. In this chapter, the generic INCI name *Cyclomethicone* will be used with the understanding that specific members of this family (D_4, D_5, D_6) alone or in various combinations are used depending upon the drying rate intended for particular formulations.

Many of the silicones used in personal care applications are substituted PDMS where some of the methyl groups have been replaced by other groups that modify the function of the silicone. For example, there are many silicones used in hair care applications that contain amine functional groups. Some of the more commonly used of there are *Amodimethicone*, *Aminopropyl Dimethicone*, and *Bis-Aminopropyl Dimethicone*. The latter name indicates the substitution occurs at each end of the siloxane polymer. Substitution of some methyl groups on PDMS with long-chain alkyl groups is denoted in the INCI nomenclature system by adding the name of the alkyl group: *Stearyl Dimethicone*, *Cetyl Dimethicone*. An example of a silicone that contains both long-chain alkyl and amine substitution is *Bis-Cetearyl Amodimethicone*.

The INCI nomenclature for SPEs used in personal care products has changed over time. When SPEs first appeared in personal care products, one INCI name, *Dimethicone Copolyol*, was used for all of these materials. As the number of different SPEs began to proliferate, the industry requested nomenclature changes that would differentiate between the various SPEs. Starting in 2001, the old name *Dimethicone Copolyol* became obsolete and all SPEs were reassigned INCI names according to the types of polymer blocks present in the SPE including a provision to identify the polyoxyalkylene substituents. So, for example, a rake type SPE with a PDMS backbone that contains side chains of polyethyleneoxide (PEO) with an average of 12 PEO repeating units is called *PEG-12 Dimethicone*. The PEG stands for polyethylene glycol, which is the INCI name for PEO.

If a copolymer of PEO and PPO is grafted onto the PDMS backbone, the new INCI name indicates the average moles of PEO and polypropylene glycol (PPG). Thus *PEG/PPG-18/18 Dimethicone* refers to a PDMS that has substituents that are random copolymers composed of an average of 18 units of PEO and 18 units of PPO. When polyoxyalkylene chains are grafted onto the ends of PDMS to make a polyether-PDMS-polyether (ABA) block copolymer, the INCI name reflects this by including the term "bis" in the name. *Bis-PEG/PPG-14/14 Dimethicone* therefore denotes an SPE composed of PDMS with a random copolymer of 14 units of PEO and 14 units of PPO grafted at each end.

While the nomenclature scheme for SPEs adopted in 2001 provides more information about the structure of an SPE than before, it still contains neither information on molecular weight nor the ratio of dimethyl siloxane units to polyoxyalkylene substituted units (D/D′ ratio). For example, *PEG-12 Dimethicone* may be water soluble if it contains a sufficient number of PEG-12 side chains. On the other hand, if D/D′ ratio is high, the molecule can be water insoluble.

9.2 SILICONES IN PERSONAL CARE FORMULATIONS

9.2.1 SILICONES IN SKIN CARE FORMULATIONS

9.2.1.1 Hand and Body Care Formulations

The first category of personal care formulations to incorporate silicones was skin-care products intended for use on the hands and body. As mentioned earlier, this began in the 1950s when PDMS (*Dimethicone*) was incorporated into these formulations because of the novel physical characteristics of this family of polymers. The properties of interest were hydrophobicity and good spreading. The former property allowed the formulation of products that produce a water-repellent film on the skin, while the latter property helps to improve the esthetics of the formulation during application, as will be explained.

To explain the benefits that silicones bring to skin care formulations, it is necessary to review what these formulations are designed to do when applied to the skin and the types of ingredients that are included to bring about the intended function. This discussion will focus on the primary function of skin care products intended for use on the hands and body, which is to ameliorate dry skin. It is beyond the scope of this chapter to delve into the causes of dry skin in detail, but suffice it to say that dry skin is a condition that exists due to lifestyle factors prevalent in developed countries, and particularly for people who live in northern climates (e.g., the United States and northern Europe). The use of soaps and detergents (some would say the excessive use of these skin cleaning agents) disrupts the physiology of the skin leading to roughness, flaking, itching, and an "ashy" appearance, which is especially noticeable among people with dark skin pigmentation. The term ashy is often used by sufferers of dry skin to describe the appearance of dry, flaking skin because the scattering of light by the flakes (actually aggregates of dead skin cells that are being shed from the skin) makes the skin appear as if ashes had been rubbed over the skin surface. Low humidity conditions that prevail during winter, especially in homes equipped with forced-air heating systems, further stress the skin and exacerbate the effects of detergent exposure.

Because skin care products are primarily intended to alleviate dry skin, their overall function can be described as "moisturization," and this term is accurate in the literal sense because essentially all moisturizers are water-based formulations so water is being applied along with all of the other ingredients in a skin care formulation. However, the application of water provides only temporary relief because it quickly evaporates and does nothing to address the conditions that cause dry skin. So skin care products are formulated with two types of ingredients that address the symptoms and underlying causes of dry skin. One type of ingredient that is included in nearly every skin care product is a humectant. Humectants are water-soluble, hygroscopic ingredients that can help to retain water on the skin surface. The other category of ingredients used in every skin care formulation is emollients. The term emollient in the context of skin care refers to something that provides a soothing or softening effect. Emollients are generally nonvolatile water-insoluble ingredients that form a coating on the skin that covers up the rough surface of dry skin and in some cases, can form a protective barrier that allows the skin to recover from environmental stresses (detergent exposure and low humidity). Silicones such as *Dimethicone* are used as emollients, typically in combination with other emollients such as *Mineral Oil*, *Petrolatum*, long-chain esters such as *Isopropyl Myristate*, or triglyceride oils such as sunflower seed oil. It is important to stress that when emollients are selected for use in skin care products, their esthetic properties are arguably as important as the physical properties because consumers expect the application of the formulation to be an enjoyable experience and the coating left behind after the formulation dries on the skin to have a pleasant feel. Silicones in general, and *Dimethicone* in particular have a pleasant feel on the skin, which is often described as smooth or silky and this is largely attributed to their spreading and lubricating properties.

PDMS is available in a wide range of viscosity grades and the most commonly used *Dimethicone* in skin care products is in the range of 300–500 centistokes (cSt). PDMS in this range of molecular weights is insoluble in most other emollients, and because of its ability to spread over other surfaces (see Chapter 1), it has the ability to mask the undesirable skin feel of other ingredients. High levels of humectants such as *Glycerin* can feel sticky and some common emollients such as *Petrolatum* have a greasy feel. Even relatively small amounts of PDMS can improve the overall esthetics of the formulation significantly. Another reason to include small amount of *Dimethicone* in a skin care formulation is to provide a defoaming effect. The emulsifiers used to stabilize skin care formulations (see Section 9.5.5) can produce a dense foam (lather) when the formulation is spread over the skin and the presence of droplets of hydrophobic PDMS effectively eliminates the formation of lather when the formulation is applied.

Since the first applications of PDMS in skin care products, the use of silicones has proliferated with the introduction of other silicones based on dimethyl siloxane polymers. Silicone gums gained popularity in the 1980s and cross-linked dimethyl siloxanes (silicone elastomers) were introduced in the late 1990s. The term silicone gum refers to dimethyl siloxane polymers with sufficiently high molecular weight as they are soft solids (gums). These siloxanes are typically silanol (–SiOH) terminated, so they have the INCI name *Dimethiconol*. Because of their high molecular weight, they form more long-lasting films on the skin because they are slower to spread to other surfaces (e.g., clothing) compared to lower viscosity PDMS. They are also perceived as being less oily on the skin compared to PDMS oils and have a pleasant silky feel. The use of silicone gums was initially limited due to difficulties in achieving a thin film on the skin, but this problem was solved by the use of volatile silicones (e.g., *Cyclomethicone*) as a carrier for the silicone gum. A solution of 15%–25% silicone gum in a volatile solvent has a sufficiently low viscosity to facilitate incorporation into a formulation and spreading over the skin. As the formulation dries, the solvent evaporates to leave behind a thin film of gum on the skin. If a non-volatile carrier (e.g., low viscosity *Dimethicone*) is used to deliver the silicone gum, the esthetic effects of the silicone can be accentuated and made more long-lasting.

9.2.1.2   Facial Care Formulations

Skin care products that are designed for use on the face are an important part of the skin care market. In terms of the intended function, there is considerable overlap with hand and body skin care formulations in the sense that facial products are also promoted as moisturizers. But actually dry skin is less of a problem for facial skin due to intrinsic factors such as the fact that facial skin has a much higher capacity to generate natural oils (sebum) that function as emollients and lifestyle factors relating to skin cleaning regimens. For example, most women in developed countries wash their hands multiple times each day, but usually wash their face only once or twice daily (e.g., to remove makeup before going to bed). And most of these women perceive their facial skin to be more delicate compared to the skin on other parts of their body. Consequently, they usually use facial cleansing formulations that are much less disruptive to skin physiology compared to soap or a liquid detergent used to wash the hands. Facial cleanser formulations are made with lower levels of detergent ingredients or with surfactants that are less irritating to the skin. Additionally, facial cleansing is often part of a multistep regimen where a moisturizer is applied immediately after cleansing to ameliorate the damaging effects of the cleanser.

Because the requirements for moisturization are less in face care formulations, a typical facial moisturizer contains lower concentrations of the emollients and humectants used in hand and body care formulations. And the emollients must be chosen so that there is no perception of an oily or greasy film after the product has dried. As mentioned earlier, the skin of the face has many more sebum glands compared to other parts of the body so women are very sensitive to oiliness and will not use a facial care product that leaves an oily or greasy residue. Another reason that facial care formulations contain different emollients at lower concentrations is that these formulations usually must provide other functions, which means that the formulations must accommodate additional ingredients that provide these functions. Sun protection has become an important benefit for facial care products since the 1980s, so most facial care formulations today contain ingredients (sunscreens) that absorb or reflect ultraviolet (UV) radiation from the sun. The face is more susceptible to the damaging effects of UV radiation because it has more exposure to the sun compared to other parts of the body. And consumers have become more concerned about the cumulative effects of sun exposure such as wrinkling and "age spots" (uneven pigmentation caused by chronic exposure to UV radiation).

In addition to sunscreens, facial care products usually contain special ingredients that are intended to provide a more youthful appearance. These "antiaging" additives include a wide variety of materials ranging from polypeptides to various plant extracts that are promoted as being effective for reversing the signs of aging. Other additives are used to mask the appearance of wrinkles by scattering visible light (often called a "soft focus" effect). So the need to accommodate a variety of special purpose ingredients in facial care formulations while providing a nongreasy/nonoily skin feel presents a challenge to the formulator. Silicone oils like *Dimethicone* are good emollients for facial care formulations because they have better esthetics compared to conventional emollients like *Mineral Oil*, but silicones also tend to produce glossy films that consumers perceive as oily skin. So the use of nonvolatile silicones as emollients in facial care products was limited until silicone elastomers were introduced in the 1990s.

The silicones that have been discussed in this chapter to this point are based on linear siloxane polymers. Silicone elastomers are a large family of materials that are characterized by the presence of cross-linking, which convert liquid silicone polymers into rubbery solids. These materials are used in a variety of consumer products from baby bottle nipples to heat-resistant cooking utensils. For these applications the silicone elastomers are designed to resist swelling when they come into contact with oils, and this requires a relatively high cross-link density. The silicone elastomers developed for personal care applications have a lower cross-link density so that they are softer and will swell in the presence of silicone oil. The development of silicone elastomers for use in personal care formulations began in Japan in the late 1980s. This work was focused on the preparation of spherical particles that were made using suspension polymerization. In this scheme, the silicone elastomer reactants are dispersed in water to form droplets and then the cross-linking reaction was initiated. The suspended elastomer particles are then spray-dried to produce silicone elastomer powder. Details regarding how silicone elastomer powders are produced are provided in Chapters 6 and 7.

Silicone elastomers are produced by cross-linking reactive linear siloxane polymers using a variety of chemical reactions. The silicone elastomers used in personal care applications are all made with the hydrosilylation reaction using a dimethyl methylhydrogen siloxane and a cross-linker with multiple vinyl groups. The dimethyl methylhydrogen siloxanes are prepared in the same way as the precursors for SPE emulsifiers although the number of reactive sites and molecular weights differ according the degree of cross-linking that is desired. Silicone elastomers that use PDMS with a vinyl group on each end as the cross-linker have been assigned the INCI name *Dimethicone/Vinyl Dimethicone Crosspolymer*, where term crosspolymer is used to indicate a cross-linked material. A similar group of silicone elastomers prepared using this chemistry was assigned the INCI name *Polysilicone-11*. Another widely used silicone elastomer is made using a dialkene (e.g., hexadiene) as the cross-linking agent. These materials have been assigned the INCI name *Dimethicone Crosspolymer*. More recently, silicone elastomers that are cross-linked with a polyalkylene oxide have been introduced. An example is *Dimethicone/Bis-Isobutyl PPG-20 Crosspolymer*, where the cross-linker is a PPO that contains an average of 20 mol of propylene oxide and is terminated at both ends with methallyl groups.

Silicone elastomer powders made via suspension polymerization are discussed in Chapters 6 and 7, so the focus here will be on the silicone elastomers that are made using a solution process. For this process, the dimethyl methylhydrogen siloxane and the cross-linker are dissolved in a solvent (e.g., *Cyclomethicone*) followed by the addition of the catalyst. The cross-linking reaction gels the contents of the reactor, and the gel is composed of silicone elastomer that is swollen by the solvent. Then the contents of the reactor are sheared using high-shear mixing blades to produce a thick paste composed of small (20–80 μ) gel particles. Because the gel particles are produced by a shearing process, they are irregularly shaped. Figure 9.1 is a micro-photograph of the swollen silicone elastomer gel particles. Silicone elastomers gel particles made by this process typically contain 10%–15% elastomer and 85%–90% solvent. The solvent can be essentially any nonpolar liquid and commercial sili-cone elastomer blends based on *Cyclomethicone, Dimethicone, Isododecane*, and *Isohexadecane* are available.

Silicone elastomers rapidly became popular ingredients in skin care formulations and especially facial care products because of their novel physical properties. They provide a soft, silky feel on the skin that is somewhat like that associated with sili-cone oils but because they are soft particles not liquids, they do not have an oily feel even when used at high concentrations in a formulation. And silicone elastomers are not glossy when spread into films. Instead, silicone elastomers produce a matte (dull) appearance due to light scattering by the silicone elastomer particles. Thus they are well-suited for masking the appearance of fine wrinkles on the face. The esthetic effects of silicone elastomers can be tailored by changing the solvent used to swell the elastomer particles. If the elastomer gel particles are swollen with a volatile

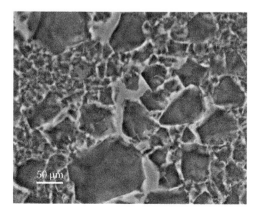

FIGURE 9.1 Microphotograph of silicone elastomer gel particles dispersed in solvent.

solvent, the skin feel that is associated with the swollen particles will gradually diminish as the solvent evaporates; if a nonvolatile solvent is used to make the elastomer gels, the skin feel of the elastomer does not change over time. The silky, velvety feel produced by silicone elastomers is more noticeable and persistent because the elastomer particles do not spontaneously spread into thin films like liquid silicones.

9.2.2 Silicones in Cosmetic Formulations

Silicones are used extensively in cosmetics for their spreading and film-forming properties. Cosmetic formulations encompass a wide variety of physical forms (liquids, solids, and powders) and compositions, but they all share a common characteristic that is to deliver various combinations of pigments and other colored materials to the skin. In the context of cosmetics, pigments are insoluble particles that are derived from minerals, the most common being *Iron Oxides* and *Titanium Dioxide* (note that for these materials the INCI names are identical to the common chemical names). The INCI name *Iron Oxides* refers to three different types of pigment: black iron oxide, red iron oxide, and yellow iron oxide, which differ in the oxidation state of the iron and degree of hydration of the iron oxide. *Titanium dioxide* is a white pigment that when combined in different proportions with the three iron oxide pigments, produces a variety of colors that cover the entire range of shades of human skin color. These pigment mixtures are used in "foundation" cosmetics that cover the skin (typically the face) to provide a uniform appearance and cover up defects (e.g., blemishes, freckles, and small wrinkles). So for example, a pigment mixture in a product intended for African Americans would contain a relatively large amount of black iron oxide and a lesser amount of titanium dioxide while a foundation formulation designed for Caucasians would contain much less black iron oxide, more red iron oxide, and more titanium dioxide.

The combination of different *Iron Oxides* and *Titanium Dioxide* can be used not only to match the various shades of human skin but also to provide other colors that fall into the category of "earth tones." Other colors such as greens, blues, and bright reds are based on organic dyes that are precipitated with inorganic salts (mordants) to form insoluble compounds called lakes. The use of lakes improves the stability of the dye in the formulation by rendering it insoluble in both water and oil, thus eliminating the tendency of the dye to "bleed" into the other ingredients and to separate from the other ingredients after the product has been applied to the skin. For the purposes of formulation, lakes can be considered insoluble particles like the oxide pigments that must be dispersed uniformly into the cosmetic formulation.

Cosmetic products must fulfill several objectives in order to be accepted by consumers. The product must provide the desired shade and appearance, when applied to the skin, have a lasting and durable color that resists rub-off or transfer, and then be easily removed, typically at the end of the day. To accomplish these objectives, the formulation must deliver the pigments and lakes uniformly to the skin in such a way that these colored ingredients become fixed on the skin after the product dries. If the pigments and lakes are not uniformly dispersed in the formulation, this can result in a blotchy, uneven appearance when the product is initially applied. If the colored particles are not fixed properly on the skin, this leads to smearing and transfer of the color to other surface (e.g., the lipstick "prints" on coffee cups). The first problem is solved by predispersing the colored particles in one of the formulation components (typically an oil or silicone fluid) using high-shear mixing equipment to ensure that the different colored particles are intimately mixed and uniformly dispersed. *Iron Oxides* together with *Titanium Dioxide* are usually mixed with an oil or silicone fluid and ground together in a ball mill or similar device. The problem of smearing and color transfer can be alleviated by minimizing the amount of oil used in the formulation or including waxes to increase the viscosity of the dried film.

The introduction of the volatile silicones (i.e., *Cyclomethicone*) into the personal care market provided formulators of cosmetics a nearly ideal material for delivering colored particles to the skin. *Cyclomethicone* has a very low surface tension and effectively wets the particles so it is a good vehicle for making the predispersions, but more importantly it also spreads very well onto the skin and then evaporates leaving a uniform colored film behind. The only comparable materials available to the formulator for this application are low molecular weight hydrocarbon solvents, but these can have a disagreeable odor and may irritate the skin. *Cyclomethicone* is essentially odorless, completely nonirritating, and does not cool the skin due to a low heat of vaporization. By replacing nonvolatile oils in color cosmetic formulations with volatile silicones, formulators were able to make much better products.

When surface-treated pigments for color cosmetics were introduced in the 1970s, this encouraged further penetration of silicones into this market. Inorganic pigments have a relatively hydrophilic surface and treatment of the surface with a hydrophobic material like silicone renders the pigment particle much more compatible with the oils and silicones used in the formulation; it also improves water resistance of the pigments on the skin. The first hydrophobic pigment treatments utilized PDMS that was blended with the pigment and heated to drive water from the particle surfaces and achieve a durable coating. A more durable surface treatment can be obtained using poly(methyl hydrogen)siloxanes (INCI name: *Methicone*) that reacts with itself and with free hydroxyl groups on the pigment surface in the presence of moisture. The resulting highly cross-linked siloxane film is very hydrophobic and durable, but this technique for coating particles has the disadvantage of producing hydrogen gas when the *Methicone* reacts with the particles, and this in turn can lead to explosions if the equipment is not properly ventilated! The most common surface treatments for pigment particles today are trialkoxyalkylsilanes (e.g., *Triethoxycaprylylsilane*) that react with the moisture on the particle surface to form a hydrophobic resinous coating.

The combination of silicone-treated pigments with *Cyclomethicone* proved to be particularly effective for delivering a uniform and durable color to the skin. Transfer resistance was further improved by including a silicone-soluble silicone resin (e.g., *Trimethylsiloxysilicate*) in the formulation. This technology was first patented by Revlon and is the basis for the Colorstay™ line of transfer-resistance color cosmetics [5].

9.2.3 Silicones in Hair Care Formulations

Silicones have been used in hair care formulations such as shampoos and conditioners for almost as long as they have been used in skin care formulations. They are useful in hair care formulations because of their ability to spread over the surface of the hair and in contrast to skin care formulations, the tendency of silicone oils to form glossy films is a positive attribute in hair care applications. Consumers associate a lustrous appearance with hair that is clean and healthy. Silicones are also good lubricants in hair care applications and make the hair easier to comb and style. Hair-care products that claim to reduce hair damage often include silicones because the ability of a silicone coating to reduce combing forces leads to less breakage of hair fibers. A detailed review of applications of silicones in hair care formulations can be found in Reference 6.

For hair conditioning applications, silicones can be delivered from either a detergent formulation (shampoo) or a hair conditioner that is typically used after shampooing that can be either rinsed out or left on. In either case, the silicone is deposited onto the surface of the hair where it provides lubrication and luster after the hair is dry. The most important lubrication effect is the reduction of interfiber friction, and this facilitates the removal of tangles as the comb is pulled though the hair. Lubrication between the comb and the hair fibers also reduces combing force. Overall, the effect is to reduce hair breakage and facilitate styling.

In order to be cost-effective, the silicone in a hair care product that is used in the shower must deposit onto hair in an aqueous environment; otherwise the silicone in the formulation will be rinsed off and go down the drain. Obtaining silicone deposition from a hair conditioning formulation is relatively easy. Conditioner formulations are usually based on quaternary ammonium compounds that have long alkyl chains (e.g., *Cetrimonium Chloride*), and these compounds are attracted to the negatively charged surfaces of the hair fibers. Silicones present in the formulation tend to deposit along with the quaternary ammonium compounds, presumably due to their affinity for the hydrophobic alkyl chains of the quaternary ammonium compounds. To further enhance silicone deposition, PDMS with strongly basic amine groups that develop a positive charge water or quaternary ammonium functional groups are used.

Obtaining silicone deposition from a detergent formulation such as a shampoo is much more difficult to achieve. Detergents are designed to emulsify and wash away oily materials, mainly sebum, from the hair surface, but the detergent ingredients also tend to prevent the deposition of silicones. It is beyond the scope of this chapter to present the various formulation schemes used to make a shampoo that provides acceptable cleaning and also achieves silicone deposition onto the hair. Several patent references are provided and these illustrate different approaches that have been used by shampoo manufacturers [7–9]. Most shampoo formulations that provide a conditioning effect utilize very high molecular weight PDMS. High molecular weight PDMS resists emulsification and the subsequent removal by detergents. The high molecular weight PDMS used in shampoos are quite viscous and therefore difficult to handle (i.e., pump into the mixing vessel), and they are also not readily dispersed into a shampoo formulation. So they are usually supplied in emulsion forms by silicone manufacturers.

9.2.4 Silicones in Antiperspirant Formulations

As explained previously in the context of cosmetic formulations, volatile silicones such as *Cyclomethicone* can be excellent vehicles for delivering active ingredients to the skin. The active ingredients in antiperspirants are salts of aluminum and zirconium (e.g., *Aluminum Chlorohydrate*, *Aluminum Zirconium Tetrachlorohydrex GLY*). When *Cyclomethicone* was first introduced to the personal care market in the mid-1970s, it was shown to the Gillette Company. They first conceived the use of this volatile silicone as a vehicle for delivering antiperspirant active ingredients to the underarm. The product that resulted, Dry Idea®, is a low-viscosity anhydrous dispersion of the powdered active in *Cyclomethicone*. A small amount of treated clay was included to increase the formulation viscosity and help keep the aluminum salts suspended, but the product still required shaking before use. The formulation is applied from a "roll-on" package where the formulation is dispensed through a narrow gap between the wall of the package and a large plastic ball that was fitted to the top of the package. Despite the inherent instability of the formulation and higher ingredient cost, the esthetics of the formulations were such a great improvement over roll-on antiperspirants on the market at the time that Dry Idea rapidly captured a large share of the antiperspirant market. In fact, it is still sold today, more than 35 years later, as this chapter is being written, and the formulation is essentially the same as when it was launched.

Anhydrous antiperspirant formulations that utilize volatile silicones as the vehicle have superior esthetics but the lack of effective thickeners for these formulations remained a problem until silicone elastomers were introduced. As discussed in Section 9.2.1.2, there are two types of silicone elastomers used in personal care applications: silicone elastomer blends, which are silicone elastomer gel particles that are swelled and dispersed in a diluent fluid, and silicone elastomer powders, which are spherical particles of silicone rubber. Both are effective thickeners for dimethyl silicones, especially *Cyclomethicone* when used at concentrations of 10%–20%. Silicone elastomer gel particles are more efficient thickeners compared to silicone elastomer powders because of greater interaction between the irregularly-shaped gel particles (see Figure 9.1), but the elastomer powders also absorb solvents and swell to provide a thickening effect. So 20 years after the first anhydrous antiperspirant formulations were developed, effective thickening technology became available in the form of silicone elastomers. Stable, nonsettling formulations with superior esthetics could be made with consistencies ranging from pastes to low-viscosity liquids (i.e., roll-on formulations). A patent that was issued in 2002 [10] teaches the use of combinations of silicone elastomer gels and powders as thickeners and stabilizers in anhydrous antiperspirants formulations. Although several commercial antiperspirant products that included silicone elastomers as thickeners were launched, these products were not successful, probably due to the high cost of the silicone elastomers.

In aqueous antiperspirants, silicones improve the esthetics of the formulation by masking the sticky feel of the active ingredients. Low viscosity *Dimethicone* (e.g., 10 cSt) is also used to reduce the white appearance of the active ingredient after the formulation has dried. Novel antiperspirant formulations based on SPE emulsifiers are discussed in Section 9.3.3.1.

Table 9.1 provides a list of major categories of silicone and the associated benefits when used in various formulations.

TABLE 9.1

Major Categories of Silicone and the Associated Benefits

Examples (INCI Names)	Benefits in Formulation

Silicone oils—Polydimethylsiloxanes (PDMS)

For skin care
- Emolliency
- Skin protection (water repellent)

- *Dimethicone*
- *Dimethiconol*

For cosmetics
- Improved spreading of hydrophobic pigments

For hair care
- Hair conditioning (reduced combing force)
- Luster (gloss)

Volatile silicones—Methyl siloxane oligomers

- *Cyclopentasiloxane*
- *Cyclotetrasiloxane*
- *Trisiloxane*
- *Caprylyl Methicone*
- *Methyl Trimethicone*

For all formulations
- Transient emolliency, spreading, and lubrication
- Vehicle (solvent) to deliver high molecular-weight silicones

PDMS with amine functional groups

- *Amodimethicone*
- *Aminopropyl Dimethicone*

For hair care
- Hair conditioning (reduced combing force)
- Luster (gloss)
- Improved deposition and durability

PDMS grafted with polyoxyalkylenes

- *PEG-12 Dimethicone,*
- *PEG/PPG-18/18 Dimethicone*
- *Cetyl PEG/PPG-10/1 Dimethicone·*
- *PEG-10 Dimethicone*

For skin care and cosmetics
- Wetting (low molecular weight)
- Stabilization of W/O emulsions (high molecular weight)

Methyl/phenyl siloxanes

- *Phenyl Trimethicone*
- *Diphenyl Dimethicone*
- *Diphenylsiloxy Phenyl Trimethicone*

For all formulations
- Emolliency
- Improved solubility in organic ingredients
- Luster (gloss)

Silicone waxes—PDMS grafted with hydrocarbon chains

- *Stearyl Dimethicone*
- *Cetyl Dimethicone*

For all formulations
- Thickening of silicone and organic oils
- Improved solubility in organic ingredients

Silicone resins—Silicates and silsesquioxanes

- *Trimethylsiloxysilicate*
- *Polymethylsilsesquioxane*

For cosmetics
- Durability (long wear)
- Anti-transfer

(Continued)

TABLE 9.1 (*Continued*)

Major Categories of Silicone and the Associated Benefits

Examples (INCI Names)	Benefits in Formulation
Silicone elastomers—Cross-linked PDMS	
• *Dimethicone Crosspolymer*	For skin care
• *Dimethicone/Vinyl Dimethicone*	• Elegant skin feel (silky, velvety)
Crosspolymer	• Thickening (W/O emulsion formulations)
• *Polysilicone-11*	For cosmetics
	• Matte appearance
	• Thickening (pigment suspension)

9.3   DISPERSING SILICONES IN PERSONAL CARE FORMULATIONS

9.3.1   DISPERSING WATER-INSOLUBLE SILICONES IN WATER-BASED FORMULATIONS

The vast majority of personal care products are aqueous formulations where water is vehicle for all the other ingredients. These formulations usually contain water-insoluble ingredients such as organic oils, silicones, waxes, and powders that are dispersed in water to form an oil-in-water (O/W) emulsion. Such ingredients are emulsified or suspended in the aqueous formulation using various surfactants and rheology control agents (thickeners) so that the formulation is stable and does not require mixing prior to use by consumers. Personal care products are typically designed for a shelf life of 3–5 years and the product must also withstand the rigors of transportation and storage, including vibration and temperature extremes without exhibiting any separation. Consequently, formulators use a wide variety of different schemes to stabilize their products, which typically utilize combinations of several emulsifiers and thickeners. Water-insoluble silicones such as PDMS are often incorporated into personal care formulations using the same emulsifiers and thickeners used for organic oils and waxes. Typically, two or more emulsifiers that accommodate a variety of different oils are used.

Thickeners are used in most personal care formulations to provide the desired consistency. Thickeners also improve emulsion stability by slowing down the creaming rate. Thickeners that provide non-Newtonian flow properties are popular in creams and lotions because they provide very high viscosity when the formulation is at rest, but exhibit shear-thinning behavior when the formulation is dispensed and applied. Viscosity reduction produced by shear forces when the formulation is rubbed onto skin or spread over the hair facilitates even application. For O/W emulsion formulations, *Carbomer* (cross-linked polyacrylic acid) and similar water-dispersible polymers (e.g., *Acrylates/C10-30 Alkyl Acrylates Crosspolymer*) form water-swollen gel particles in the aqueous phase and these provide the desired shear-thinning behavior. In anhydrous formulations, especially formulations that contain large amount of silicone, the gel particles formed when silicone elastomers are swollen with solvent function in essentially the same way as the water-swollen gel particles formed by cross-linked acrylate polymers. Because thickeners provide added stability, aqueous O/W

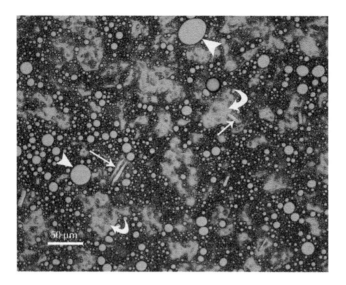

FIGURE 9.2   Microphotograph of a commercial sunscreen formulation.

emulsion formulations are usually crude emulsions with a wide range of oil droplet sizes. If thickeners were not included, this emulsion would exhibit instability (creaming) and would not be stable enough for a commercial product. Figure 9.2 shows such a formulation at high magnification. In the figure, oil droplets are indicated by large arrowheads while the curved arrows point out the gel particles of thickening additives. The straight arrows indicate other solid particles composed of wax ingredients.

SPEs with the proper balance between the hydrophilic polyether groups and the hydrophobic dimethyl siloxane units of the copolymer can be used to aid the preparation of oil-in-water (O/W) emulsions. Many commercial SPEs have the appropriate hydrophilic/hydrophobic balance to effectively disperse silicones into an aqueous formulation. However, rarely is a silicone emulsifier used for this purpose. Selection of an emulsifier package is largely empirical, based on formulation stability, performance attributes, and cost. Silicone surfactants are generally more costly than organic surfactants so the overwhelming majority of O/W emulsion formulations on the market are made using organic surfactants. The only exceptions are formulations where a silicone emulsifier provides an additional benefit (e.g., improved skin feel). In a few cases, silicone surfactants enhance the emulsion stability when specific silicones are emulsified. It was shown that emulsions of D_5 made using conventional organic surfactants lacked long-term stability whereas the emulsion prepared using a combination of a silicone polyether and an ethoxylated lauryl sulfate, were stable [11]. The reason was unexplained.

9.3.2   DISPERSIONS OF HIGH MOLECULAR WEIGHT SILICONES

As previously mentioned, high molecular weight, hence high viscosity PDMS is used in many shampoo formulations, as they generally result in a greater amount of deposition than lower molecular weight silicones. In skin care formulations, high viscosity PDMS provides a pleasant skin feel without any of the oiliness associated with low viscosity silicone oils. Highly viscous silicones are, however, difficult to transfer into a formulation batch and to disperse into an aqueous formulation. Highly intense shear is needed in order to disperse the very viscous silicone oil into small droplets. Many silicone manufacturers help their customers avoid these difficulties by providing them with the high molecular weight silicones in emulsion form. These emulsions can be then added directly to the customer's aqueous formulation. The silicone emulsion is usually added near the end of the process for the finished products, when the emulsification of other oils in the formulation has been completed and the formulation is near room temperature. When using pre-emulsified silicones, it is important to ensure that the emulsifiers used in the silicone emulsion are compatible with other emulsifiers that are used in the formulation. The general rule is that cationic emulsifiers are compatible with other cationic emulsifiers and the same holds for anionic emulsifiers. Mixing anionic and cationic emulsifiers in the same formulation usually leads to the formation of insoluble complexes and sometimes catastrophic emulsion failure (separation). Nonionic emulsifiers are generally compatible with all other emulsifier types.

Various techniques for preparing silicone emulsions are covered in Chapters 2 through 4. One convenient and popular approach for synthesizing high molecular weight PDMS is to use emulsion or suspension polymerization. This route avoids the problems of handling such viscous materials and has the added advantage of producing the PDMS in the form of an emulsion that can be incorporated easily into an aqueous shampoo formulation. By properly designing the polymerization process, silicone emulsions with an optimum droplet size for deposition can be made.

9.3.3   INVERSE (W/O) EMULSION FORMULATIONS STABILIZED BY SILICONE EMULSIFIERS

9.3.3.1   Background and Early History

This section will cover the technology for preparing W/O, or inverse emulsion formulations where an aqueous phase is dispersed into an oil phase composed of silicone as well as other cosmetic oils. The vast majority of personal care emulsion formulations are O/W emulsions, as discussed in Section 9.3.1. W/O emulsions were quite rare in the personal care market until the development of SPE emulsifiers designed to stabilize these emulsions in the late 1970s. W/O emulsion formulations were rare partly because of the limited selection of organic emulsifiers that were capable of stabilizing these emulsions, but primarily due to esthetic reasons. The organic emulsifiers that were developed for W/O emulsion formulations generally required that the formulation contain at least 30% by weight of oil and the formulations usually contain large amounts of waxy ingredients in the oil phase to thicken and stabilize the emulsion. Such formulations are perceived as being "heavy" and greasy by consumers. There are a few moisturizing creams that are W/O emulsions, for example, Eucerin® cream; the esthetics of these formulations are only tolerated by a small group of consumers who suffer from severely dry skin. The introduction of volatile silicone emollients (i.e., *Cyclomethicone*) allowed formulators to contemplate for the first time inverse emulsion formulations that would have much better esthetics. Volatile silicones have a light, pleasant feel during application and then evaporate after application, leaving no residue on the skin or hair. However, production of stable W/O emulsions with a high proportion of silicone in the continuous phase required the development of new emulsifiers.

The ability to produce stable W/O emulsion formulations with good esthetics based on volatile silicone allowed manufacturers to apply this technology in a variety of products where W/O emulsions have inherent advantages over O/W emulsion formulations. W/O emulsion formulations are well-suited for delivering water-soluble ingredients that have a sticky or slimy feel on the skin. Because these ingredients are contained in the internal phase of the emulsion, the esthetics of the silicone external phase dominate the formulation esthetics when the formulation is applied and spread over the skin. The emulsifiers used to stabilize W/O emulsions are water-insoluble so once the formulation dries, the resulting film is very resistant to removal by water. This is useful for cosmetics and waterproof sunscreens.

As explained in Section 9.3.1, the use of SPEs for making O/W emulsion formulations for personal care applications is uncommon. For these formulations, SPEs provide no critical roles in stabilizing the emulsion but are added mostly for sensory benefits. In contrast, stable W/O emulsions require the use of an emulsifier where the hydrophobic part of the emulsifier is composed of silicone. This can be explained by the principal as expressed in Bancroft's rule [12], which states that stable emulsions can only be produced when the emulsifier is more wettable by the continuous phase than by the dispersed phase. The dimethyl siloxane portion of SPEs meets this criterion for W/O emulsions where silicone is the principal component of the continuous phase. Organic emulsifiers based on hydrocarbons do not sufficiently stabilize water-in-silicone emulsions. The introduction of SPE emulsifiers designed for making water-in-silicone emulsions enabled this type of formulation to be made for the first time, and they remain the most popular choice for this purpose. W/O emulsion formulations based on silicone were completely unknown to the industry when this technology was introduced and it enabled many commercially successful products.

One of the first silicone surfactants for producing water-in-silicone emulsions was developed in the laboratory of Joseph Keil at Dow Corning. His work was prompted by the growing interest in silicone-based formulations due to the introduction of volatile silicones, which were first used as vehicles for antiperspirant formulations (see Section 9.2.4). Up until that time (ca. 1978), silicones had been used mostly as additives in personal care formulations at relatively low use levels (below 5%). Anhydrous antiperspirant formulations where *Cyclomethicone* was the vehicle would have as much as 60%–70% silicone in the formulation. This got the attention of the silicone industry because it represented an opportunity to sell much larger volumes of silicone.

The active ingredients used as antiperspirant are aluminum or aluminum/zirconium salts. Antiperspirant roll-ons that were on the market in the mid-1970s, when Gillette was developing Dry Idea, were based on aqueous solutions of the aluminum salts with a small amount of thickener to provide the proper consistency for dispensing the product. Such formulations were cheap and effective but had terrible esthetics because the salt solution felt wet and sticky on the skin. It was during this time that Keil working with Ronald Gee conceived the idea of making an emulsion of the aqueous salt solution in *Cyclomethicone*. Such emulsions are stabilized by a high molecular weight SPE emulsifier that was designed using the cohesive energy ratio (CER) calculation that is based on three-dimensional solubility parameters [13]. Their inventions were disclosed in two patents granted in late 1970s [14,15].

In an antiperspirant W/O emulsion formulation where the salt solution is the dispersed phase, the sticky feel is masked by the feel of the *Cyclomethicone* in the continuous phase. When the formulation dries, both the water and the volatile silicone evaporate, leaving behind the active ingredient on the skin. Further refinement of W/O emulsion antiperspirant formulations led to the development of clear gel formulations. These were achieved by matching the refractive index of the aqueous antiperspirant salt solution to that of the silicone phase. The refractive index of the aqueous phase can be adjusted by adding water-soluble glycols; the refractive index of the silicone phase can be adjusted by adding *Phenyl Trimethicone*, which has a higher refractive index than *Cyclomethicone*. An example of clear gel antiperspirant formulations is given in the formulary (Section 9.5.4).

One of the first SPE emulsifiers for W/O emulsion formulations was assigned the INCI name *PEG/PPG-18/18 Dimethicone*. Due to the high viscosity of the SPE, a consequence of its high molecular weight, the emulsifier was sold in the form of a 10% dispersion in *Cyclomethicone*. It was much too viscous to handle unless it was dispersed into a solvent. Work continued on new silicone emulsifiers and it was found that adding long-chain alkyl groups to the backbone of the SPE emulsifier broadened the range of oils that could be included in the oil phase. A patent disclosing the use of these multifunctional emulsifiers to prepare water-in-mineral oil emulsions was granted in 1985 [16]. This development work eventually led to the commercial W/O emulsifier that was assigned the INCI name, *Lauryl PEG/PPG-18/18 Methicone*. Note the use of the term methicone in this INCI, which indicates that this emulsifier does not contain any dimethyl siloxane. The properties and applications of these SPEs with alkyl groups are discussed by Dahms and Zombeck [17].

To achieve faster drying times for W/O emulsions based on volatile silicone, ethanol can be added to the formulation. The formulation dries faster because ethanol evaporates faster than *Cyclomethicone*. In the course of developing such formulations based on the new SPE emulsifier, it was discovered that substantial amounts of ethanol could be included in the dispersed phase. In an O/W emulsion stabilized by an organic surfactant, a significant level of ethanol can destabilize the emulsion because the emulsifier becomes solubilized in the water-ethanol solution and desorbs from the oil/water interface. In a W/O emulsion stabilized by a silicone surfactant, on the other hand, the ethanol that is present in the internal aqueous phase has no detrimental effect to the emulsion stability, since the type of silicone surfactants that are used to stabilize W/O emulsions have a high molecular weight siloxane portion, which renders the molecule insoluble in a water–ethanol phase. A patent disclosing these emulsions was granted in 1982 [18].

As mentioned previously, the development work for new SPE W/O emulsifiers was done during a period of growing interest by the personal care industry in the use silicones, particularly *Cyclomethicone*. The personal care industry is in many respects quite conservative with regard to new formulation technology and it took many years for the use of volatile silicones to become widespread. And it was not until the benefits of *Cyclomethicone* were fully realized by formulators that they were ready to adopt the new emulsion technology using SPE emulsifiers. Nevertheless, silicone manufacturers continued to invest in the development of this technology and it eventually became a commercial success. An important part of the work was to develop recommendations for potential customers in equipment selection and process conditions to scale up the technology commercially.

Today, the two largest commercial applications for this technology are the clear antiperspirant gels mentioned earlier, and W/O emulsion color cosmetics (e.g., foundations). The development of W/O cosmetic emulsion formulations led to the use of silicone-soluble film-formers that were incorporated in the silicone continuous phase to provide transfer resistance and result in a longer-lasting effect on the skin (see Section 9.2.2).

9.3.3.2   Recent Developments in Silicone Emulsifiers for W/O Emulsions

While high molecular weight SPEs based on polyalkylene oxides are still the most widely used emulsifiers for W/O emulsion formulations in personal care applications, other types of emulsifiers have been introduced that utilize different hydrophilic groups. This development has been driven in part by the "greening" of the personal care industry that has increased interest in ingredients based on natural materials. Since the polyethers used in conventional SPEs are derived from petrochemicals, alternative starting materials such as glycerol (derived from triglycerides produced by plants) are attractive to personal care product manufacturers. Shin-Etsu, for instance, has introduced a number of W/O emulsifiers where the hydrophilic groups grafted to the PDMS backbone are polyethers made by condensing 3 moles of glycerin. An allyl group on the polyglycerin serves to attach the polyether to the PDMS backbone using hydrosilylation.

The W/O silicone emulsifiers based on polyglycerin often contain dimethyl siloxane branches that are attached using a terminal vinyl group on the PDMS branch. Shin-Etsu claims that the PDMS side chains provide improved performance. The INCI name for these emulsifiers is *Polyglyceryl-3 Polydimethylsiloxyethyl Dimethicone*, where polydimethylsiloxyethyl denotes the PDMS branches. Shin-Etsu also introduced a variant with C_{12} alkyl groups that allow for stabilization of emulsions that contain organic oils (*Lauryl Polyglyceryl-3 Polydimethylsiloxyethyl Dimethicone*). As mentioned previously, many of the W/O silicone emulsifiers are high molecular weight polymers and are therefore quite viscous. To facilitate handling, they are often sold as dispersions in low molecular weight silicone carriers. These new emulsifiers from Shin-Etsu are sold without any carrier, and yet, they are not especially viscous (<1000 cSt).

9.3.4   PROCESSING REQUIREMENTS

The preparation of W/O formulations requires different techniques compared to O/W formulations. To understand the differences, it is helpful to briefly review the general approach for making O/W emulsion formulations. I will refer to the basic moisturizing lotion formulation given the formulary (Section 9.5.5). Formulations of this type usually include ingredients that are solids at room temperature so the emulsification is done at an elevated temperature to keep all of the ingredients in the liquid state. For this lotion, the emulsification is carried out at 70°C. The oil phase ingredients and the aqueous phase ingredients are first weighed out in separate vessels, heated to 70°C, and mixed until and the ingredients in each phase are combined. Then the oil phase is slowly poured into the aqueous phase while mixing. The emulsifier for this formulation is a soap (triethanolamine stearate) that forms in situ when the stearic acid in the oil phase comes into contact with the triethanolamine in the aqueous phase. A coemulsifier, *Cetyl Alcohol* is used, which when combined with the soap and upon cooling, forms a lamellar phase that substantially thickens the emulsion. Once the emulsification of the oil phase is completed (essentially as soon as all of the oil phase has been added), the hot emulsion is mixed for a few minutes to ensure homogeneity, and then the batch is cooled. If a polymeric thickener is used, it is usually predispersed in water at room temperature and added after the oil phase. After the batch has cooled to about 40°C–45°C, the temperature sensitive ingredients (fragrance, preservative) are mixed in and cooling is continued until the batch temperature reaches 25°C–30°C. This type of emulsion does not require intense shear; a typical agitator for doing this on the lab scale is a marine propeller placed near the bottom of the mixing vessel at about one fourth of the distance from the bottom to the top of the vessel. The agitator is turned at sufficient speed to produce rapid turnover (top-to-bottom mixing). As the emulsion thickens, the mixer speed is increased to maintain turnover. For thicker emulsions, some type of side-scraping agitator is often added to the mixer shaft, particularly if a cooling jacket is used. This is because the batch thickens first near the walls of the mixing vessel where cooling takes place and this thick material must be moved towards the center of the batch to allow the rest of the batch to cool efficiently.

In summary, the basic approach for an O/W emulsion formulation includes, (1) hot processing to melt the solid ingredients, (2) addition of the oil phase to the aqueous phase with low shear but good circulation of the batch, and (3) provisions to maintain circulation of the batch as it thickens upon cooling. In contrast, W/O emulsions made with SPE emulsifiers do not generally require heating but require the water phase be added to the oil phase instead of the reverse. Also W/O emulsions typically require more intense shear to prepare and good bottom-to-top circulation is critical. This is especially true for production equipment that is configured to load ingredients near the bottom of the vessel.

Another important difference between an O/W and W/O emulsion are the factors that control the viscosity of the emulsion. Most O/W emulsion formulations include thickeners by way of using lamellar phase-forming coemulsifiers or water-soluble polymeric thickeners (e.g., *Carbomer*), but the viscosity of W/O emulsion formulations depend mostly on the volume ratio between the two phases and the droplet size of the dispersed (aqueous) phase. To illustrate this, a series of simple emulsions were made using different ratios of aqueous phase to oil phase. The aqueous phase was composed of 1 part sodium chloride and 99 parts of deionized water. The oil phase was composed of 1 part of a commercial W/O SPE emulsifier (10% *PEG/PPG-18/18 Dimethicone* in *Cyclopentasiloxane*) and 29 parts of *Cyclopentasiloxane*. The emulsions were made using the laboratory mixer

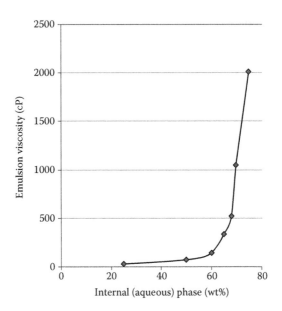

FIGURE 9.3   Viscosity of a water-in-silicone emulsion as a function of weight percent aqueous internal phase.

configuration described here. Figure 9.3 shows a plot of viscosity versus weight percent of aqueous phase. As the weight percent of aqueous phase rises above 50%, emulsion viscosity begins to increase rapidly. This is a consequence of resistance to flow caused by increasing interactions between droplets of aqueous phase as the number of droplets increase. Like O/W emulsions, W/O emulsion must be thick enough to prevent sedimentation of the dispersed droplets. Two schemes for thickening are generally used. First, viscosity for a given formulation can be maximized by reducing the average droplet size, but this approach requires additional processing equipment. The second approach is to increase the volume of the dispersed (aqueous) phase, so typical W/O emulsion formulations for personal care applications have more than 50% by volume of dispersed phase and often the ratio is in the range of 70–80%.

Given the requirements for making W/O emulsions, a typical laboratory setup for making these formulations includes a mixer shaft equipped with multiple agitators. An agitator with blades perpendicular to the plane of rotation (i.e., a turbine agitator) is fixed to the bottom of the mixer shaft and above that one or more agitators with tilted blades (or marine-propeller-style agitators) are used to produce top-to-bottom mixing in the batch. It is important to ensure that the aqueous phase is rapidly incorporated into the batch as it is added so the mixer speed should be set to produce a vortex at the top of the batch that pulls aqueous phase into the batch as it is added. It has been found that the rate of addition rate of the aqueous phase per se is less important than ensuring that the rate is not so rapid that aqueous phase accumulates at the top of the batch. The emulsion will thicken as the aqueous phase is added, so it may be necessary to increase the mixer speed to maintain the vortex. The turbulent mixing produced by the turbine agitator is usually sufficient to produce stable emulsions, but particle size can be further reduced (and viscosity maximized) by using a rotor–stator mixing device such as a Silverson mixer (Silverson Machines, Inc., East Longmeadow, MA) as a secondary mixing step.

A problem that occurs with commercial-scale production of W/O emulsions is that the volume of the oil phase is small relative to the batch size so that the agitator may not effectively mix the batch in the early stage of aqueous phase addition because the liquid level is too low. This is less of a problem for laboratory production because the mixing shaft can be lowered very close to the bottom of the mixing vessel, but this cannot usually be done with commercial-scale production equipment. If the production vessel is equipped with a side sweeping mixer, this can usually be used to provide sufficient mixing until the center mixer is covered. High-shear mixing to reduce particle size after all the aqueous phase has been incorporated is often accomplished with a separate rotor–stator mixing head that is installed beside the main mixer shaft. The other mixers are used in conjunction with the rotor–stator mixer to provide the circulation needed to ensure that all parts of the batch are moved to the rotor–stator mixing head, which is usually small compared to the size of the vessel. An alternative to using a rotor–stator mixer in the vessel is pass the batch through inline homogenizer or other high-shear mixer as the batch is pumped from the mixing vessel.

9.4   ELASTOMERIC SILICONE EMULSIFIERS AND DISPERSION AIDS

With the development of cross-linked silicones for personal care industry, it was a logical step to develop silicone elastomers with polar functionality that could be used as emulsifiers. Both Dow Corning and Shin-Etsu pursued this line of development. Dow Corning introduced *PEG-12 Dimethicone Crosspolymer* and Shin-Etsu launched *PEG-10 Dimethicone Crosspolymer*. To prepare these materials, some of the silicon hydride (SiH) sites on the dimethyl methylhydrogen siloxane are used to attach allyl-terminated PEO and the remainder of the SiH sites is used for cross-linking. Figure 9.4 shows a schematic of a functionalized silicone elastomer gel particle swollen with solvent.

The most important variables affecting the final properties of these functional elastomers are cross-link density and the amount of hydrophilic groups grafted. These hydrophilically modified silicone elastomers can be used as emulsifiers and also thickeners in anhydrous formulations. They were studied for compatibility with tocopherol, also known as vitamin E, an attractive ingredient for skin care due to its antioxidant properties [19]. It was shown that silicone elastomers that were modified with PEG-12 or PEG/PPG-18/18 groups could be used as emulsifiers to prepare W/O emulsions as well as multiple emulsions (water-in-silicone-in-water). The same article discusses the rheological effects of silicone elastomers with PEG-12 functionality in anhydrous formulations based on volatile silicones. It was found that a PEG-12-modified silicone elastomer with low cross-link density was particularly effective for reducing syneresis in liquid anhydrous antiperspirant formulation. Syneresis is a type of formulation instability where silicone separates from the bulk of the formulation.

The use of silicone elastomers as delivery systems for active ingredients in skin care and sunscreen products has been explored. Many active ingredients that are of interest

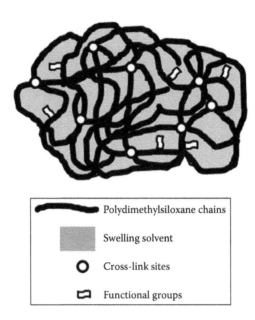

Polydimethylsiloxane chains

Swelling solvent

O   Cross-link sites

◻   Functional groups

FIGURE 9.4   Schematic of a functionalized silicone elastomer gel particle swollen with solvent.

for antiaging skin care products are water soluble and relatively unstable (e.g., *Ascorbic Acid*). Silicone elastomers that contain polar functionality within their cross-linked structure are therefore suitable vehicles to incorporate these active ingredients into water-based formulations. By isolating the active inside silicone elastomer gel particles, interaction with water is minimized and therefore stability of the active is prolonged. Shin-Etsu was the first to introduce a silicone elastomer that utilized PEO as the cross-linker (*PEG-10/Lauryl Dimethicone Crosspolymer, PEG-15/Lauryl Dimethicone Crosspolymer*). These elastomers are cross-linked with *bis*-allyl terminated PEOs.

Sunscreen actives are relatively polar organic compounds and they are incompatible with silicone elastomers composed of solely dimethyl siloxane. Silicone elastomers modified with hydrophilic groups increase the compatibility for sunscreen formulations. The improved compatibility increases the UV protection of the formulation because thicker films can be spread to the skin without segregation into uneven films. The sunscreen actives are trapped within the silicone elastomer gels. Dow Corning introduced silicone elastomers that are cross-linked with *bis*-methallyl terminated PPO. The formulary includes a clear sunscreen gel that demonstrates the compatibility of this type of silicone elastomer (see Section 9.5.7).

9.5   FORMULARY

9.5.1   INTRODUCTION

In this section, a variety of personal care formulations are presented. These formulations are intended to illustrate the many ways that silicone dispersions can be formulated. These prototypes are stable under the ambient laboratory conditions. More rigorous testing may be needed if the formulation is to be made into a commercial product to ensure the product has adequate stability.

The ingredients in the following are referred to by their INCI names. Suppliers for these ingredients can be easily found by searching the Internet. Ingredients that share the same INCI name, but are sold by different suppliers under different trade names are in most cases interchangeable. However in other cases, especially for polymers, the same INCI name can lead to ingredients that are quite different in properties from each other. For example, the degree of substitution and molecular weight are not reflected in the INCI name but can critically affect material properties and therefore performance. For these ingredients, the particular trade name and supplier are noted.

Anhydrous formulations require very little if any preservative, so for these formulations, no preservative is included. For water-containing formulations, a preservative is included. The preservative concentrations indicated are sufficient to prevent microbial growth during study in the laboratory. Any formulations being considered for commercial use will require further microbial testing in accordance with standards for consumer use. None of the formulations include a fragrance. The requirements for selecting an appropriate fragrance for a particular formulation vary greatly depending upon which types of consumers are being targeted.

9.5.2 ANHYDROUS ROLL-ON ANTIPERSPIRANT

This formulation (Table 9.2) is modeled on Dry Idea, which was the first commercial product that used *Cyclomethicone* (volatile silicone) as the vehicle for the active ingredients. It contains *Aluminum Zirconium Tetrachlorohydrex GLY* (antiperspirant active) that is dispersed in cyclomethicone. Clay that has been treated with a quaternary ammonium compound (*Quaternium-18 Hectorite*) is used as a thickener.

9.5.2.1 Procedure
Disperse component 2 into component 1 using a high-shear mixer, then add component 3 and continue mixing until a homogeneous dispersion is obtained.

9.5.2.2 Variations of This Formulation
Virtually any volatile silicone can be used in place of the *Cyclomethicone* used as the vehicle (D_4, D_5, *Trisiloxane*, or combinations). To reduce the white appearance of

---

TABLE 9.2

Ingredients for the Anhydrous Roll-On Antiperspirant

Component	INCI Name	Weight (%)
1.	*Cyclomethicone*	65.0
2.	*Cyclomethicone (and) Quaternium-18 Hectorite (and) SD Alcohol 40*	15.0
3.	*Aluminum Zirconium Tetrachlorohydrex GLY*	20.0

TABLE 9.3

Ingredients for the Anhydrous Antiperspirant Stick

Component	INCI Name	Weight (%)
1.	*Hydrogenated Castor Oil*	5.0
2.	*Stearyl Alcohol*	16.0
3.	*Cyclopentasiloxane*	43.0
4.	*Phenyl Trimethicone*	9.0
5.	*Aluminum Zirconium Tetrachlorohydrex GLY*	25.0
6.	*Talc*	1.0

the antiperspirant salt on the skin, 5%–10% of the volatile silicone can be replaced with a low viscosity nonvolatile silicone (e.g., Dimethicone, 10 cSt).

9.5.3 ANHYDROUS ANTIPERSPIRANT STICK

Table 9.3 shows a solid suspension antiperspirant formulation commonly known as a stick. The antiperspirant salt is dispersed in a molten mixture of volatile silicone and waxes, poured into a suitable mold and then cooled.

9.5.3.1 Procedure

Load all of the components except for component 3 into a mixing vessel that can be heated. Heat and stir the components to 80°C–85°C until the waxes (components 1 and 2) have melted completely and then slowly add component 3 while still applying heat to maintain the temperature of the batch. When all of component 3 has been added, pour the molten mixture into molds and cool rapidly to minimize settling of components 5 and 6. For larger batches, it is useful to use a mixing vessel that is equipped with a condenser to return the volatile components to the batch.

9.5.4 CLEAR ANTIPERSPIRANT GEL

This is a W/O emulsion that is stabilized with an SPE emulsifier. To allow the formulation to dry after application, the silicone phase contains a low molecular weight, volatile PDMS and a hydrocarbon solvent. This formulation (Table 9.4) has better esthetics than a conventional (O/W) emulsion because the antiperspirant salt is in the internal (aqueous) phase and the stickiness of the antiperspirant salt solution is masked by the silicone in the external phase. A transparent emulsion is achieved by carefully matching the refractive indices of the two phases.

9.5.4.1 Procedure

Load components 1–3 (silicone phase) into a mixing vessel that is large enough to hold the entire batch and mix until uniform. A dual mixer as described in Section 9.3.4 is recommended. In a separate vessel, combine the ingredients for the aqueous

TABLE 9.4
Ingredients for the Clear Antiperspirant Gel

Component	INCI Name	Weight (%)
1.	*Dimethicone (and) PEG/PPG-18/18 Dimethicone*[a]	3.0
2.	*Dimethicone (2 cSt)*	6.0
3.	*Isohexadecane*	7.0
4.	*Aluminum Sesquichlorohydrate*	32.0
5.	*Water*	24.5
6.	*Ethanol*	3.0
7.	*Propylene Glycol*	24.5

[a] This formulation was developed using Dow Corning® ES-5226 DM Formulation Aid.

phase (components 4–7) and stir until a clear solution is obtained. Measure the refractive index of both phases. The refractive index of both phases should be close to 1.4105. If necessary adjust the refractive index of the aqueous phase to match that of the silicone phase to within 0.0003. Add water to raise the refractive index or propylene glycol to lower the refractive index. Slowly add the aqueous phase to the silicone phase while keeping the mixer speed high enough to quickly incorporate the aqueous phase as it is added. The emulsion will thicken as the aqueous phase is added so the mixer speed will need to be increased. After all of the aqueous phase has been added, mix for 5–10 minutes. To achieve a gel consistency, the formulation must be homogenized using a high-shear mixer such as a Silverson.

9.5.5 BASIC MOISTURIZING LOTION (O/W EMULSION)

This formulation (Table 9.5) illustrates the use of an anionic emulsifier (triethanolamine stearate) to stabilize an emulsion made with both silicone and organic oils. The emulsifier is formed in situ as the oil phase is added to the water phase. *Cetyl Alcohol* and *Carbomer* are used as thickeners.

9.5.5.1 Procedure

Load the components for the oil phase (components 1–5) into a mixing vessel. Heat the oil phase to ~70°C and mix gently until the solid components (4, 5) are melted. In a separate container that is large enough to hold the entire batch, load components 6–8 and heat them to ~70°C. Start the mixer and after the *Triethanolamine* and *Glycerin* are dissolved, add the molten oil phase. This formulation does not require a high-shear mixer; a marine propeller agitator is sufficient. The mixer speed should be set to provide turbulent mixing with good circulation but not so fast that there is splashing. If the mixer speed is too fast, air will become entrained in the batch and these bubbles are difficult to remove once the lotion has thickened. After all the oil phase has been added, mix for another 5 minutes and then begin to cool the batch by removing the source of heat. The cooling process can be accelerated by immersing

TABLE 9.5
Ingredients for the Simple Moisturizing Lotion

Component	INCI Name	Weight (%)
1.	Dimethicone (350 cSt)	3.0
2.	Mineral Oil	1.5
3.	Petrolatum	1.0
4.	Stearic Acid (triple pressed)	3.0
5.	Cetyl Alcohol	1.0
6.	Water	74.0
7.	Glycerin	5.0
8.	Triethanolamine	1.3
9.	Carbomer (1% dispersion in water)	10.0
10.	DMDM Hydantoin	0.2

the mixing vessel in a cold water bath. The emulsion will thicken near the walls of the vessel first, especially if a cooling bath is used. This thick layer should be scraped off periodically to facilitate cooling of the center of the batch.

The *Carbomer* dispersion should be prepared ahead of time by slowly sifting *Carbomer* powder into cold water, while stirring at a speed rapid enough to create a vortex that draws the powder into the water. Avoid generating foam by stirring too fast. If the *Carbomer* powder is added too fast, lumps will form that are very difficult to disperse. The *Carbomer* dispersion will be translucent because *Carbomer* will not become fully hydrated until it is neutralized.

When the temperature of the batch has dropped below about 45°C, add the preservative (component 9) and the *Carbomer* dispersion. The batch will thicken further and the mixer speed will need to be increased to maintain good circulation. Continue mixing until the batch reaches room temperature. For these types of emulsions that are heated in an open mixing vessel, it is best to use a tared vessel so water that has been lost to evaporation can be weighed when the formulation is finished. The lost water is typically added back at the end of the process.

9.5.5.2 Variations of This Formulation

This is a versatile emulsion that can accommodate a variety of different oils and other ingredients such as botanical extracts. To make a cream, the amount of oil and *Cetyl Alcohol* can be increased.

9.5.6 Oil-Free W/O Emulsion Sunscreen with Titanium Dioxide

The formulation given in Table 9.6 has a very light feel after application, because it contains such a large proportion of volatile ingredients and therefore leaves a very small amount of residue on the skin after drying. The *Aluminum Starch Octenylsuccinate* is a hydrophobic powder that provides a pleasant dry silky feel. This formulation contains two UV absorbers (sunscreen actives). One is a water

TABLE 9.6

Ingredients for the Oil-Free Sunscreen

Component	INCI Name	Weight (%)
1.	*Titanium Dioxide*[a]	3.0
2.	*Cyclopentasiloxane*	19.5
3.	*Cyclopentasiloxane (and) PEG/PPG-18/18 Dimethicone*[b]	7.5
4.	*Aluminum Starch Octenylsuccinate*	5.0
5.	*Phenylbenzimidazole Sulfonic Acid*[c]	2.0
6.	*Sodium Hydroxide (10% aqueous solution)*	~2.7
7.	*Polysorbate-20*	0.2
8.	*Water*	q.s. to 65.0%

[a] A triethoxycaprylylsilane-treated, attenuation (sunscreen) grade of titanium dioxide was used to develop this formulation.

[b] Dow Corning® 5225C Formulation Aid was used to develop this formulation.

[c] Also known as Ensulizole for sunscreen product labeling in the United States.

soluble salt (*Sodium Phenylbenzimidazole Sulfonate*), and the other is a powder (*Titanium Dioxide*) which is included in the silicone phase. *Phenylbenzimidazole Sulfonic Acid* is not soluble in water, but forms the soluble salt when mixed with sodium hydroxide.

9.5.6.1 Procedure

This formulation requires the mixer configuration described in Section 9.3.4. Load the silicone phase (components 1–4) into the mixing vessel that is large enough to contain the entire batch and mix until uniform. Prepare the aqueous phase by loading ~50% of the batch weight of deionized water and add the *Sodium Hydroxide* solution, followed by the *Phenylbenzimidazole Sulfonic Acid*. Stir until all of the acid is dissolved and, if necessary add more sodium hydroxide solution to bring the pH of the solution up to about pH 7. Add component 7 and then enough deionized water to bring the total weight of the aqueous phase up to 65% of the total batch weight.

Slowly add the aqueous phase to the mixing vessel with sufficient agitation to rapidly incorporate it into the batch as it is added. The batch will turn white and thicken as the emulsion is formed. After all of the aqueous phase has been added, mix at high speed for 10–15 minutes. Passing the batch through a high-shear mixing device (e.g., inline homogenizer) will maximize the viscosity of the formulation.

9.5.7 CLEAR ANHYDROUS SUNSCREEN GEL

This formulation (Table 9.7) illustrates the use of a silicone elastomer that is cross-linked with polypropylene oxide. This polar content in the silicone elastomer increases compatibility with the sunscreens. This formulation has an estimated Sun Protection Factor (SPF) of about 15.

TABLE 9.7

Ingredients for the Clear Anhydrous Sunscreen Gel

Component	INCI Name	Weight (%)
1.	*Isododecane (and) Dimethicone/Bis-Isobutyl PPG-20 Crosspolymer*[a]	43.0
2.	*Ethylhexyl Methoxycinnamate*[b]	7.5
3.	*Ethylhexyl Salicylate*[c]	5.0
4.	*Caprylic/Capric Triglyceride*	5.5
5.	*Dicaprylyl Ether*	7.0
6.	*Butyl Methoxydibenzoylmethane*[d]	2.5
7.	*Caprylyl Methicone*	12.3
8.	*Phenyl Trimethicone*	4.0
9.	*Dimethicone* (2 cSt)	10.0
10.	*Isododecane*	3.0
11.	*Silica Silylate*[e]	0.2

[a] This formulation was developed using Dow Corning® EL-8050 ID silicone organic elastomer blend.

[b] Also known as octinoxate for sunscreen product labeling in the United States.

[c] Also known as octisalate for sunscreen product labeling in the United States.

[d] Also known as avobenzone for sunscreen product labeling in the United States.

[e] For maximum clarity, a small particle-size grade should be used. This formulation was developed using Dow Corning® VM-2220 aerogel fine particles.

9.5.7.1 Procedure

Combine components 2–9 in a mixing vessel. Component 6 is a crystalline solid that must be completely dissolved in order to maximize stability and clarity. Heat components 2°C–9°C to ~50°C and stir until a clear solution is obtained. Heat component 1°C to ~50°C and combine with the rest of the ingredients, then cool to room temperature. Combine components 10 and 11 and stir into the rest of the batch. The *Isododecane* in component 1 is volatile, so the heating time should be minimized. On a larger scale, the vessel used to heat component 1 can be equipped with a condenser to prevent the solvent from escaping.

9.5.8 LIQUID FOUNDATION (COLOR COSMETIC)

The formulation shown in Table 9.8 delivers a mixture of pigments (*Iron Oxides, Titanium Dioxide*) that are predispersed in silicone prior to combining them with the rest of the silicone phase. Silicone elastomer is included in the formulation to help suspend the pigment particles in the silicone phase and provide a smooth silky feel when the formulation is applied. The precise amounts of the different *Iron Oxides* are important to produce a color that is within the normal range for human skin. A pigment grade of *Titanium Dioxide* with a relatively large particle size is normally used in foundation formulations. This formulation contains relatively large amounts of *Titanium Dioxide* and *Zinc Oxide* nanoparticles that are surface-treated to reduce

TABLE 9.8
Ingredients for the Liquid Foundation

Component	INCI Name	Weight (%)
1.	*Cyclopentasiloxane and Dimethicone/Vinyl Dimethicone Crosspolymer*[a]	25.00
2.	*Lauryl PEG-9 Polydimethylsiloxyethyl Dimethicone*[b]	1.00
3.	*Dimethicone (50 cSt)*	2.40
4.	*Cyclopentasiloxane*	10.72
5.	*Cyclopentasiloxane and Dimethicone/Vinyl Dimethicone Crosspolymer*[a]	0.800
6.	*Cyclopentasiloxane*	1.998
7.	*Iron Oxides (Yellow)*[c]	0.311
8.	*Iron Oxides (Red)*[c]	0.137
9.	*Iron Oxides (Black)*[c]	0.134
10.	*Water*	40.00
11.	*Glycerin*	5.00
12.	*Sodium Chloride*	2.00
13.	*Titanium Dioxide*[d]	5.00
14.	*Zinc Oxide*[e]	5.00
15.	*Alumina*[f]	0.50

[a] A silicone elastomer blend with about 9% elastomer was used to develop this formulation. The amount of silicone elastomer blend will need to be adjusted if the elastomer content is substantially higher or lower than 9%.

[b] This formulation was developed using KF-6038, supplied by Shin-Etsu.

[c] Triethoxycaprylylsilane-treated iron oxides were used.

[d] This formulation was developed with an attenuation (sunscreen) grade of titanium dioxide (A40-TIO2-D57 from Kobo Products). The particle size range is 30–50 nm.

[e] This formulation was developed with an attenuation (sunscreen) grade of zinc oxide (ZNO-USP1-MS3 from Kobo Products). The primary particle size is 100 nm.

[f] A fumed alumina, SpectrAL® PC-401 from Cabot was used to develop this formulation.

their photocatalytic activity so that they are suitable for use as UV filters to provide protection from the sun. The small particle size of *Titanium Dioxide* and *Zinc Oxide* used here provide much less whitening than pigment grade ingredients.

9.5.8.1   Procedure

This formulation requires the mixer configuration described in Section 9.3.4. Load the silicone phase (components 1–4) into a mixing vessel that is large enough to contain the entire batch and mix until uniform. Note that the silicone elastomer blend and *Cyclopentasiloxane* used for this formulation have been divided between the silicone phase and the pigment dispersion (components 5–9). The amount of silicone elastomer blend and *Cyclopentasiloxane* in the pigment dispersion have been set to provide a suitable viscosity for dispersing the pigments into the silicone while providing a viscosity high enough to help suspend the pigment particles after they have been dispersed.

The pigment dispersion should be made in a separate mixer that is equipped for high-shear mixing in order to combine and disperse the pigments so that they provide a uniform color. For larger batches, a ball mill is useful for preparing the pigment dispersion. Combine the silicone phase and the pigment dispersion and mix until uniform. Prepare the aqueous phase (components 11–13) in a separate vessel and mix to dissolve the salt and glycerin. Slowly add the aqueous phase to the mixing vessel with sufficient agitation to rapidly incorporate it into the batch as it is added. The batch will thicken as the emulsion is formed. After all of the aqueous phase has been added, mix at high speed for 10–15 minutes. Add the remaining ingredients (powders) to the batch and continue mixing until the batch is homogeneous. Passing the batch through a high-shear mixing device (e.g., inline homogenizer) will maximize the viscosity of the formulation and ensure uniform dispersion of the TiO_2 and ZnO.

9.5.8.2 Variations of This Formulation

This formulation will provide a modest level of protection from the sun (estimated SPF of 10–12). The addition of ensulizole to the water phase (see Section 9.5.6) will provide a higher level of sun protection.

REFERENCES

1. http://shop.avon.com/product.aspx?newdept=&s=AV_GGL_PLA&c=iProspect& otc=03490980_BathBody_handCream&bnd=&pf_id=41894&level1_id=300&level2_ id=303&pdept_id=343&dept_id=334 (Internet search on October 19, 2015).
2. Owen, M.J. *CHEMTECH, 11,* 1981, 288 (also available from Dow Corning as publication 01-3078-01).
3. Warrick, E.L. *Forty Years of Firsts: The Recollections of a Dow Corning Pioneer,* McGraw-Hill, New York, 1990, p. 246.
4. Hill, R.M. (ed.). *Silicone Surfactants,* Marcel Dekker, New York, 1999, p. 138.
5. Salvatore, J.B. et al. US 5837223, 1998.
6. Berthiaume, M.D. *Silicones in Hair Care,* SCC Monograph Series, Society of Cosmetic Chemists, New York, 1997.
7. Adams, G.P. US 3950510, 1976.
8. Fieler, G.M., Stacy, L.V. US 7728457, 1988.
9. Bolich, R.E., Williams, T.B. US 4788006, 1988.
10. Fecht, C.M., Starch, M.S. US 6406684, 2002.
11. Feng, Q.J. et al. US 2007/0190012 A1, 2007.
12. Bancroft, W.D. *J. Phys. Chem., 17,* 1913, 501–519.
13. Beerbower, A., Hill, M. *McCutcheon's Detergents and Emulsifiers,* Allured Publishing Company, Carol Stream, IL, 1971.
14. Gee, R.P., Keil, J.W. US 4122029, 1978.
15. Keil, J.W. US 4268499, 1981.
16. Keil, J.W. US 4532132, 1985.
17. Dahms, G.D., Zombeck, A. *Cosmetics and Toiletries, 101,* 1995, 91.
18. Starch, M.S. US 4311695, 1982.
19. Starch, M.S., Fiori, J.E., Lin, Z., *J. Cosmet. Sci., 54,* 2003, 193–205.

10 Silane-Based Water Repellents for Inorganic Construction Materials

Jean-Paul Lecomte and Oliver Weichold

CONTENTS

10.1 INTRODUCTION

Concrete based on ordinary Portland cement (OPC) accounts for a large portion of the building materials. For instance, 70% of the building materials used in 2000 in the European Union construction industry are concrete and cement based [1]. Limited availability of some raw materials, increased requirements to recycle material from destroyed buildings, cost of rehabilitation, and cost of constructing new buildings drive to the need to increase service life of buildings and infrastructures. This chapter provides a review of silane- and siloxane-based technologies employed in the industry to increase the durability of cement and concrete materials.

The cement matrix found in mortar or concrete forms during the hydration of initially unhydrated cement clinker. The cement matrix has a porous structure with pore sizes ranging from nanometers up to millimeters. In the set (hardened) cement matrix, pores are interconnected and form a "continuum" throughout the matrix

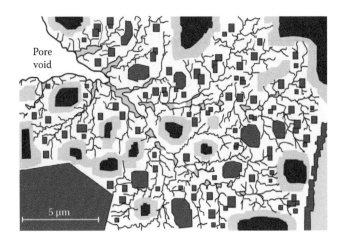

Pore
void

5 µm

FIGURE 10.1 Schematic representation of the cement matrix after setting. Black, portland-ite; dark blue, un-hydrated cement particles; yellow, CSH gel formed upon cement hydration; light blue, water absorbed into the pores.

(see Figure 10.1). As the surface of a cement-based material is easily wet by water (the contact angle between water and the cement matrix surface is low), this leads to capillary absorption of water into the interconnected pore system [2].

Absorption of water by a cement matrix can lead to both physical and chemical attacks. Freeze-thaw cycles can lead to cracks in the concrete cover as freezing causes water in the pores to expand, thereby exerting pressure on the cement matrix. Many water-soluble substances, from either the environment or man-made products, can penetrate into the porous structure of the cement matrix along with water and subsequently cause damage (see Figure 10.2). The so-called "sulfate attack," for example, refers to cracking or spalling due to the formation of ettringite from water-soluble sulfate salts. Soluble chlorides from sea water or deicing salts can be washed into the concrete pores and increase the corrosion rate of the steel reinforcement.

In order to increase service life of concrete structures, protective measures against water ingress are applied (see Figure 10.3). Surface treatment is commonly used to protect new or old concrete against water penetration. Coating made of a film-forming binder (a polymer dispersion), filler, and optionally pigment can be applied on the surface of the concrete. This leads to a "mechanical" protection of the surface as a coherent film is formed between the concrete surface and the external environment. Sealers or "pore blockers" are used to block the pores by the accumulation of fine solid particles within the cement matrix pores. The use of either coating or pore blockers can help reduce the ingress of water, but unfortunately results in a decrease in the water vapor permeability of the treated material. Silanes and siloxanes act differently as they chemically modify the surface of the concrete pores and decrease its affinity for water (Figure 10.4). Silanes impregnate the surface by penetrating into the pores and reacting at their surface. Alternatively, water repellents based on silanes can be used as "admixtures" or "integral water repellents (IWR)," which are

FIGURE 10.2 *Top:* structural concrete exposed to deicing salts showing damage of the concrete cover and corrosion of the reinforcement bars. *Bottom:* water droplets placed on an autoclaved, aerated, light concrete treated with a silane/siloxane water repellent showing decreased wetting of water on the concrete.

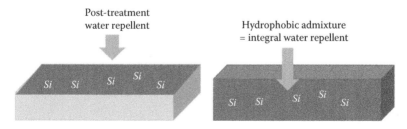

FIGURE 10.3 Schematic illustration of using post-treatment water repellent and IWR.

added during the preparation of the cement-based materials (see Figure 10.3). This leads to the presence of the water repellent in the bulk of the cement-based materials. Water repellents based on silanes can therefore be post-applied on an existing surface or mixed into a mortar/concrete formulation. Both application methods will be discussed.

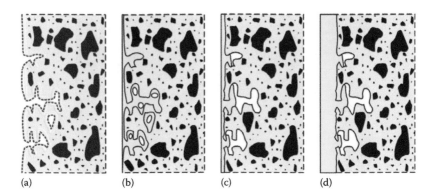

(a)   (b)   (c)   (d)

FIGURE 10.4 Schematic representation of different surface treatment: (a) impregnation, no film; (b) pore blocking, pores are filled up; (c) thin film formation; (d) thick film formation (render/stucco).

Silanes and also siloxanes and silicone resins have now become a well-known class of hydrophobers used in the construction industry. They are used as post-treatment [3–5] or admixtures in non-load-bearing concrete [6]. In the following, the reactivity of silanes is described, with focus given to that under the alkaline conditions. Preparation of and use of emulsions and powders of silanes will also be described.

10.2   MECHANISM OF ACTION AND REACTION OF SILANE

The chemistry of alkoxy silanes is governed by the properties of the Si–O bond (Figure 10.5). Due to the differences in electronegativity (Si = 1.74 and O = 3.50 in the Allred–Rochow scheme [7]), the bond is highly polarized in such a way that the silicon bears a partial positive charge δ^+ and the oxygen a partial negative charge δ^- [8, p. 152]. This is called the inductive effect and oxygen atoms are termed −I substituents. The ionic character of the bond is approximately 54% [9]. However, unlike atoms from the second row of the periodic table, silicon, being located in the third row, possesses unoccupied d-orbitals, which allows violation of the octet rule. Consequently electron pairs formed with neighboring atoms (α-position) are stabilized giving rise to a second resonance structure (Figure 10.5). This is called the mesomeric effect, and the OR groups are termed +M substituents. Although the

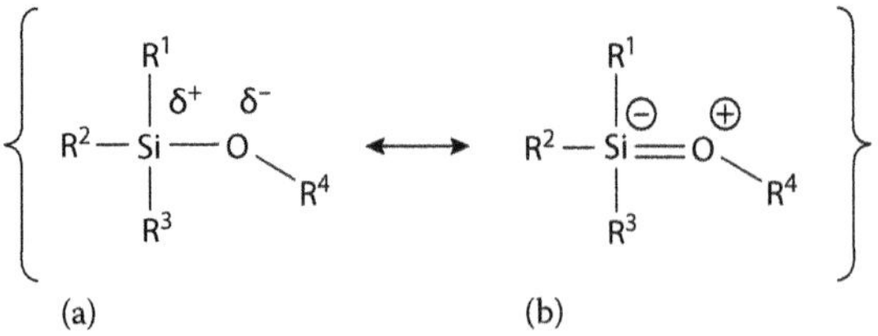

$$
\left\{
\begin{array}{c}
R^1 \\
| \quad \delta^+ \ \delta^- \\
R^2 - Si - O \\
| \qquad\quad R^4 \\
R^3
\end{array}
\qquad \longleftrightarrow \qquad
\begin{array}{c}
R^1 \\
| \quad \ominus \ \oplus \\
R^2 - Si = O \\
| \qquad\quad R^4 \\
R^3
\end{array}
\right\}
$$

(a)   (b)

FIGURE 10.5 Contributing resonance structures of the Si–O bond in trialkylalkoxy silanes: single bond with partial charges (a) and double bond with full charge separation (b).

contribution of structure b in Figure 10.5 is low, the presence of a double bond share in the resonance hybrid is one of the reasons for the extraordinary large Si–O–R bond angles, 120°–150° [10].

Due to the distribution of partial charges, the Si–O bond is prone to both nucleophilic and electrophilic attack. Electrophiles—the simplest ones being protons, for example, in acidic hydrolysis—attack the oxygen atom. Electrophilic attack at the oxygen increases the partial positive charge on the silicon atom and makes it even more susceptible to nucleophilic attack by, for example, water, which completes the acid-catalyzed hydrolysis reaction to form silanol and alcohol. The competing E1-type mechanism leading to the formation of R_3Si^+ fragments is believed not to occur in solution due to the high rate of nucleophlic attack at the silicon [11].

Although acid-catalyzed hydrolysis of alkoxy silanes is much more frequently encountered in literature than basic hydrolysis, the former is not separately discussed in this review, since the substrates under consideration exhibit a highly alkaline environment. However, observations from acid-catalyzed hydrolysis are used as comparison where necessary. Before discussing the reactions and reactivities of alkyl trialkoxysilanes, it should be noted that the established compounds used as hydrophobing agents on building materials as well as all common trialkoxysilanes are immiscible with water. In order to elucidate the reaction mechanisms and reactivities, the transformations are either investigated in solvents or in water/solvent mixtures. Although it can be safely assumed that the reaction mechanism or the reactivities in solution and at interface are similar, the rates of reaction when the reaction is confined at the interface of the pores can be expected to differ from those in solution.

10.2.1 HYDROLYSIS OF ALKYL TRIALKOXYSILANES IN BASIC MEDIA

Nucleophiles such as OH$^-$ ions directly attack the silicon atom. While at carbon centers, such reactions proceed through a concerted S_N2 mechanism, that is, the existing bond is broken in a way the new bond is formed, the presence of unoccupied d-orbitals at the silicon allow the formation of pentacoordinate intermediates. However, the hydrolysis in alkaline media is not limited to OH$^-$ attack only but subject to a general base catalysis [12]. Any species acting as base accelerates the reaction by assisting the deprotonation of water in transition state (TS) 1 (Figure 10.6).

All reactions in Figure 10.6 are equilibria and the extent of hydrolysis depends on the free enthalpy associated with the reaction. Shifting the equilibrium to one side requires a driving force such as subsequent reactions with favorable free enthalpy or changing concentrations of reaction partners, for example, the evaporation of the formed alcohol.

The reactivity of alkoxy silane is controlled by the two electronic effects—I/+M (vide supra) and steric effects of the alkyl and alkoxy substituents. Under base catalysis, the reactivity decreases in the order $CH_3Si(OCH_3)_3$ > $(CH_3)_2Si(OCH_3)_2$ > $(CH_3)_3SiOCH3$ [13,14]. This indicates that the –I effect of the OR groups prevails over the +I of the alkyl groups and the +M of the OR groups. Hydrolysis of alkoxy silanes with more than one OR group proceeds in a stepwise manner [15]. A great difference between the rates of the first and the second step was found; hydrolysis of the first OR group is significantly faster [16]. However, for $PhSi(OCH_2CH_2OCH_3)_3$,

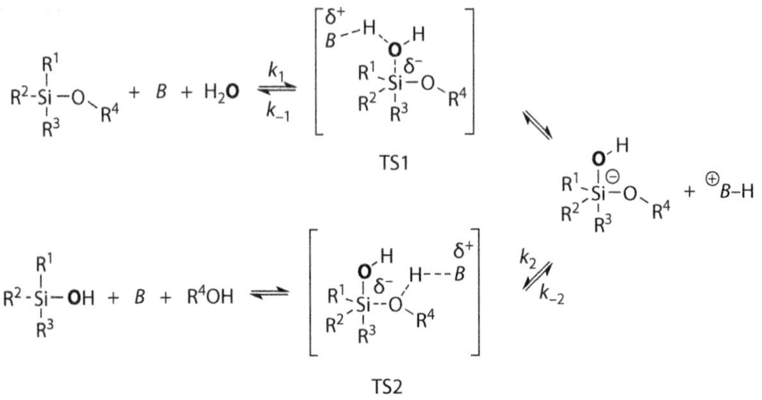

FIGURE 10.6 Mechanism for the general base-catalyzed hydrolysis of alkoxy silanes. *B* indicates any species acting as base. (From Osterholtz, F.D. and Pohl, E.R., *J. Adhes. Sci. Technol.*, 6, 127, 1992.)

the rate increases with every step; this is probably due to a reduction of the steric demand at the silicon. Within a series of alkoxy substituents, the rate of hydrolysis decreases with increasing steric demand, in the order $CH_3O > CH_3CH_2O >$ $(CH_3)_3CO$ [15]. The alkyl substituents have profound effects on the rate of hydrolysis. Generally, substituents close to the reaction center increase the steric hindrance, so that bulky substitutents slow down the reaction rate. For a series of alkyl substituents, the half-life of the hydrolysis increases in the order CH_3 (6.8 min) $< CH_3CH_2$ (25 min) $< CH_3CH_2CH_2$ (58 min) $<$ cyclo-C_6H_{11} (340 min) that correlates with Taft's steric parameter [17]. In addition, plain alkyl substituents increase the electron density at silicon, thus increasing the energy of the transition state resulting in lower rates of hydrolysis. Electron-withdrawing substituents have the opposite effect and $ClCH_2Si(OCH_2CH_2OCH_3)_3$ was found to hydrolyze 1600 times faster than $(CH_3)_2CHSi(OCH_2CH_2OCH_3)_3$, despite having the same steric demand [12].

The rates of hydrolysis also strongly depend on the pH of the surrounding medium [16]. Starting at highly acidic pH values, the rates decrease to reach a minimum at a pH value of approximately 6.5, after which they increase again (Figure 10.7).

The hydrolysis of $PhSi(OCH_2CH_2OCH_3)_3$ was found to become pH independent at pH values > 10, indicating that at high pH, a change in the rate-determining step occurs [16]. Quenching experiments at low temperatures revealed that at higher pH values, the hydrolysis consists of two stages, with a slow second stage whose rate is pH independent. It was concluded that deprotonation of the formed silanol follows hydrolysis and the resulting anion **d′** is inert to the nucleophilic attack by hydroxide anions. Consequently, structures **e′** and **f′** can only be obtained by starting from the corresponding alkoxy silanols **d** and **e**, but not from the trialkoxysilane **c**. Based on these findings, a modified scheme for the hydrolysis of alkoxy silanes was proposed (Figure 10.8).

It should be noted that in the presence of water, the reaction of **c** with hydroxide leads directly to **d′** without the formation of **d** due to the difference in pK_a values of **d**, water, and ethanol.

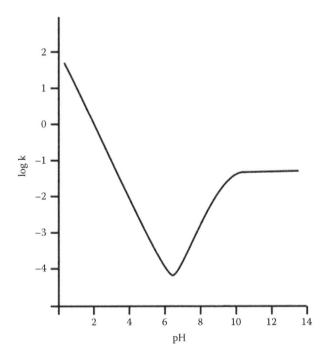

FIGURE 10.7   The rates of hydrolysis strongly depend on the pH of the surrounding medium. The Hydrolysis of PhSi(OCH$_2$CH$_2$OCH$_3$)$_3$ becomes pH independent at pH values > 10. (From McNeil, K.J. et al., *J. Am. Chem. Soc.*, 102, 1859, 1980.)

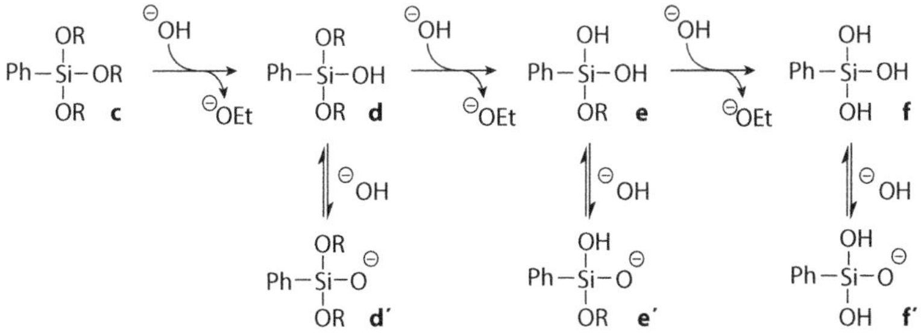

FIGURE 10.8   Stepwise hydrolysis of phenyl trialkoxysilanes in basic media. (From McNeil, K.J. et al., *J. Am. Chem. Soc.*, 102, 1859, 1980.)

10.2.2 CONDENSATION AND REACTION WITH SURFACES

Alkylalkoxysilanes that are kept in high purity water with low ionic strength are stable for weeks and months provided that the container is not made of glass [15]. Traces of ions, for example, from the "tap water," acids, or bases can drastically accelerate the hydrolysis and condensation. Since the condensation reaction is catalyzed by species that also catalyze the hydrolysis of alkoxysilanes, hydrolysis and condensation reactions cannot be separated under practical conditions. Early literature describing the hydrolysis of alkoxysilanes in high concentrations (1–2 mol/L) reported the direct formation of oligo- and polymeric condensation products without observing the monomeric silanols [18,19]. Alkaline hydrolysis was found to produce higher molecular weight and higher branched products than acid catalysis [20]. Thorough kinetic investigations corroborated these findings, and today the understanding is that acid catalysis causes fast hydrolysis and the formation of silanols, which undergo slow gelation [15]. This is supported by findings that silanols prepared and kept under mild and strictly neutral conditions can persist for extended periods of time even in the absence of bulky substituents [21,22]. In contrast, base catalysis is prone to high rates of condensation and fast gelation [15], and the rate was found to depend only on the hydroxide concentration and not on the counter cation [23]. Protic solvents form hydrogen bonds with the nucleophilic Si–O⁻ group that reduces the electron density at the oxygen and hampers its ability to attack further molecules. As a result, protic solvents retard base catalyzed condensation [20]. The reactivities are summarized in an overall model (Figure 10.9) that was originally devised for tetraalkoxysilanes [24] and later modified to represent the reaction path for trialkoxysilanes [15].

Figure 10.9 contains two types of reactions: pure hydrolysis producing silanols (horizontal arrows) and pure condensation of silanols to siloxanes (vertical arrows). Both reactions are mainly observed at acidic pH values. The diagonal arrows describe the replacement of an alkoxy group by a silyloxy group according to Figure 10.10, which is most frequently observed in alkaline media.

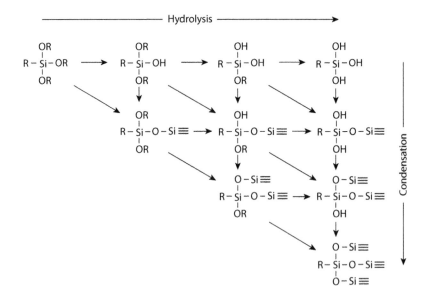

FIGURE 10.9 Reaction pathway from monomolecular alkyl trialkoxysilanes to high molecular weight condensation products. Diagonal arrows indicate direct base-catalyzed condensation.

FIGURE 10.10   Substitution reaction of an alkoxy group by a silyloxy group in alkaline media.

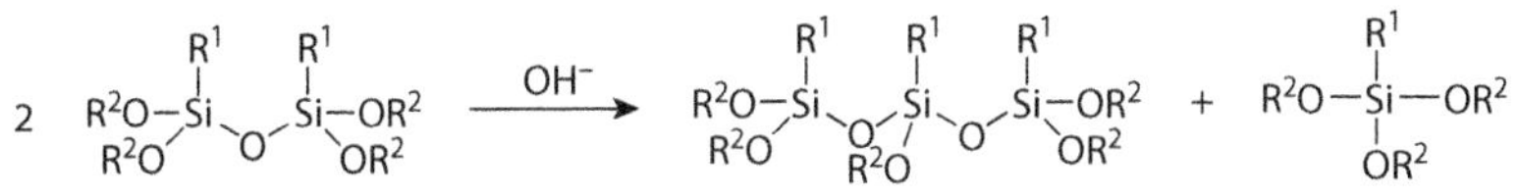

FIGURE 10.11   Redistribution of siloxane bonds.

As mentioned in the description of Figure 10.6, all reactions are equilibria, and in fact, alcoholysis and hydrolysis ($R^4 = H$) of the siloxane bond occur even in basic media. Between pH 3 and 8, the rate of hydrolysis increases by more than three orders of magnitude and appears to reach a plateau at a pH of approximately 10 [25]. Alcoholysis in basic media causes a redistribution of siloxane bonds and may be the reason for unhydrolyzed monomers being found after gelation of the mixture [20,26]. As a result, high and low molecular weight species are maximized at the expense of intermediate molecular weights (Figure 10.11).

Heterogeneous reactions of alkoxy silanes with surfaces generally resemble condensation reactions in homogeneous media. All untreated siliceous materials contain Si–OH bonds on the surface that in the presence of water or an alkaline pore solution partially deprotonate forming Si–O⁻ groups. Both surface silanols and silanolates can attack alkoxy silanes in a reaction similar to Figure 10.10 forming surface-bound siloxanes. For surface-bound monolayers prepared from dilute solutions under controlled conditions, a threshold temperature exists above which closely packed monolayers with low surface energy cannot be obtained [27–29]. However, the physical meaning of this temperature is still under discussion. Some authors report a temperature range from 0°C for alkylalkoxysilanes containing alkyl chain of C_{10} to 38°C for C_{22} with a slope of 3.5°C per additional CH_2 group [27,28]. Others report 0°C for a C_{16} chain with a slope of 5°C per CH_2 group [29]. Since common hydrophobing agents are applied from highly concentrated solutions, the formation of monolayers is unlikely and the significance of the threshold temperature in practice may be irrelevant. In any case, the threshold temperature for alkylalkoxysilane containing C_8 in the alkyl group, which are common hydrophobing agents, would be below application temperature.

Another discovery more relevant to field applications was that the formation of ordered domains requires an optimum level of water on the solid surface [30,31]. Since the reactive alkylsilanes are immiscible with water, they adsorb at the air–water interface with the alkyl tails pointing to the air, and the alkoxy groups react under the alkaline condition to form polycondensates. These polycondensates initially float on the surface of the liquid layer in a random motion, during which they either get trapped by surface silanols or silanolates through covalent or hydrogen bonding or collide with each other to form larger structures [29,32]. On the contrary, relatively dry substrates trap reactive molecules more efficiently and preferentially, covalently resulting in less ordered structures on the surface.

Based on studies over the past several decades, the actions of hydrophobing agents on porous inorganic building materials can be understood as the following. Upon application, the liquid silane enters the pores by capillary suction. Under almost all conditions, the pore walls are covered with a molecularly thin layer of water, with which the silane is immiscible. Initial hydrolysis of the silane is supposed to occur exclusively at the air–water interface, forming amphiphilic silanols or silanolates that stick to the air–water interface and improve compatibility between the water layer and the silane. The hydrolysis reaction consumes water and liberates low molecular weight alcohol, which is miscible with both the silane and water and further improves compatibility between water and silane phases.

Meanwhile, condensation occurs, which generates a water molecule that compensates for the water consumed during hydration only in acidic media. On alkaline substrates such as concrete, condensation proceeds according to Figure 10.12 without the formation of water. That is, the hydrolysis and condensation of alkoxy silanes on concrete ultimately deprives the pore system of water. While the water layer is still there, the oligomers float until they are anchored to the surface by reactive groups. Since condensation and reaction with the surface under alkaline conditions with a limited amount of water at the interface are incomplete, the anchored oligomers contain reactive groups and the oligomers can grow further. Finally, the pore walls are covered with a hydrophobic gel (Figure 10.12).

Hydrolysis, condensation, and anchoring to the surfaces occur while the liquid is progressing through the pore system. However, the rates of these processes strongly depend on the type of silane and the local environment of the pore system. This is exemplified by comparing the amount of ingress of two common hydrophobing agents, *n*-octyltriethoxysilane and *i*-butyltriethoxysilane using single-sided nuclear

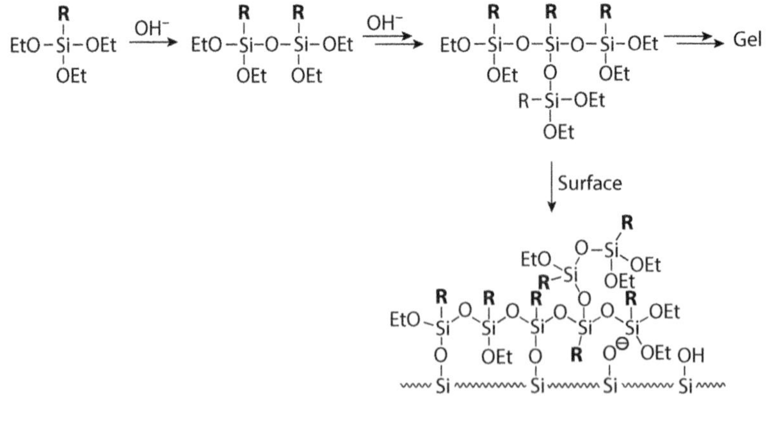

FIGURE 10.12 Alkaline hydrolysis and condensation in the presence of a surface leading to the formation of a surface-bound gel.

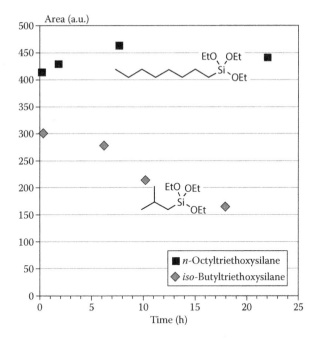

FIGURE 10.13 Quantity proportional to the silane volume-fraction in concrete (shown as signal area) as function of ingress time recorded by single-sided NMR. (From Antons, U. and Weichold, O., *J. Infrast. Syst.*, submitted.) The area plotted on the ordinate is a measure proportional to the silane volume content.

magnetic resonance (Figure 10.13). For *n*-octyltriethoxy silane, the signal intensity proportional to the silane volume fraction inside the concrete appears to remain constant over the first day of ingress during which the silane penetrates 6.5 mm deep into the structure. For *i*-butyltriethoxy silane, the signal continuously decreases during the first day even though penetration of the silane reaches a depth of 7.5 mm [33]. Thus, the ingress of *n*-octyltriethoxy silane into the pores involves only transport of the silane while the ingress of *i*-butyltriethoxy silane involves simultaneous transport and reaction of the silane.

10.3 FORMULATION OF SILANE-BASED WATER REPELLENTS

10.3.1 Siloxanes, Silicone Resins, and Alkoxy Silanes

This section describes the use of silicon-based hydrophobers with a special focus on silanes. Besides alkyl trialkoxysilanes, other silicon-based active materials are also used in water repellents. Figure 10.14 illustrates the structure of polydimethylsiloxane, alkyl trialkoxysilane, and silicone resin.

Silicone is a generic term used to describe polymers based on a siloxane backbone comprising of the Si–O repeat unit. Polydimethylsiloxanes (PDMS) are the most common siloxanes used worldwide. They are available as low or high viscosity fluids.

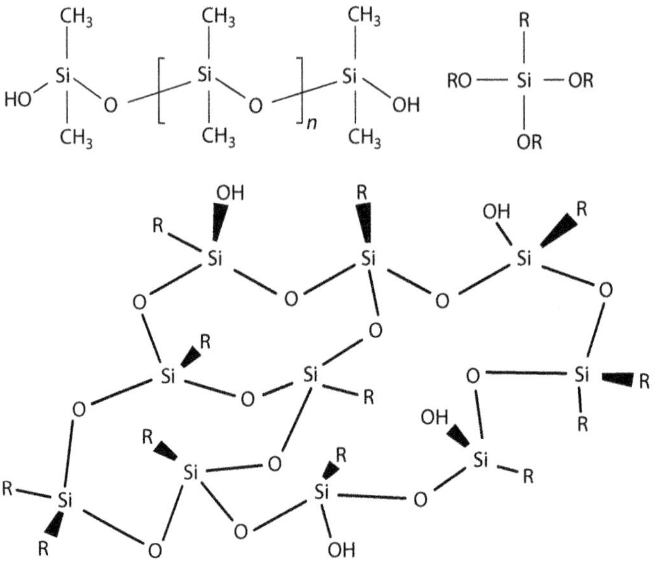

FIGURE 10.14 Structure of PDMS, alkyl trialkoxysilane, and silicone resin. R can be ethyl, methyl, phenyl, or octyl group.

Terminated with silanol groups (as drawn in Figure 10.14), they are reactive. The low surface tension, better resistance to UV radiation as compared to organic polymers, and high gas permeability are properties of great benefit in hydrophobic treatment.

Silicone resins are produced by controlled hydrolysis and condensation reactions of silanes.

Alkyl trialkoxysilanes are low viscosity liquids that can be used as they are, without dilution, to post-treat building materials. However, alkyl trialkoxysilanes are often used at a lower concentration in water repellent formulations. Such water repellent formulations with low active content can be obtained by dilution of the silane in a solvent or emulsified into water. Some applications such as dry mortar formulations require the silane to be formulated as fine free-flowing powders.

10.3.2   EMULSION OF SILANE

As post-treatment water repellent, silane must often be diluted. Dilution in various solvents such as aliphatic alcohols or hydrocarbons was practiced till regulations on the use of volatile organic content limited their use and lead to the need to develop water-based systems. As an IWR, silane needs to be effectively mixed and homogeneously distributed into mortar or concrete mixes. This can be achieved only if interfacial tension between silane and water is decreased significantly.

It has been reported that functionalization of alkyl trialkoxysilane with amine groups results in the silane to be readily dilutable [34]. Emulsions of silanes were then developed. Oil-in-water emulsions of silane are easy to dilute in water and this enables the preparation of water-based water repellents. Oil-in-water emulsion of silane is also easier to mix homogeneously in a mortar or concrete mixture than the neat silane.

The presence of a long alkyl chain and three bulky leaving groups (ethoxy) confers a low solubility in water and low rate of hydrolysis to octyltriethoxysilane. These characteristics makes octyltriethoxysilane the preferred silane for the preparation of oil-in-water emulsion. Emulsions made of alkyl trialkoxysilanes that are either faster to hydrolyze or more soluble in water are indeed not stable.

DePasquale disclosed the preparation of shelf stable oil-in-water emulsions of alkyl trialkoxysilanes with non-ionic surfactants (alkylethoxylates) of HLB value between 4 and 15 in the early 1990s [35]. Numerous formulations of shelf-stable emulsions of alkyl trialkoxysilanes have been described since then. Improved stability of the emulsion is obtained by buffering the continuous phase (aqueous), since a stable neutral pH reduces hydrolysis rate of silanes [36]. Anionic [37], alkyl-aminealkoxylate [38], or mixture of nonionic emulsifiers [39] can also be used for the preparation of silane emulsions. Mixture of silane, siloxane, and silicone resin can be emulsified as well [40].

Silane-and siloxane-based water repellents can be formulated such as to have a gel or cream consistency. This enables the user to apply a thick layer of water repellent on vertical substrate without having the water repellent running off or slumping. This also provides enough time for the water repellent to completely absorb into the construction material. "Gel" of silane can be formulated by mixing an inorganic thickener, such as bentonite or montmorillonite, into the neat liquid silane [41]. After application of the gel on the construction material and letting it absorb into the substrate. "Gel" or "cream" of silane can also be formulated by preparing an oil-in-water emulsion containing a high silane content [42]. Water-in-oil emulsions can be formulated such as to have the same cream or gel consistency [43].

More recently, oil-in-water microencapsulation of silanes has been used to hydrophobe porous construction materials [44].

10.3.3 Powdered Silane

The continuous need to improve quality and consistency of cement-based mortars prepared on the job site has led to the development of so-called dry mixes. Dry mixes are preformulated, mostly cement-based mortars, which are premixed in a manufacturing equipment, and shipped as "ready-to-use" mixtures (ready to be mixed with water). These mortar formulations are designed for specific applications, like masonry mortar, tile adhesive, grouts, render, or skim coats [45]. Some applications require the finished mortar to be protected against water penetration. This is the case for tile grouts, render for façade, skim coat, and top coat for insulation façade system.

Mortar with low absorption of water can be obtained by the addition of powdered IWR into the dry mix formulation. Historically, oleochemicals and metal soaps were used for that purpose [46]. Cement-based dry mixes that contain oleochemicals such as calcium stearate or zinc stearate are difficult to mix with water. Reduction of water absorption by the finished mortar has been reported to decrease as a function of weathering. These limitations lead to the development of new powdered hydrophobic additives based on silicone chemistry [47,48].

For 10–15 years, silanes and siloxanes have been included in the formulation of new powdered additives that can be used as "IWR" in dry mix formulations. Different formulations and processes for the conversion of liquid silane or siloxane into powders have been described.

Neat silane or emulsion of silane can be sprayed on pyrogenic or precipitated silica [49]. If the silica used has a high surface area of at least 50 m²/g, it can adsorb enough liquid (here the neat silane or an emulsion of silane). After adsorption on silica, a fine, free-flowing powder is obtained, which can be used as powdered IWR.

Atomization or spray drying of emulsion is a very common process used to convert emulsion into powder. Atomization of emulsion of silane is now currently practiced to prepare powdered IWR (e.g., Reference 50). Powder containing both silanes and oleochemicals (fatty acids or fatty acid esters) can be prepared by atomization [51]. Octyltriethoxysilane is mostly used for the preparation of powdered IWR based on silane used in cement-based dry mix. Powders based on silane bearing a shorter alkyl chain (from methyl to butyl) and obtained by atomization can be added as IWR for mortar with low cement content (i.e., gypsum) [52].

Size enlargement or so-called granulation process has been used to convert silane/siloxane mix into fine free-flowing powder [47,53,54]. Powders obtained by such process are illustrated in Figure 10.15.

10.4 APPLICATION OF SILANE-BASED WATER REPELLENTS

10.4.1 POST-TREATMENT WATER REPELLENT

Deterioration of steel-reinforced concrete can be accelerated by the ingress of chloride ions, which can be found in deicing salts or in sea water. Chlorides accelerate corrosion of steel bars (see Figure 10.16). As corroded steel occupies more space than the original steel, it leads to cracking of cement matrix, spalling or delamination of the concrete cover [55]. Sodium or calcium chloride are solid salts that are soluble in water. Solubilized in water, they can be transported by water into porous construction materials, such as concrete. Ingress of chloride into concrete is initiated by the absorption of chloride-containing water. After evaporation of water, the salts remain inside the pore system, leading to increasing concentration of chloride as a function of exposure time [56]. Ingress of chloride ions into concrete structure can continue by further absorption of chloride-rich water or by diffusion of chloride ions from high to low concentration regions.

Extensive laboratory tests have demonstrated that impregnation of concrete surfaces with silanes can effectively minimize water penetration, minimize chloride penetration, and postpone and reduce the rate of corrosion of steel reinforcements. Silane needs to penetrate deep enough into the concrete to effectively protect concrete against water penetration and chloride ingress [57–59]. "Depth of penetration" of silane and therefore depth of the impregnation layer depends however on the concrete mix or water–cement ratio (for a given concrete mix) [60,61].

TM3000_0029   2015/09/21 11:49 N   1 mm

FIGURE 10.15   Micrographs of silicone hydrophobic powder obtained by size enlargement process.

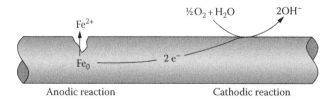

FIGURE 10.16   Mechanism of corrosion of reinforcement steel bars.

Numerous studies describe the reduction of chloride ingress; thanks to the treatment of concrete surface with silane. The method used to initiate and accelerate chloride ingress may have an impact on the results of the study. Ponding the surface of concrete specimens with sodium chloride solution requires longer testing period, but enables proper testing of the effect of the concrete impregnation on water absorption (and chloride ingress) by *capillary* action [62]. Soaking of concrete specimens in chloride solution and application of an electrical potential accelerates *migration* of chloride by diffusion. Different studies using different concrete mixes, different water-repellent formulations (and application dosages), and different methods to promote chloride ingress were described [63–66].

Penetration of chlorides can be slowed down in *concrete* in which cracks are created, if surface impregnation with silane is carried out *before* crack formation and deep enough [67]. Deep impregnation of concrete surface with silane *after* cracks formation might still be effective to delay chloride-induced corrosion of steel reinforcement [68]. Impregnation of concrete already submitted to moderate chloride ingress with silane helps to avoid further chloride ingress [69].

Accelerated aging of concrete treated with silane was carried out by exposure to UV radiation, thermal shocks, carbonation, or moistening cycles [3,70,73]. Experimental data shows good durability of the hydrophobic treatment, meaning that water ingress in concrete is still much reduced after accelerated aging. Some long-term field exposures of concrete treated with silane have shown a residual protective effect against water ingress even after 20 years of service [4,71].

Carbonation of concrete based on Portland cement has no influence on silane-based hydrophobic treatment (applied before carbonation). Carbonation of concrete based on blast furnace cement [84] has on the contrary, a major impact on the efficiency of the hydrophobic treatment to reduce water penetration [72,73].

Models were developed for the prediction of service life of reinforced concrete structures. They estimate and use in the calculation an initiation period that is the exposure time required till chlorides concentration close to the reinforcing steel reaches a value, which can significantly increase the rate of corrosion [4,74]. Some lab data showed that treatment of concrete surface with silane is also an effective method to decrease surface attack by sulfate [75,76]. Impregnation of the surface of concrete showing sign of alkali-silica-reaction (ASR) with silane was shown also to stop concrete expansion [77] thanks to a reduction of water ingress.

The reduction of water penetration in construction materials can have other positive impacts. Excessive penetration of water in walls negatively impacts the energy consumption of buildings (see Figure 10.17). Absorption of water by construction materials increases their thermal conductivity but also leads to cooling due to the evaporation of water. A model house was built to assess the saving of heating energy that can be achieved by treating the exterior clay-brick walls with an emulsion of silane [78]. The treatment of walls with a water repellent minimizes the increase of thermal conductivity and the cooling due to the water evaporation [79].

Colonization of walls by microorganisms can lead to aesthetic alteration of façade. Microalgae, fungi, lichens, and mosses can be found on façade surfaces in advanced cases of colonization. Experimental data have shown that impregnation of mortar with water repellent formulation that contains a silane could notably slow down the

FIGURE 10.17   Aerated light concrete house (10 × 5 × 10 cm) observed with an infrared camera. The house has been half treated with a diluted silane/siloxane-based water repellent emulsion, cured for 2 days, then soaked for 2 hours followed by air dry for 30 minutes at room temperature.

progression of colonization. Surface treatment with water repellents based on silane helps to keep the surface dryer for longer period of time, reducing the availability of water, which is a major parameter of algal growth [81].

10.4.2   IWR

It is only recently that silane, siloxane, or silicone resin started to be used as "IWR" in mortar and concrete, although this potential application was described much earlier [82,83]. Experimental results have shown that only deep silane impregnation of concrete surface can reliably delay chloride penetration. Addition of silane-based IWR into the fresh mortar/concrete mix would be another way to protect the construction material throughout the volume instead of protecting the surface only.

Cement-based mortar or concrete slurries are water-based mixtures. As silane or siloxane has a high interfacial tension with water, they are not easily mixed or homogeneously dispersed into mortar or concrete mixtures. In order to be easily mixed within mortar or concrete, interfacial tension between individual silane/siloxane droplets and the cement-based slurry needs to be minimized. This can be achieved by formulating silanes or siloxanes as oil-in-water emulsion [40] or powder [47].

OPC, which is the cement mainly used in mortar or concrete, is a hydraulic binder. This means it undergoes hydration reactions when mixed with water. Upon time, reaction products of the hydration reaction are creating a matrix that leads, over time, to the setting and build up of a sturdy matrix [84]. It may therefore be perceived as counter intuitive to use a water repellent in mortar or concrete mix as they might impact the hydration reaction of OPC. Emulsion or powder of silane was however successfully used as IWR in cement-based formulation for the last 10–15 years.

As most of the different type of admixtures used in mortar or concrete [85], emulsion or powder of silane used as IWR has some impact on the behavior of the fresh mortar/concrete mix or properties of the final mortar/concrete [86]. Addition of emulsion or powder of silane into mortar or concrete mixes was shown to

- Strongly reduce capillary water absorption of set mortar/concrete [6] even after accelerated aging [87,88]
- Increase slump [89]
- Slightly delay evolution of heat generated by cement hydration, leading to increased setting time [90,91]
- Reduce slightly compressive strength, especially when the emulsion or powder is overdosed [6]

The impact of emulsion or powder of silane in mortar/concrete depends on their addition level in the mix. When overdosed, emulsion of silane can decrease mechanical properties. The right compromise between the desired reduction of water penetration and the overall concrete/mortar properties has to be determined.

Addition of IWR based on silane in steel-reinforced concrete was studied [92]. The reduction of water penetration in a concrete matrix modified with an IWR decreases the ingress of chloride [6,93]. Addition of IWR can improve the resistance to the corrosion of galvanized steel reinforcement in concrete specimens (even with crack or with high water/cement ratio) exposed to wet-dry cycles in an aqueous chloride solution [94,95]. The reduction of capillary absorption can lead to a situation where the pores are not saturated with water, even under immersion. This can increase the diffusion of oxygen [96] and increase the rate of corrosion of steel embedded into concrete specimens with cracks [97].

Addition of IWR based on silane in concrete can create an effective barrier to capillary suction of water, even if the concrete specimens have been strained until multiple cracks are formed [98]. Secondary efflorescence can help to visually assess the impact of IWR on the absorption and migration of water into a mortar/concrete specimen. If a mortar block is placed in contact of a concentrated salt solution, capillary absorption of the salt solution takes place. After migration of the salt solution through the pores of the cement matrix and its evaporation at the surface of the mortar specimen, crystallization of the salt occurs. Figure 10.18 shows secondary efflorescence at the surface of a reference mortar block placed into contact with a concentrated salt solution. When the mortar block is modified with a powdered, silane-based IWR, secondary efflorescence is strongly minimized, despite the fact that this specific block was strained till cracks were formed [99].

Emulsions of silane were used as IWR in different types of concretes. Mechanical properties and durability of concrete made with recycled aggregates are lower than concrete made with "natural aggregate" concrete (when the same mix proportion is used) probably due to the higher porosity of concrete made with recycled concrete. Addition of emulsion of silane as IWR was shown to improve durability of concrete made with recycled aggregates [89]. An emulsion of silane

FIGURE 10.18 Illustration of salt transport through EN 196-1 mortar with no additive (top image) or with 0.25% silicone hydrophobic powder vs. dry mix weight (cement + sand) (bottom image). Insert: schematic representation of the experimental set up. (From Lecomte, J.P., Personal communication, 2015.)

was also tested as IWR for pervious concrete. Significant reduction of capillary water absorption was also observed [100].

Addition of silane emulsion into cement-based mixtures (mortar/concrete) has a tendency to decrease mechanical properties of the final material, especially when overdosed. A cocktail of different admixtures was used to obtain the same final strength as the reference, self-compacting concrete mixture [101].

Effectiveness and durability of the hydrophobic treatment provided by powdered silanes and metallic soaps were compared in mortars. Powders based on silane are as effective as conventional metal soaps to effectively reduce water penetration within modified mortars [102,103]. Durability of the protection against water ingress after different accelerated aging was assessed in mortar. Addition of powders of silane was shown to provide longer protection against water ingress than addition of metal soaps.

Emulsion or powder of silane was used as IWR in limestone cement or natural hydraulic lime pozzolana-lime mortars [104–106]. Addition of silane/siloxane provides good water-repellent effectiveness without strongly influencing the setting and hydration of the binder. Integral admixtures based on silane were also successfully used into rammed earth building [107].

10.4.3 INFLUENCE OF IWR ON CEMENT HYDRATION PROCESSES

Influence of silane-based IWR on the cement hydration processes and the final cement-matrix microstructure was studied. The main hydration reaction were shown to take place as "normal" in the presence of emulsion or powder of silane [87,91]. Evolution of heat is however slightly delayed and a slight decrease of mechanical properties is observed [90,104]. From the experimental data, the following model trying to explain the interactions/reactions of emulsions or powders of silane with cement has been proposed [90]:

- Droplets of silane (provided by powder or emulsion of silane) break or coalesce in the high pH environment of a freshly mixed mortar or concrete slurry.
- Silanes react with silanol-rich "phase" (C3S, CSH) and partially treat their surface.
- The reaction of silane at the surface of nonhydrated cement particles most presumably delay the hydration reactions as suggested by the delay of heat evolution in the presence of silane emulsion.
- Even if delayed, the main hydration reactions are taking place, leading to the formation of a cement matrix with almost unmodified porosity. The small reduction of cement hydration leads to some reduction of the final mechanical properties.

10.5 METHODS USED TO ASSESS SILANE-BASED WATER REPELLENTS

The reduction of water ingress induced by the application of water repellent based on silane or siloxane at the surface of a construction material depends on the type and quantity of water repellent and on the treated substrate. The uptake of the water repellent, its penetration depth and its reactivity will all impact the effectiveness of the reduction of water ingress and the durability of the hydrophobic treatment. The effectiveness of the surface treatment can be assessed by measuring different properties [108]:

- The depth of penetration
- The water transport (or said differently, water absorption)
- The rate of chloride penetration

- The water vapor permeability/diffusion through a treated material
- The rate of carbonation of cement-based material
- The rate of corrosion of steel reinforcement
- The resistance of surface treatments to weathering

These methods will not be described here as they were reviewed previously.

The penetration depth of silane into mortar/concrete depends on different factors such as the porosity and suction capacity of the treated material, type of water repellent (Mw, reactivity), type of carrier (organic solvent, emulsion, microemulsion), and application conditions.

Depth of penetration of silane was shown to be a key parameter for an effective protection against chloride ingress into treated concrete. Depth of penetration can be easily determined by fracturing a piece of treated mortar/concrete apart and then immerging the two parts in water. The layer treated with silane do not absorb water and can be easily distinguished from the untreated bulk, which becomes darker when wet (see Figure 10.19).

Although effective, this is a destructive method that does not provide a lot of information. Different analytical methods were then used to measure the impregnation layer, to characterize the dynamics of water ingress or to determine the chemical species produced upon impregnation.

FIGURE 10.19 Mortar block $5 \times 5 \times 5 cm^3$ impregnated for 30 s in octyltriethoxysilane and left to react for several weeks. Block was then cut and pull apart. This half block was sanded, immersed in water for 10 min and wiped off before picture was taken.

Fourier transform infrared spectroscopy (FTIR) was used to characterize depth of penetration of alkyl trialkoxysilanes into concrete. Characteristics vibration bands of C–H groups were measured as a function of depth of penetration by removing step after step, thin layers of 0.3 mm of treated concrete [109]. Attenuated total reflection (ATR) used in conjunction with FTIR was used to characterize kinetics of the reaction of alkoxysilane on a SiO_2-funtionalized germanium crystal, used as a model surface [110]. This method was complemented by atomic force microscopy (AFM) to understand the formation and stability of different siloxane films formed on functionalized germanium crystals.

Release of ethanol or methanol and the evolution of the surface tension of water in which concrete powder was added were measured in order to characterize the reaction of alkoxysilane in the presence of concrete powder [111,112]. Distribution of the condensation products of silane formed upon percolation of silane through a chromatography column packed with concrete powder was studied by FT-IR spectroscopy [113]. Soxhlet-extractions of concrete treated with silane were carried out to study attachment of silane to the C-S-H gel [114].

Characterization of the reaction products of alkyl trialkoxysilanes applied on concrete was studied by time-of-flight mass spectrometry (TOF/MS) combined with electrospray ionization (ESI) and matrix assisted laser desorption ionization (MALDI) [115,116]. The combined use of ESI-TOF/MS and MALDI-TOF/MS enabled to characterize alkyl trialkoxysilane and their reaction products as intact species formed under alkaline conditions, similar to the condition found in cementitious materials [117].

Neutron radiography imaging was used to visualize and characterize water movement in cement-based materials (concrete or concrete with cracks) [118–121]. It was confirmed that neutron radiography can be used to quantify the rate of diffusion and distribution of moisture with both spatial and time resolution. The method was, for example, used to characterize the dynamics of water ingress in precracked, steel-reinforced concrete, treated or not with silanes [122]. This showed that water ingress through cracks can be strongly minimized if the concrete surface is properly impregnated with silane. Neutron radiography that provides a strong contrast between wet and dry regions of porous materials saturated with water was compared to x-rays tomography, which proved to be a very useful method to examine the internal structure of natural building stone [123].

Single-side nuclear magnetic resonance (NMR) was used as a versatile and nondestructive method for the determination of the presence or movement of liquids in concrete [72,124,125,126]. Single-sided NMR uses a sender/receiver set located at the same side of the sample, making it adapted for the testing of small lab samples (see Figure 10.20) as well as real structures. As the method can only detect liquids, the evaluation of the depth of the hydrophobic layer and of its performance is indirect. The method is used to detect the presence (or absence of water) at certain depths of the sample. The method can be used as well to detect silane ingress into concrete as long as silanes are still liquid [126]. Acoustic microscopy was used as a nondestructive method to measure very thin depth of penetration of water repellent applied on clay-based construction materials [127]. The penetration depth of a water repellent as thin as 20 μm was calculated, based on measurements of the rate of transmission of high frequencies (175 MHz) ultrasonic wave in the reference and the treated clay bricks.

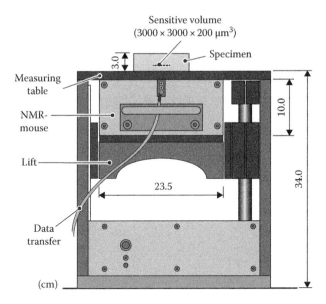

FIGURE 10.20   Schematics of the NMR MOUSE® PM25 used for the nondestructive assessment of moisture profiles in porous building materials. (From Antons, U. et al., *Restoration of Buildings and Monuments*, 20(5), 299, 2014.)

Molecular modeling in combination with experimental data were also used to calculate distribution and oligomerization of alkyl trialkoxysilane in cement matrices treated with silane [128,129]. There is also a need by users of emulsion of silane to effectively measure the silane content into the emulsion. Weighing an aliquot of an emulsion before and after heating, till complete evaporation of the water phase is the method conventionally used to measure solid content of an emulsion. Although the evaporation rate of octyltriethoxysilane (the most commonly used silane for the preparation of emulsion of silane) is low, it can still be stripped off during this heating stage. Evaporation of both the continuous phase and part of the silane leads to wrong evaluation of the silane content. A procedure using a microwave oven as heating method was proven to be adequate to measure "solid content" of emulsion of silane [52].

10.6   PERSPECTIVES

Impregnation with silicon-based water repellents is a proven technology to protect cementitious materials against ingress of water. The pressure of new regulations to limit emission of volatile organic content will drive to the use of impregnation materials that release lower level of volatiles (such as ethanol or methanol). Silicon hydrides are chemical species that could be used to prepare or formulate alternatives to alkoxysilanes [130].

Combination of Si-based water repellents with photocatalytic nanoparticles is being explored as masonry remediation [131]. Applied on a wall, such formulations are able to reduce water penetration of treated wall and initiate photocatalytic degradation of the pollutants in contact with the treated surface.

IWR based on silane or silicone resin are increasingly used in real life. Addition of these IWR into a mortar or concrete matrix may impact the mechanical properties. IWR that would not have impact on the cement hydration processes and on the cement matrix mechanical properties could be used in load-bearing concrete or light/foamed concrete.

ACKNOWLEDGMENTS

We thank Dave Selley for suggesting that we write this review. We also thank David Pierre, Frederick Gubbels, and Yihan Liu for reviewing the chapter and providing fruitful suggestions for improvement.

REFERENCES

1. Van Gemert, D.; Knapen, E. (2010) Contribution of C-PC to sustainable construction procedures, *Proceeding to ICPIC 2010, 13th International Congress on Polymers in Concrete*, Funchal, Portugal, pp. 27–36.
2. Martys, N.; Ferraris, C. (1997) Capillary transport in mortars and concrete, *Cement and Concrete Research*, 27(5): 747–760.
3. Büttner, T.; Raupach, M. (2008) Durability of hydrophobic treatments on concrete, results from laboratory tests, *Proceeding of Hydrophobe V, Fifth International Conference on Water Repellent Treatment of Building Materials*, Aedificatio Publishers: Freiburg, Germany, pp. 329–340.
4. Schueremans, L.; Van Gemert, D.; Friedel, M.; Giessler-Blank, S. (2008) Durability of water repellents in a marine environment, *Proceeding of Hydrophobe V, Fifth International Conference on Water Repellent Treatment of Building Materials*, Aedificatio Publishers: Freiburg, Germany, pp. 357–368.
5. Lecomte, J.-P.; Selley, D.; Mcauliffe, T.; Spaeth, V. (2010) Hydrophobic protection of fibre reinforced cement boards with silicon-based materials, *Proceeding of the International Inorganic-Bonded Fiber Composites Conference (IIBCC)*, Aalborg, Denmark, pp. 246–255.
6. Zhao, T.; Wittmann, F.; Jiang, R.; Li, W. (2011) Application of silane-based compounds for the production of IWR concrete, *Proceeding of Hydrophobe VI, Sixth International Conference on Water Repellent Treatment of Building Materials*, Aedificatio Publishers: Freiburg, Germany, pp. 137–144.
7. Allred, A.L.; Rochow, E.G. (1958) A scale of electronegativity based on electrostatic force, *Journal of Inorganic and Nuclear Chemistry*, 5(4): 264.
8. Holleman, A.F. (1985) Lehrbuch der anorganischen Chemie, Aufl., de Gruyter: Berlin, Germany, pp. 91–100.
9. Pauling, L. (1960) *Nature of the Chemical Bond*, Cornell University Press: Ithaca, NY, pp. 88–107.
10. Grigoras, S.; Lane Thomas, H. (1989) *Silicon-Based Polymer Science*, Vol. 224, American Chemical Society: Washington, DC, pp. 125–144.
11. Corey, J.Y. (1975) Generation of a silicenium ion in solution, *Journal of the American Chemical Society*, 97: 3237.
12. Osterholtz, F.D.; Pohl, E.R. (1992) Kinetics of the hydrolysis and condensation of organofunctional alkoxysilanes: A review, *Journal of Adhesion Science and Technology*, 6: 127–149.

13. Schmidt, H.; Scholze, H.Z.; Kaiser, A. (1984) Principles of hydrolysis and condensation reaction of alkoxysilanes, *Journal of Non-Crystalline Solids*, 63: 1–11.
14. Smith, K.A. (1986) A study of the hydrolysis of methoxysilanes in a two-phase system, *The Journal of Organic Chemistry*, 51: 3827–3830.
15. Arkles, B.; Steinmetz, J.R.; Zazyczny, J.; Metha, P. (1992) Factors contributing to the stability of alkoxysilanes in aqueous solution, *Journal of Adhesion Science and Technology*, 6: 193–206.
16. McNeil, K.J.; DiCapri, J.A.; Walsh, D.A.; Pratt, R.F. (1980) Kinetics and mechanism of hydrolysis of a silicate triester, tris (2-methoxyethoxy) phenylsilane, *Journal of the American Chemical Society*, 102: 1859–1865.
17. Taft, R.W. Jr. (1956) *Steric Effects in Organic Chemistry*, M.S. Newman (Ed.), John Wiley: New York.
18. Sprung, M.M.; Guenther, F.O. (1955) The partial hydrolysis of methyltriethoxysilane, *Journal of the American Chemical Society*, 77: 3990–3996.
19. Sprung, M.M.; Guenther, F.O. (1958) The hydrolysis of n-amyltriethoxysilane and phenyltriethoxysilane, *Journal of Polymer Science*, 28: 17–33.
20. Brinker, C.J. (1988) Hydrolysis and condensation of silicates: Effects on structure, *Journal of Non-Crystalline Solids*, 100: 31–50.
21. Adam, W.; Mello, R.; Curci, R. (1990) O-atom insertion into Si-H bonds by dioxiranes: A stereospecific and direct conversion of silanes into silanol, *Angewandte Chemie International Edition in English*, 29: 890–891.
22. Adam, W.; Mitchell, C.M.; Saha-Möller, C.R.; Weichold, O. (1999) Host–Guest Chemistry in a Urea Matrix: Catalytic and Selective Oxidation of Triorganosilanes to the Corresponding Silanols by Methyltrioxorhenium and the Urea/Hydrogen Peroxide Adduct, *Journal of the American Chemical Society*, 121: 2097–2103.
23. Grubb, W.T. (1954) A rate study of the silanol condensation reaction at 25°C in alcoholic solvents, *Journal of the American Chemical Society*, 76: 3408–3414.
24. Kay, B.D.; Assink, R.A. (1988) Sol-gel kinetics: II. Chemical speciation modeling, *Journal of Non-Crystalline Solids*, 104: 112–122.
25. Iler, R.K. (1979) *The Chemistry of Silica*, Wiley: New York.
26. Brinker, C.J.; Keefer, K.D.; Schaefer, D.W.; Assink, R.A.; Kay, B.D.; Ashley, C.S. (1984) Sol-gel transition in simple silicates II, *Journal of Non-Crystalline Solids*, 63: 45–59.
27. Brzoska, J.B.; Ben Anouz, I.; Rondelez, F. (1992) Evidence of a transition temperature for the optimum deposition of grafted monolayer coatings, *Nature*, 360: 719–721.
28. Brzoska, J.B.; Ben Anouz, I.; Rondelez, F. (1994) Silanization of solid substrates: A step toward reproducibility, *Langmuir*, 10: 4367–4373.
29. Imura, K.-I.; Nakajima, Y.; Kato, T. (2000) A study on structures and formation mechanisms of self-assembled monolayers of n-alkyltrichlorosilanes using infrared spectroscopy and atomic force microscopy, *Thin Solid Films*, 379: 230–239.
30. Angst, D.L.; Simmons, G.W. (1991) Moisture absorption characteristics of organosiloxane self-assembled monolayers, *Langmuir*, 7: 2236–2242.
31. McGovern, M.E.; Kallury, K.M.R.; Thomson M. (1994) Role of solvent on the silanization of glass with octadecyltrichlorosilane, *Langmuir*, 10: 3607–3614.
32. Silberzan, P.; Lèger, L.; Aussere, D.; Benattar, J.J. (1991) Silanation of silica surfaces. A new method of constructing pure or mixed monolayers, *Langmuir*, 7: 1647–1651.
33. Antons, U.; Weichold, O. Assessing the performance of hydrophobing agents on concrete using non-destructive single-sided nuclear magnetic resonance, *Journal of Infrastructure Systems*, submitted.
34. Cuthbert, R.; Plueddemann, E. (1991) Masonry water repellent composition, U.S. patent 5,051,129.
35. DePasquale, R.; Wilson, M. (1991) Aqueous systems containing silanes for rendering masonry surface water repellent, U.S. patent 4,648,904.

36. Wilson, M. (1991) Buffered silane emulsions having low volatile organic compounds when cured, European patent EP 0 412 515.

37. Göbel, T.; Michel, R.; Alff, H.; Karl, J. (1992) Wässrige, organosiliciumverbindungen enthaltende Emulsionen zum Imprägnieren von anorganischen Materiallen, insbesondere von Baumateriallen, European patent EP 0 538 555.

38. Kebao, R.; Kagi, D. (1995) Aqueous compositions for the water repellent treatment of masonry, Patent WO 1995/22580.

39. Fisher, P.; Gearhart, R. (2000) Stable, constant particle size, aqueous emulsions of nonpolar silanes suitable for use in water repellence applications, Patent WO 00 34206.

40. Gierke, M.; Vidal, C.; Wilson, M. (1993) Organosilicon emulsions for rendering porous substrates water repellent, European Patent EP 0606 671.

41. Folke, E.; Karlson, J. (1995) Method for the protective treatment of mineral material structures, treatment composition intended for performing of the method and use thereof, Patent WO 1995/025706.

42. Mayer, H.; König-Lumer, I; Hausberger, A. (1998) Wässrige Cremes von Organosiliciumverbindungen für die Hydrophobierung von Baustoffen, European patent EP 0 819 665.

43. Janning, F. (2001) Verfahren und system zur Herstellung einer Fassadfencreme, European patent EP 1070689.

44. Campeol, F.; Galeone, F.; Lecomte, J.-P.; Marteaux, L.; Sarrazin, M.-J.; Zimmerman, B. (2013) Water repellent organosilicon materials, Patent WO 2013/166280.

45. Leopolder, F. (September 2011) The global drymix mortar industry: Development, trends, opportunities and risks, *The Indian Concrete Journal*, 2011: 7–16.

46. Li, W.; Wittmann, F.H.; Jiang, R.; Zhao, T.; Wolfseher R. (2011) Metal soaps for the production of IWR concrete, *Hydrophobe VI, Sixth International Conference on Water Repellent Treatment of Building Materials*, Aedificatio Publishers: Freiburg, Germany, pp. 145–154.

47. Butler, D.; Thomas, D. (1997) Cementitious materials, European patent EP 0811584.

48. Roos, M.; König, F.; Stadmüller, S.; Weyershausen, B. (2008) Evolution of silicone based water repellents for modern building protection, *Proceedings of Hydrophobe V, Fifth International Conference on Water Repellent Treatment of Building Materials*, Aedificatio Publishers: Freiburg, Germany, pp. 3–16.

49. Mayer, H.; König-Lumer, I.; Hausberger, A.; Hager, R. (2001) Building composition which comprise hydrophobicizing powders comprising organosilicon compounds, US patent 6,267,423.

50. Koelliker, R. (2004) Redispersible material, process for producing and using the same, and aqueous system containing this redispersible material, US patent US patent 6,812,293.

51. Bastelberger, T.; Härzchel, R.; Jodbauer, F. (2011) Hydrophobic additive, US patent 7,972,424.

52. Aberle, T.; Pustovgar, A.; Vallée, F.; Emmenegger, P.; Schaub, D. (2010) Powder to hydrophobise and its use, Patent WO 2010/052201.

53. Lecomte, J.-P.; Stammer, A.; Campeol, F.; Thibaut, M. (2008) Cementitious material, Patent WO 2008/062018.

54. Lecomte, J.-P.; Thibaut, M.; Stammer, A. (2007) Encapsulated silicone/silane mix enhances water resistance of mortars, *European Coating Journal*, 12: 46–51.

55. Neville, A. (1995) Chloride attack of reinforced concrete: An overview, *Materials and Structures*, 28: 63–70.

56. Meijers, S.; Bijen, J.; de Borst, R.; Fraaij, A. (2005) Computational results of a model for chloride ingress in concrete including convection, drying-wetting cycles and carbonatation, *Materials and Structures*, 38: 145–154.

57. Gerdes, A.; Meier, S.; Wittmann, F. (1998) A new application technology for water repellent surface treatment, *Proceeding of Hydrophobe II, Second International Conference on Water Repellent Treatment of Building Materials*, Aedificatio Publishers: Freiburg, Germany, pp. 217–230.

58. Gerdes, A.; Wittmann, F. (2001) Requirements for the application of water repellent treatments in practice, *Proceeding of Hydrophobe III, Third International Conference on Water Repellent Treatment of Building Materials*, Aedificatio Publishers: Freiburg, Germany, pp. 155–168.

59. Gerdes, A.; Wittmann, F. (2001) Decisive factors for the penetration of silicon-organic compounds into surface near zones of concrete, *Proceedings of Hydrophobe III, Third International Conference on Water Repellent Treatment of Building Materials*, Aedificatio Publishers: Freiburg, Germany, pp. 111–122.

60. Wittmann, M.; Huang, Z.; Gerdes, A. (2005) Application of water repellent treatments for the protection of "offshore" constructions, *Proceedings of Hydrophobe IV, Fourth International Conference on Water Repellent Treatment of Building Materials*, Aedificatio Publishers: Freiburg, Germany, pp. 145–158.

61. Wittmann, F.; Zhang, P.; Zhao, T. (2014) Characteristics of the surface neat water repellent layer, *Proceedings of Hydrophobe VII, Seventh International Conference on Water Repellent Treatment of Building Materials*, Aedificatio Publishers: Freiburg, Germany, pp. 11–18.

62. Donadio, M.; Marazzani, B. (2014) Comparison of three methods to determine the active content of aqueous silane or siloxane based hydrophobic impregnation agents, *Proceedings of Hydrophobe VII, Seventh International Conference on Water Repellent Treatment of Building Materials*, Aedificatio Publishers: Freiburg, Germany, pp. 105–112.

63. McCarthy, M.; Giannakou, A.; Jones, M. (2004) Comparative performances of chloride attenuating and corrosion inhibiting systems for reinforced concrete, *Materials and Structures*, 37: 671–679.

64. Medeiros, M.; Helene, P. (2008) Efficacy of surface hydrophobic agents in reducing water and chloride ion penetration in concrete, *Materials and Structures*, 41: 59–71.

65. Medeiros, M.; Helene, P. (2009) Surface treatment of reinforced concrete in marine environment: Influence on chloride diffusion coefficient and capillary water absorption, *Construction and Building Materials*, 23: 1476–1484.

66. Wittmann, F. (2007) Effective chloride barrier for reinforced concrete structures in order to extend the service-life, *Advances in Construction Materials*, C. Grosse (Ed.), Springer: Berlin, Germany, pp. 427–437.

67. Wittmann, F.; Guo, P.; Zhao, T. (2008) Influence of cracks on the efficiency of surface impregnation of concrete, *Proceedings of Hydrophobe V, Fifth International Conference on Water Repellent Treatment of Building Materials*, Aedificatio Publishers: Freiburg, Germany, pp. 287–298.

68. Dai, J.-G.; Akira, Y.; Wittmann, F.; Yokota, H.; Zhang, P. (2010) Water repellent surface impregnation for extension of service life of reinforced concrete structures in marine environments: The role of cracks, *Cement and Concrete Composites*, 32: 101–109.

69. Zhao, T.; Zhu, G.; Wittmann, F.; Li, W. (2008) On surface impregnation of chloride contaminated cement based materials, *Proceedings of Hydrophobe V, Fifth International Conference on Water Repellent Treatment of Building Materials*, Aedificatio Publishers: Freiburg, Germany, pp. 311–326.

70. Courard, L.; Lucquiaud, V.; Gérard, O.; Handy, M.; Michel, F.; Aggoun, S.; Cousture, A. (2014) Evaluation of the durability of hydrophobic treatments on concrete architectural heritage, *Proceedings of Hydrophobe VII, Seventh International Conference on Water Repellent Treatment of Building Materials*, Aedificatio Publishers: Freiburg, Germany, pp. 29–38.

71. Christodoulou, C.; Goodier, C.; Austin, S.; Webb, S.; Webb, J.; Glass, G. (2013) Long term performance of surface impregnation of reinforced concrete structures with silane, *Construction and Building Materials*, 48: 708–716.

72. Antons, U.; Weichold, O.; Raupach, M. (2013) Measuring the effectiveness of hydrophobic layers using a non-destructive method, *Advanced Materials Research*, 687: 298–302.

73. Antons, U.; Raupach, M.; Weichold, O. (2014) Influences on the hydrophobicity of concrete surfaces treated with alkyl trialkoxysilanes, *Restoration of Buildings and Monuments*, 20(6): 405–412.

74. Liang, M.; Wang, K.; Liang, C. (1999) Service life prediction of reinforced concrete structures, *Cement and Concrete Research*, 29: 1411–1418.

75. Suleiman, A.; Soliman, A.; Nehdi, M. (2014) Effect of surface treatment on durability of concrete exposed to physical sulfate attack, *Construction and Building Materials*, 73: 674–681.

76. Donadio, M.; Schuerch, H.; Marazanni, B. (2014) Concrete durability improvement in presence of sulphates using a silane based hydrophobic impregnating agents, *Proceedings of Hydrophobe VII, Seventh International Conference on Water Repellent Treatment of Building Materials*, Aedificatio Publishers: Freiburg, Germany, pp. 95–104.

77. Bérubé, M.-A.; Chouinard, D.; Pigeon, M.; Frenette, J.; Rivest, M.; Vézina, D. (2002) Effectiveness of sealers in counteracting alkali-silica reaction in highway median barriers exposed to wetting and drying, freezing and thawing, and deicing salt, *Canadian Journal of Civil Engineering*, 29: 329–337.

78. McMullen, J.; Zhang, Z.; Rirsch, E.; Dhakal, H.; Bennet, N. (2011) Brick and mortar treatment by cream emulsion for improved water repellence and thermal insulation, *Energy and Buildings*, 43: 1560–1565.

79. McMullen, J.; Zhang, Z.; Radulovic, J.; Herodotou, C.; Totomis, M.; Dhakal, H.; Bennet, N. (2012) Titanium dioxide and zinc oxide nano-particulate enhanced oil-in-water (O/W) façade emulsions for improved masonry thermal insulation and protection, *Energy and Buildings*, 52: 86–92.

80. Lecomte, J.P. (2015) Personal communication.

81. Martinez, T.; Bertron, A.; Escadeillas, G.; Ringot, E. (2014) Algal growth inhibition on cement mortar: Efficiency of water repellent and photocatalytic treatments under UV/VIS illumination, *International Biodeterioration and Biodegradation*, 89: 115–125.

82. Carlson, B.; Hartlein, R. (1965) Use of silane as concrete additives, US patent 3,190,762.

83. Pühringer, J. (1981) Process for the preparation of a mortar and the product thereby obtained, Patent WO 81/01703.

84. Hewlett, P. (Ed.) (1998) *Lea's Chemistry of Cement and Concrete*, Elsevier Butterworth Heinemann: Oxford, U.K.

85. Cheung, J.; Jeknavorian, A., Roberts, L.; Silva, L. (2011) Impact of admixtures on the hydration kinetics of Portland cement, *Cement and Concrete Research*, 41: 1289–1309.

86. Tittarelli, F.; Moriconi, G. (2009) Effectiveness of surface or bulk hydrophobic treatments in cementitious materials, *Protection of Historical Buildings, PROHITECH 09*, Mazzolani, F.M. (Ed.), Taylor & Francis Group: London, U.K., pp. 1071–1075.

87. Spaeth, V.; Lecomte, J.-P.; Delplancke, M.-P.; Orlowsky, J.; Büttner, T. (2013) Impact of silane and siloxane based hydrophobic powder on cement based mortar, *Advanced Materials Research*, 687: 100–106.

88. Spaeth, V.; Lecomte, J.-P.; Delplancke-Ogletree, M.-P. (2014) IWR based materials: Impact of aging on cement microstructure and performance, *Proceedings of Hydrophobe VII, Seventh International Conference on Water Repellent Treatment of Building Materials*, Aedificatio Publishers: Freiburg, Germany, pp. 57–66.

89. Zhu, Y.-G.; Kou, S.-C.; Poon, C.-S.; Dai, J.-G..; Li, Q.-Y. (2013) Influence of silane-based water repellent on the durability properties of recycled aggregate concrete, *Cement and Concrete Composites*, 35: 32–38.
90. Milenkovic, N.; Staquet, S.; Lecomte, J.-P.; Pierre, C.; Delplancke M.-P. (2014) Non ionic silane emulsion as IWR—Impact on cement hydration process, *Proceedings of Hydrophobe VII, Seventh International Conference on Water Repellent Treatment of Building Materials*, Aedificatio Publishers: Freiburg, Germany, pp. 47–56.
91. Spaeth, V.; Delplacke-Ogletree, M.-P.; Lecomte, J.-P. (2010) Development of cement materials by incorporation of WR powder additives, hydration process, microstructure development and durability, Restoration of buildings and monuments, *Bauinstandsetzen und Baudenkmalpflege*, 16(4/5): 1–10.
92. Tittarelli, F.; Moriconi, G. (2006) Efficiency of traditional and innovative protection methods against corrosion, *Measuring, Monitoring and Modeling Concrete Properties*, M.S. Konsta-Gdouts (Ed.), Springer: Dordrecht, the Netherlands, pp. 545–555.
93. Zhang, P.; Wittmann, F.; Villmann, B.; Zhao, T.; Slowik, V. (2008) Moisture diffusion in and capillary suction of IWR cement-based materials, *Proceedings of Hydrophobe V, Fifth International Conference on Water Repellent Treatment of Building Materials*, Aedificatio Publishers: Freiburg, Germany, 273–285.
94. Tittarelli, F.; Moriconi, G. (2010) The effect of silane-based hydrophobic admixture on corrosion of galvanized reinforcing steel in concrete, *Corrosion Science*, 52: 2958–2963.
95. Tittarelli, F.; Moriconi, G. (2011) Comparison between surface and bulk hydrophobic treatment against corrosion of galvanized reinforcing steel in concrete, *Cement and Concrete Research*, 41: 606–614.
96. Tittarelli, F. (2009) Oxygen diffusion through hydrophobic cement-based materials, *Cement and Concrete Research*, 39: 924–928.
97. Tittarelli, F.; Moriconi, G. (2008) The effect of silane-based hydrophobic admixture on corrosion of reinforcing steel in concrete, *Cement and Concrete Research*, 38: 1354–1357.
98. Sahmaran, M.; Li, V. (2009) Influence of microcracking on water absorption and sorptivity of ECC, *Materials and Structures/Materiaux et Constructions*, 42(5): 593–560.
99. Lecomte, J.-P.; Llado, D.; Salavati, S.; Rodrigues, G.; Riberiro, M. (2013) Mortar protection. New silicone resin-based hydrophobic powder for the dry mix market, *European Coating Journal*, 12: 88–91.
100. Tittarelli, F.; Carsana, M.; Ruello, M. (2014) Effect of hydrophobic admixture and recycled aggregate on physical-mechanical properties and durability aspects of no-fines concrete, *Construction and Building Materials*, 66: 30–37.
101. Corinaldesi, V. (2012) Combined effect of expansive, shrinkage reducing and hydrophobic admixtures for durable self compacting concrete, *Construction and Building Materials*, 36: 758–776.
102. Lanzon, M.; Garcia-Ruiz, P. (2008) Effectiveness and durability evaluation of rendering mortars made with metallic soaps and powdered silicone, *Construction and Building Materials*, 22: 2308–2315.
103. Lanzon, M.; Garcia-Ruiz, P. (2009) Evaluation of capillary water absorption in rendering mortars made with powdered waterproofing additives, *Construction and Building Materials*, 23: 3287–3291.
104. Falchi, L.; Zendri, E.; Müller, U.; Fontana, P. (2015) The influence of water-repellent admixtures on the behaviour and the effectiveness of Portland limestone cement mortars, *Cement and Concrete Composites*, 59: 107–118.
105. Falchi, L.; Müller, U.; Fontana, P.; Izzo, F.; Zendri, E. (2013) Influence and effectiveness of water-repellent admixtures on pozzolana-lime mortars for restoration application, *Construction and Building Materials*, 49: 272–280.

106. Falchi, L.; Varin, C.; Toscano, G.; Zendri, E. (2015) Statistical analysis of the physical properties and durability of water-repellent mortars made with limestone cement, natural hydraulic lime and pozzolana-lime, *Construction and Building Materials*, 78: 260–270.

107. Kebao, R.; Kai, D. (2012) Integral admixtures and surface treatment for modern earth buildings, *Modern Earth Buildings in Materials, Engineering, Constructions and Applications*, Woodhead Publishing Series in Energy, Woodhead Publishing Limited: Cambridge, U.K., pp. 256–280.

108. Basheer, P.; Basheer, L.; Cleland, D.; Long, A. (1997) Surface treatments for concrete: Assessment methods and reported performance, *Construction and Building Materials*, 11(7–8): 413–429.

109. Gerdes, A. (1995) Assessment of water repellent treatments by the application of FT-IR spectroscopy, *Proceedings of Hydrophobe I, First International Conference on Water Repellent Treatment of Building Materials*, Aedificatio Publishers: Freiburg, Germany, pp. 12.1–12.9.

110. Glowacky, J.; Heissler, S.; Boese, M.; Leiste, H.; Koker, T.; Faubel, W.; Gerdes, A.; Müller, H. (2008) Investigation of siloxane film formation on funtionalized germanium crystals by atomic force microscopy and FTIR-ATR spectroscopy, *Proceedings of Hydrophobe V, Fifth International Conference on Water Repellent Treatment of Building Materials*, Aedificatio Publishers: Freiburg, Germany, pp. 219–232.

111. Oehmichen, D.; Gerdes, A.; Wefer-Roehl, A. (2008) Reactive transport of silanes in cement based materials, *Proceedings of Hydrophobe V, Fifth International Conference on Water Repellent Treatment of Building Materials*, Aedificatio Publishers: Freiburg, Germany, pp. 205–218.

112. Oehmichen, D.S.; Gerdes, A.; Wefer-Roehl, A. (2007) *GDCh-Monographie*, 37: 165–174.

113. Gerdes, A.; Oehmichen, D.; Preindl, B.; Nüesch, R. (2005) Chemical reactivity of silanes in cement-based materials, *Proceedings of Hydrophobe IV, Fourth International Conference on Water Repellent Treatment of Building Materials*, Aedificatio Publishers: Freiburg, Germany, pp. 47–58.

114. Glowacky, J.; Gerdes, A.; Rüesch, R. (2005) Bonding of silanes on CSH-gel, *Proceedings of Hydrophobe IV, Fourth International Conference on Water Repellent Treatment of Building Materials*, Aedificatio Publishers: Freiburg, Germany, pp. 39–78.

115. Herb, H.; Gerdes, A. (2008) TOF/MS for characterization of silicone based water repellents, *Proceedings of Hydrophobe V, Fifth International Conference on Water Repellent Treatment of Building Materials*, Aedificatio Publishers: Freiburg, Germany, pp. 197–204.

116. Herb, H.; Brenner-Weiss, G.; Gerdes, A. (2011) Characterization of degradation of silicon-based water repellents by Maldi-TOF/MS, *Proceedings of Hydrophobe VI, Sixth International Conference on Water Repellent Treatment of Building Materials*, Aedificatio Publishers: Freiburg, Germany, pp. 61–68.

117. Herb, H.; Gerdes, A.; Brenner-Weiss, G. (2015) Characterization of silane based hydrophobic admixtures in concrete using TOF-MS, *Cement and Concrete Research*, 70: 77–82.

118. Kanematsu, M.; Maruyama, I.; Noguchi, T.; Iikura, H.; Tsuchiya, N. (2009) Quantification of water penetration into concrete through cracks by neutron radiography, *Nuclear Instruments and Methods in Physics Research A*, 605: 154–158.

119. Zhang, P.; Wittmann, F.; Zhao, T.; Lehmann, E.; Tian, L.; Vontobel, P. (2010) Observation and quantification of water penetration into strain hardening cement-based composites (SHCC) with multiple cracks by means of neutron radiography, *Nuclear Instruments and Methods in Physics Research A*, 650: 414–420.

120. Zhang, P.; Wittmann, F.; Zhao, T.; Lehmann, E.; Jin, Z. (May 2010) Visualisation and quantification of water movement in porous cement-based materials by real time thermal neutron radiography: Theoretical analysis and experimental study, *Science China Technological Sciences*, 53: 1198–1207.

121. Zhang, P.; Wittmann, F.; Zhao, T.; Lehmann, E. (2010) Neutron imaging of water penetration into cracked steel reinforced concrete, *Physica B*, 405: 1866–1871.

122. Zhang, P.; Wittmann, F.; Zhao, T-J.; Lehmann, E.; Vontrobel, P. (2011) Neutron radiography, a powerful method to determine time-dependent moisture distributions in concrete, *Nuclear Engineering and Design*, 241: 4758–4766.

123. Cnudde, V.; Masschaele, B.; Vlassenbroeck, J.; Dierick, M.; De Witte, Y.; Lehmann, E.; Van Hoorebeke, L.; Jacobs, P. (2008) X-rays and neutrons used for the visualization of oligomeric siloxanes, *Proceedings of Hydrophobe V, Fifth International Conference on Water Repellent Treatment of Building Materials*, Aedificatio Publishers: Freiburg, Germany, pp. 31–42.

124. Raubach, M.; Büttner, T. (2009) Hydrophobic treatments on concrete—Evaluation of the durability and non-destructive testing, *Concrete Repair, Rehabilitation and Retrofitting II*, Alexander, M.G. et al. (Eds.), Taylor & Francis Group: London, U.K.

125. Antons, U.; Orlowsky, J.; Raupach, M. (2012) A non-destructive test method for the performance of hydrophobic treatments, *Proceedings of the Third International Conference on Concrete Repair Rehabilitation and Retrofitting (ICCRRR)*, Cape Town, South Africa, September 3–5, 2012, CRC Press/Taylor & Francis Group: London, U.K.

126. Antons, U.; Raupach, M.; Weichold, O. (2014) Einsatzmöglichkeiten der NMR-MOUSE in der Bauforschung-aktueller Stand der Arbeiten am ibac, *Restoration of Buildings and Monuments*, 20(5): 299–310.

127. Matziaris, K.; Stefanidou, M.; Karagiannis, G. (2011) Impregnation and superhydrophobicity of coated porous low-fired clay building materials, *Progress in Organic Coatings*, 72: 181–192.

128. Suessmuth, J.; Gerdes, A. (2008) Computional chemistry to investigate the chemical behavior of silanes and CSH-gel, *Proceedings of Hydrophobe V, Fifth International Conference on Water Repellent Treatment of Building Materials*, Aedificatio Publishers: Freiburg, Germany, pp. 233–244.

129. Suessmuth, J.; Weidler, P.; Gerdes, A. (2011) Combination of quantum-mechanical and experimental investigations for the development of a model of the size and distribution of silane oligomers on the pore surface of mineral based material, *Proceedings of Hydrophobe VI, Sixth International Conference on Water Repellent Treatment of Building Materials*, Aedificatio Publishers: Freiburg, Germany, pp. 69–80.

130. Stammer, A. (2014) Silicon hydrides as water repellents, a lower VOC alternative to alkoxysilanes and alkoxyfunctional polysiloxane, *Proceedings of Hydrophobe VII, Seventh International Conference on Water Repellent Treatment of Building Materials*, Aedificatio Publishers: Freiburg, Germany, pp. 19–28.

131. MacMullen, J.; Radulovic, J.; Zhang, Z.; Dhakal, HN.; Daniels, L.; Elford, J.; Leost, M-A.; Bennett, N. (2013) Masonry remediation and protection by aqueous silane/silopxane macroemulsions incorporating colloidal titanium dioxide and zinc oxide nanoparticulates: Mechanisms, performance and benefits, *Construction and Building Materials*, 49: 93–100.

11 Application of Silicones in the Oil and Gas Industry

Randal M. Hill, Siwar Trabelsi, and Gianna Pietrangeli

CONTENTS

11.1 INTRODUCTION: A BRIEF REVIEW OF SILICONE MATERIALS

Silicones are materials built from organosiloxane subunits. The most common example is polydimethylsiloxane (PDMS), also called silicone oil or silicone fluid. PDMS varies from 0.65 cSt silicone fluid to silicone gums. Other examples are copolymers of PDMS with hydrocarbon or fluorocarbon blocks. An important class of copolymers where A and B are two different polymer blocks of polyalkylene oxide with PDMS [1]. Another important category is silicone resins, cage-like molecules such as the polyoligosilsesquioxanes [2]. The final category is silanes (small-molecules that incorporate one or more silicon atoms), which are reactive toward metal-oxide surfaces [3,4]. The particular molecular structures associated with these and other details regarding their synthesis and properties can be found in many good references including Brook [5] and the other chapters in this book. All of these different types of silicones have found application in the oil and gas industry.

The different types of silicones have different properties, but most are useful because they have useful surface properties—low surface energy, hydrophobicity, or high surface activity. Only silicone and fluorocarbon materials have useful surface activity toward the hydrocarbon–air interface. Their surface-active properties provide the motivation for most of their uses in the oil and gas industry. A most useful discussion of the key properties of silicones, especially their surface-active properties is provided by Owen [6]. The unique properties of silicones are potentially of great value in oil and gas applications, including their low surface energy and their remarkable ability to promote wetting and spreading [6,7]. The challenge to making use of them is often delivering a finely divided form of the material to the desired interface—they must usually be made into a dispersion of some sort to be used effectively. Hence the emphasis of this book is on silicone dispersions.

The parts of the oil and gas industry covered by this review are exploration, well completion, and production. Aspects of petroleum refining and fuels that make use of silicones will not be addressed. Well operations begin with drilling a well into a subterranean formation [8]. The well may be vertical or include long horizontal runs. When drilling and cementing are completed, most wells these days in North America are hydraulically fractured to increase hydrocarbon production. The fracturing operation consists of pumping a mixture of water, polymer, and sand into the well under very high pressure to create fractures within the formation. There are many aspects of these operations that require addition of chemicals including surfactants, biocides, polymers, corrosion inhibitors, scale, and clay-control agents. And there are many opportunities for silicones to be used advantageously throughout these operations.

The use of silicone materials has been discussed for a wide range of different applications in oil and gas including defoaming [9], demulsification [10], hydrophobing proppant, and rock surfaces [11–14]. Each of these and several others will be discussed in more detail here.

11.2 APPLICATIONS

11.2.1 ANTIFOAMS AND DEFOAMERS

A review on defoamers as applied to all applications including oil and gas is given by Owen [15]. The topic is covered in monographs on oilfield chemicals such as chapter 21 of Fink [16] and chapter 12 of Kelland [17].

The foaming process consists of entrainment of a gas (air) into a liquid accompanied by the adsorption of surface-active agents (surfactants) at bubble interfaces thus lowering the interfacial tension of the system. Foam stability is controlled mainly by mechanisms deriving from the adsorption process itself: film elasticity [18], viscous surface layer formation [19], gas diffusion [20], and electrical double-layer repulsion specific for aqueous foam stabilized by an ionic surfactant [21].

By far, the most studied and understood are aqueous foams. In comparison, few fundamental studies exist, and relatively little is known about stability mechanisms in nonaqueous foams. The latter are used in cosmetics and in manufacturing of polymer foams. However, in crude oil production, foam is detrimental and undesirable. Foaming problems commonly occur in the following processes—crude oil production, gas sweetening (amine), and gas dehydration (glycol) [22].

11.2.1.1 Nonaqueous Foam (Petroleum Foams)

The primary use of silicones in the oil and gas industry is foam control during oil and gas separation, crude oil distillation, propane deasphalting, and thermal-cracking processes. Polydimethylsiloxanes (PDMS), fluorosilicones, and silicone glycols are the primary silicones used in the oil industry [9]. Their unique properties have made them a high performing material for antifoam application [23]. Defoamers are materials that break existing foam whereas antifoams prevent the formation of foam. Antifoams are usually good defoamers. However, some defoamers are not good antifoams. Silicones are effective as both antifoams and defoamers. They have a high permeability to gases [24] and thus will remove resistance to gas diffusion between bubbles, which helps destabilize foams. To the best of our knowledge, antifoam compounds—blends of silicone fluid and hydrophobic silica—are not used to control nonaqueous petroleum foam.

To function as an antifoam, the following requirements have to be met:

- Must be insoluble in the foaming medium under specific conditions
- Must have a lower surface tension than the foaming medium
- Has to be rapidly dispersed in the foaming medium
- Must be chemically inert

Most oil wells produce a combination of brine and crude oil, and many also produce some amount of natural gas and/or CO_2. When this mixture is depressurized upon passing through the choke valve, the process may result in emulsions, aqueous or nonaqueous foams, or a combination of emulsion and foam. The brine and gas must be separated before the oil is transported to a refinery, so the emulsions and foams must be broken. When surfactants are used in a well treatment operation, they may contribute to both foaming and emulsion formation, thus creating serious problems in downstream processing of the oil. To overcome foaming problems, either a mechanical foam separator (an engineering solution) or a combination of mechanical device and addition of chemicals must be used.

Many factors can contribute to petroleum foam formation and stability once it is generated, such as naturally occurring surfactants; asphaltenes and resins [25,26], short-chain carboxylic acids, and phenols of molecular weight ≤400 [27]. Foam stability is also due to certain physical properties such as surface tension, surface and bulk viscosity [26], and film elasticity [28]. Surface tension [29] plays a key role in foam stability but other factors are also important. Dynamic surface tension has been reported as being particularly important for foam stability [30].

Due to the complexity of crude oil chemistries, it is not possible to fully discern the effect of individual crude oil components on foam stability. Petroleum foams usually contain water and solid particles (clay, sand, wax, salt, and precipitated paraffin) [31,32] as well as different chemicals introduced during drilling or completion operations (corrosion inhibitors, biocides, demulsifiers etc.) which make them even more complex. Petroleum foams are one of the most abundant and complex nonaqueous foams. A better understanding of their properties could lead to the tailoring of a more efficient chemical method to control crude oil foaming.

Petroleum foams present a more difficult problem in high producing wells with high gas–oil ratio (GOR) and high pressures and temperatures. In the case of light oils, foams are very unstable due to very fast bubbles coalescence, but when it comes to heavy oils, stable oil foams containing dispersed bubbles are formed.

Insolubility of the antifoam in the continuous phase is essential. This condition is met for many hydrocarbon and silicone materials in aqueous foams. However, in petroleum foams, the solubility of silicone in the crude oil varies with molecular weight and substitution [33] and depends on the gas and oil ratio as well as crude oil and gas type. High molecular weight silicones (PDMS) have proven their efficiency and are the preferred material for these applications [34]. They are often dissolved in a suitable solvent such as limonene or xylene before injection. Silicones have low surface and interfacial tension and can become profoamers and stabilize foam when they are used at relatively low concentrations below their solubility limit. Once saturation is attained, the silicone can become a separate phase and act as antifoam since it is no longer soluble in oil [35]. Callaghan et al. [36] also showed that PDMS with low viscosity loses efficiency and becomes a profoamer when the percentage of gas is increased, while PDMS with higher viscosities maintains efficiency even at high gas/oil ratio.

It has also been reported [30] that silicone antifoams greatly decrease both dilatational elasticity and viscosity of crude oil foam systems, which leads to a significant reduction in foam stability. This collected data allowed differentiation between profoaming and antifoaming and therefore optimization of dosage.

Fluorosilicones are highly effective for the treatment of particularly difficult crude oils and aggressive environments (sour crude treatment system) because they present greater resistance to chemical attack and low solubilization [37,38]. They are generally not soluble in organic solvents, which is the basis for their effectiveness as crude oil defoamers.

After a suitable selection of the silicone material, the required concentration of active silicone is usually in the range of 1–10 ppm of PDMS and 0.5–2 ppm of fluorosilicone. This will depend of course on the crude oil characteristics, the residence time, and process conditions [39].

Other types of antifoams are also used in the oil and gas industry, including polyglycols, phosphate esters, and sulfonated compounds [40]. However, they are generally much less effective than silicones due to their higher solubility in oil and their interaction with other completion chemicals [9]. They are usually used at higher levels, which range between 100 and 1000 ppm of bulk fluid.

The simplest method to characterize petroleum foams is the Bickerman method [34,41]. This test creates foams by flowing gas through the oil sample and measuring two different parameters: foaming index and average foam lifetime. The disadvantage of this method is that it does not consider high temperature and pressure conditions.

Consequently, alternative methods that take into account these conditions have been developed [42]. Two new techniques have been also developed: foam and entrained air test (FEAT) and FEAT-II methods [43]. The FEAT method consists of a fluid recirculation loop and a temperature controlled foam generation column to entrain air in the fluid. The column is filled with the fluid, and density is continuously recorded once recirculation begins. The effect of antifoams can be investigated at any time by injection into the foam column. FEAT-II is used for nonaqueous foams and is able to simulate high temperature and pressure conditions.

Untreated crude oil is first saturated with gas for at least 4 hours in a pressurized tank operating up to 225°C and 3000 psi and continuously stirred to simulate a "live oil" system. Upon depressurizing, "live oil" travels to the foam tank, and its density and flow rate are measured using a mass flow meter. The instrument is also equipped with a receiving foam tank to monitor foam height. Other completion fluids and water may be mixed with the crude oil before placing it into the pressurized tank. The antifoam is injected into the fluid while entering the foam tank. Fluid density, foamability, and foam knockdown time is used to evaluate the performance of the antifoam. This technique has been used to evaluate three PDMS defoamers with viscosity of 1,000, 12,000, 60,000 cST for triphase oil/water/gas separators. PDMS 60,000 was highly efficient in defoaming the fluid sample reducing foamability from 29.5% to about 2.5% followed by PDMS 12,000. PDMS 1000 exhibited a poor performance and a final foamability of about 8%.

11.2.1.2 Aqueous Foams

Some examples of aqueous foams in the oil and gas industry include drilling, acidizing and fracturing fluid foams, gas blocking and diverting foams, and gas mobility control foams (see Section 11.2.6). The products most commonly used to control aqueous foams are silicone antifoam compounds—blends of silicone fluids with hydrophobic silica. This type of antifoam compound is most effective if it is prepared as an emulsion for better dispersion in water-based systems. The synergistic performance of the mixture of oil and particles was reported [15]. The particles will be located at the interface between the oil droplet and the liquid and act as asperities to facilitate penetration through the pseudoemulsion film.

11.2.1.3 Cementing

The use of silicones in cementing is primarily foam control. But because the application is so different from the foaming problems discussed earlier, it is best dealt with separately.

After drilling an oil/gas well, casing (large, steel pipe) is put into the ground. The space that remains between the casing and the formation (the annulus) is then filled with cement by pumping cement slurry through the casing and back up the annulus as primary cementing. Repair or remedial cementing jobs are performed after the primary job usually as part of a secondary cementing.

The first aim of cementing is to seal off formations to prevent fluids from one formation migrating up or down the hole and polluting the fluids in another formation, prevent leak off through formation areas, fix as well as reinforce the casing and protect it against corrosion, help prevent blowouts, and finally plug old wells (abandonment). Cement additives include accelerators, lightweight and heavyweight additives, retarders, lost circulation and fluid loss additives, dispersant, gas control additives, and defoamers.

Defoamers are used in cementing application to remove air entrainment in cement slurries that may occur during the mixing process. Excessive foam can lead to cavitation during mixing and thus loss of hydrostatic pressure during pumping. Additionally, slurry foaming can cause an underestimation of slurry density downhole. This is due to the entrapped air being compressed in downhole pressure conditions and increasing the density of the slurry over the lb/gal weight measurement made at the surface. Defoamers that are commonly used in cementing applications include fatty alcohols, sulfonated oils, and silicone emulsions [44]. Liquid or solid defoamers are added in small amount ranging from 0.1% to 0.3% by weight of cement (BWOC) to remove foam during the mixing of cement slurries [45].

Bava et al. [45,46] compare the performance of silicone and nonsilicone products as defoamers and antifoamers on cement slurries using two different techniques: FEAT and blender foam test. Results indicated that nonsilicones outperformed silicones as antifoamers and defoamers especially in slurry formulations containing polyvinyl alcohol (PVA) as fluid loss additive and latex as a gas mitigation control at brine conditions.

Silicone chemicals have been used in cement applications other than as defoamer agents. In US8030253 [47], Halliburton disclosed a foamed cement comprising swellable particles such as fluorosilicone and silicone particles among others. The swelling might be up to 50% of the original size at downhole conditions. The particles will swell when contacted with oil to inhibit flow through the crack and/or microannulus, preventing and/or reducing the loss of zonal isolation. These oil-swellable particles may be present in the cement in an amount up to about 27% BWOC.

In EP0315243 [48], Pumptech N.V. disclosed a new cement slurry for the cementing of oil and gas wells with good fluid-loss control. The cement slurry is an oil-in-water emulsion in which the oil phase is a silicone oil. The emulsion has an oil–water ratio of 5%–50% by weight.

Well abandonment represents a major cost issue for most operators. Oil and gas wells that can no longer be used must be plugged to prevent oil and gas from migrating uphole and contaminating surface aquifers. Conventional well abandonments consist in plugging the well by mechanical and cement plugs in the wellbore. Novel silicone rubbers (and gels)/Portland cement plugging materials have been developed by Shell International E&P [49]. Room temperature vulcanizing (RTV) silicone rubbers and gels showed a superior performance to most alternative thermosetting resins, a high stability at very high temperature, chemical inertness as well as withstanding extremely high differential pressures. Trial well abandonment using RTV silicone rubber/cement composite as a primary plug was successful. A second trial using the same sealing agent for a gas shutoff application was also successful.

11.2.2 DEMULSIFICATION

The demulsification process is similar to the defoaming process. In this case, a liquid phase needs to be separated from another liquid phase through interface

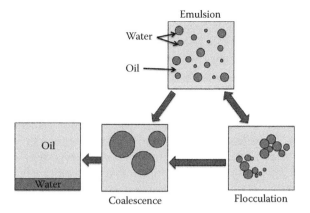

FIGURE 11.1 Schematic of flocculation and coalescence process. (From Pietrangeli, G.A. et al., Treatment of water in heavy crude oil emulsions with innovative microemulsion fluids, *SPE Heavy and Extra Heavy Oil Conference—Latin America*, Society of Petroleum Engineers, Medellin, CO, 2014.)

destabilization, droplet flocculation, and coalescence [39,50,51]. Figure 11.1 shows a schematic of the process.

Silicones are used as demulsifiers in the following operations [52,53]:

Demulsification and antifoaming in gas/oil/water separators
Dispersion and drag reduction of asphaltenes
Dehazing of diesel fuel (including biodiesel)
Enhanced oil recovery
Separation of waste oil and sludge
Dehazing of lubrication oils
Bitumen extraction from oil sands
Steam-assisted gravity drainage
Preventing emulsion formation

One of the important requirements to successfully use a silicone for demulsification is to ensure that it is delivered to the emulsion problem in an appropriately dispersed state. This may be achieved by using the silicone dissolved in a solvent such as limonene or xylene, using an emulsion of the silicone, or a water-dispersible form such as a silicone polyether surfactant [52,54].

Water is almost always produced from the well along with crude oil, often in the form of water-in-oil emulsions (W/O), where the crude oil is the external/continuous phase and the water is the internal/dispersed phase. The origin of the emulsified water is diverse; it could be production water, completion brine, drilling fluid, injection water, flowback water, or a combination of some or all of these [50,55]. The amount of emulsified brine produced along with crude oil varies from less than 1 vol% to about 60 vol%, with heavier oils often containing larger amounts than light oils. Crude oil physical characteristics are affected by the presence of water.

Viscosity and flow patterns are very noticeable when water is present. Medium crude oils could behave as a heavy or extraheavy crude oil if the water content is high enough [50,56].

W/O emulsions form because of turbulent flow as a mixture of brine and crude oil flows into the wellbore, into the choke at the wellhead, and during surface handling [50,51]. Surface-active substances are naturally present in crude oil and help to form and stabilize the W/O emulsions. These include naphthenic acids, asphaltenes and resins, inorganic solids, polymers, and/or dispersed particulate solids [51,52,55]. Asphaltenes can reduce the interfacial tension between the crude oil and the water to low values, but their role in stabilizing W/O emulsions also involves interactions with resins, and formation of aggregates that accumulate at the oil–water interface [56,57]. In addition to surface-active substances in the crude oil, additives used during drilling, completion, hydraulic fracturing, and acidizing operations could also potentially produce emulsions when they come into contact with some crude oils [58].

Other operations that could increase the possibility of emulsion formation include (1) water injection to maintain reservoir pressure, (2) casing leaks and perforated production pipe that can lead to the migration of water into an oil-producing zone, and (3) gas lift, where agitation in the produced liquid is introduced, favoring emulsification of water in the crude oil, etc. [59]. Figure 11.2 shows a microscopic view of small water droplets in crude oil samples.

Emulsified water must be separated before further transportation and refining of crude oil. Before pumping through a pipeline, and especially before refining, crude oil must have a low water content, typically less than 0.05–2 vol%, preferably lower than 1 vol%. The high salt content encountered in the emulsified water causes several problems downstream, such as corrosion during refining processes, scaling in pipes, disruption of distillation processes, and a greater energy requirement to pump the more viscous emulsion, just to name a few [52,53,55,60].

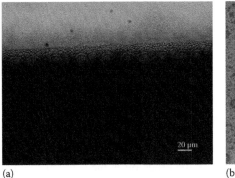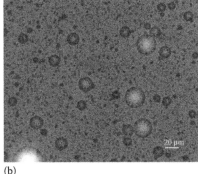

(a)   (b)

FIGURE 11.2 Water droplets present in a crude oil sample. (a) shows a neat emulsified crude oil sample in a sealed thin/flat glass capillary. (b) shows the water droplets after crude oil sample is diluted with xylene. (From Pietrangeli, G.A. et al., Treatment of water in heavy crude oil emulsions with innovative microemulsion fluids, *SPE Heavy and Extra Heavy Oil Conference—Latin America*, Society of Petroleum Engineers, Medellin, CO, 2014.)

W/O stable emulsions not only create flow assurance issues, they also increase surface treatment costs for pipelines, separation facilities, and processing plants. The disposal of the water, salt, and residual oil separated from the crude oil before the refining process can be costly and affect the value of the operator's asset [50,55,61,62].

The problem of emulsified water in crude oil (or dehydration) is dealt with (a) using additives put into well-completion fluids (termed "upstream"), (b) with engineering solutions at the surface, and (c) with additives put into the oil after production (b and c are termed "downstream"). There are many good reviews of the general problem, including chapter 22 of *Oil Field Chemicals* [16], and chapter 11 of *Production Chemicals for the Oil and Gas Industry* [17]. It is often observed that additives used upstream are effective in a wide range of conditions, while very specific formulations tailored almost well-by-well may be required in downstream applications.

Tailored formulations are created based on the physicochemical properties of the emulsion components. The idea is to create a formulation which is close to the optimum formulation, where the drops essentially coalesce upon contact [60,62–64]. Demulsifiers should reach the oil–water interface and concentrate there. Once at the interface, the demulsifier displaces the stabilizer amphiphilic molecules offering little resistance to coalescence [39,61]. Demulsifiers accelerate the flocculation and enhance the film drainage for faster coalesce, by inverting the interfacial tension gradient between the inner and outer film; Figure 11.3 shows the process in more detail. Due to the increasing demand of environmentally safe chemistry, silicone products have been proposed as demulsifiers more often. The introduction of these new molecules seeks a less toxic, safer, and more efficient demulsifier [53,61,66].

Demulsifiers are typically injected into the well, into the crude oil stream, at the separation equipment, or at any other suitable points [52]. Despite the large number of demulsifiers available on the market, it is not possible to break all crude oil emulsions rapidly, safely, and efficiently with small quantities of a single product [53]— there is no single demulsifier that is universally effective.

The performance of demulsifiers can be measured and studied using different techniques including water–oil ratio, adsorption kinetics, interfacial elasticity, rheology, microscopy, and turbidity [55,60,65]. The selection of the demulsifier should take into account the appearance of the separated water, which may be turbid if the demulsifier is hydrophilic and able to form oil/water emulsions. Fine particles, such as clays, silicas, iron oxides, and precipitated salts could enhance emulsion stability,

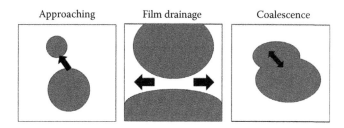

FIGURE 11.3 Emulsion destabilization process.

particularly when modified through the adsorption of resins and asphaltenes to render them strongly, interfacially active [57]. These types of emulsions are called Pickering emulsions [60,63,66]. Certain demulsifiers may change the wettability of these particles, thus aggravating the emulsion problem.

Common demulsifiers include polyols, ethoxylated phenolic/formaldehyde resins, ethoxylated phenols, alkoxylated amines, polyamine derivatives, sulfonic acid salts, etc. The common doses range from 1 ppm up to 500 ppm and will depend on emulsion stability, fluid properties, and environmental conditions. Any error in dosage would increase the stability of the emulsion. Other optional ingredients that may be added to the demulsifying composition include activators, retarders, accelerators, processing additives, reinforcing materials, wetting agents, anticorrosion additives, hydrogen sulfide scavengers, biocides, etc. [52,54,56,60].

Use of silicones to demulsify water-in-crude oil was claimed early [62] and is discussed in many publications [10,54,57,61,64–67]. Low-to-medium molecular weight PDMS is soluble in crude oil and is not an effective demulsifier [39]. We are not aware of resins and silanes being used for demulsification. Organomodified silicones, such as silicone polyethers work well. They are water dispersible and do not suffer from solubility issues even at lower molecular weights. In this application, organomodified silicones are copolymers with a PDMS backbone. Those chains can be polyether, polyethylene oxide, polypropylene oxide, amines, and other organic groups [39,53,65].

Gotz and Gerd [62] proposed the use of polyoxyalkylene–polysiloxane mixed block copolymers as emulsion breakers for crude oil/water emulsions, where the polyoxyalkylene blocks consist of a mixture of ethylene oxide and propylene oxide units in a weight ratio of 40:60 to 100:0 and the polysiloxane blocks contain three to five silicon atoms per block. Daniel-David et al. [64] evaluated the demulsification properties of polyalkylene oxide modified PDMS chains with a molecular weight of 2100. They found that the modified silicone product was an effective demulsifier.

Daniel-David et al. [61] studied the elasticity properties and demulsification effectiveness of the different silicone copolymers. They have found that the absence of ethylene oxide moieties leads to a poor demulsification. They have also concluded that destabilization of the droplet is enhanced by the interaction of the natural surface-active molecules and the copolymers.

Dalmazzone and Noik [66] discussed silicones for demulsification of water in crude oil emulsions. This paper gives detailed procedures for demulsification studies, including the formulation of reproducible water-in-crude oil emulsions, their characterization, and the demulsification bottle test in order to determine the correlation between demulsifier and coalescence.

Dalmazzone and Noik [51] tested a family of silicone demulsifiers. Two silicone derivatives were found efficient for breaking emulsions in paraffinic and asphaltenic crude oils. The silicones (vs the other chemistries they tested) were insensitive to salinity. Demulsification efficacy increased strongly with temperature. The authors concluded that the classical mechanisms of sedimentation, flocculation, and coalescence are operating in demulsification of crude oil, but it was not clear in what way the silicone materials are influencing these phenomena. On the other hand, the viscosity of crude oil played a key role in the demulsifier dosage, with greater viscosities requiring larger dosages of treatment.

In other research, Noik et al. [54] studied the use of polyalkylene oxide-modified PDMS with different siloxane units and compared them with a polyglycol and a sulfosuccinate product. It was concluded that organic demulsifiers offered the largest separation, while silicone-based products gave faster separation but did not separate as much water. Mixtures of silicone-based molecules and organic-based molecules offer the best water separation with minimum residence time.

Due to the low rotational barrier in the Si–O bond (<0.2 kcal/mol) and long bond length (0.165 nm), the siloxane chain is more flexible than an organic carbon chain (rotational barrier of 3.6 kcal/mol, bond length of 0.154 nm). Similar to paraffins, silicones exhibit a low surface energy (about 21 mN/m), which has been attributed to the methyl groups of the PDMS molecules.

Phukan et al. [65] observed that organomodified silicones gave a faster demulsification process than common organic molecules. They also established that synergy between silicone and organic demulsifiers resulted in a faster action and lower dosage.

In another paper, this group found that formulations based on polysiloxane copolymers promote separation of water from crude oil even at concentrations of a few tens of ppm and discussed the mechanism by which the copolymers function [10]. They proposed the use of polyalkylene oxide-modified poly(dimethylsiloxane) chains as a demulsifier. The amphiphilic copolymer absorbs at the oil–water interface to disrupt the network of asphaltenes aggregates and induce emulsion breakup.

Le Follotec et al. [67] used triblock copolymers in the form of pEOx—pDMSy—pEOx as demulsifiers. For values of x similar to the value of y and no larger than three times y, demulsification was observed. In contrast, if x was six times y, demulsification was not observed. The higher the affinity of the copolymer for the oil, the less water separation was obtained.

Polyorganosiloxanes have several advantages over other conventional demulsifiers. They have a seawater biodegradability of at least 18% in 28 days, when tested according to the Organization for Economic Co-operation and Development (OECD) 306 guidelines. They improved biodegradability compared to silicone polyethers. Crude oil that has been treated with polyorganosiloxane can contain less than 0.5% by weight of water after separation of the water phase from the oil phase, lower treatment rates are possible with the addition of polyorganosiloxane molecules to the demulsifier package [52,53].

11.2.3 Drilling Muds

Drilling fluids or muds, as they are commonly named, are fluids created and used with specific purposes and functions using suspended solids in a liquid [68,69]. They are used during exploration and production operations for oil and gas wells [70]. They are complex fluids that contain several components and additives. The components, type, and amount of additives are based on the drilling method and the type of reservoir to be drilled [71].

Drilling fluids transport cuttings from the wellbore to the surface, thus providing geological information about the rock being drilled. They cool and lubricate the drill bit, support the drill pipe and the drill bit, provide wellbore integrity and stabilization preventing collapse, regulate the formation pressure to avoid unwanted fractures, and minimize formation fluid invasion by sealing the rock. An ideal drilling fluid would be efficient, low-cost, and low maintenance. The flow properties and ability to transport cuttings will depend on the drilling fluid's density, viscosity, and velocity through the annulus.

Generally, drilling fluids are non-Newtonian, viscoelastic fluids, which exhibit thixotropic behavior. Viscoelastic fluids have an elastic component or modulus (G') and a viscous component or modulus (G''). At very low stress values, G' values are greater than G'' values for thixotropic fluids. Typically, for an oil-based mud, the ratio G'/G'' is close to 3.5 [72]. This characteristic allows the mud to hold the cuttings if the mud is under static conditions. The gel strength of some muds increases with time after agitation has ceased; these types of fluids are known as thixotropic. Furthermore, if the mud is subjected to a constant shear rate, its viscosity decreases with time as its gel structure is destroyed, until equilibrium is reached [69].

Water-based drilling fluids or water-based muds (WBMs) contain solid particles suspended in brine or water [69]. They are environmentally more friendly and are usually less expensive. The cuttings generated from the drilling operation can be easily disposed of when they have been in contact with WBM [71].

Although water-based drilling fluids are the most common fluid system used worldwide during drilling operations, certain circumstances require the use of oil- or synthetic oil-based muds (OBMs or SBMs). OBMs are brine-in-oil emulsions stabilized with emulsifiers and solids, also called inverted emulsions. Diesel, mineral oils (paraffins, isoparaffins, and cyclic and branched alkanes), and synthetic oils (esters and olefins) are used as base fluids.

These types of drilling fluid are thermally stable and desirable for high temperature applications. They offer a unique shale stability and imbibition for highly active shales, create thinner filter cakes, and reduce the risk of stuck pipe and bit balling. These drilling fluids can be reused and treated multiple times because they are resistant to chemical and solids contamination. They are flexible, available at low densities, and the rate of penetration is higher when they are used. They do not need the addition of lubricants due to their oily nature, which reduces torque and drag. They can deal with natural gas hydrates better than WBMs. They can support the shale formation, and molecules of the oil do not penetrate into organic and nonorganic pores under the capillary pressure. Additionally, these fluids are not corrosive.

Silicone materials can be used in drilling muds as emulsifiers. They provide a stable emulsion while including an option of weighting agents. Romenesko and Schiefer [73] disclosed a method to prepare water-in-oil inverted emulsions for use as drilling fluids using a silicone-based polymer containing both polar and alkyl groups.

Donatelli and Keil [74] disclosed an improved silicone emulsifier composition useful for preparing thermally stable, solids-free inverted emulsions for the drilling operation in deep wells at high densities. The improved molecule is a terpolymer, a silicone backbone with both alkyl groups and PO/EO groups pendant to the chain, where the PO could be equal to zero and the alkyl group has between 6 and 18 carbon atoms. Figure 11.4 shows a representation of structure of the improved silicone emulsifier.

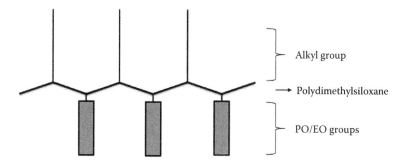

FIGURE 11.4 Schematic representation of the improved silicone emulsifier described in U.S. Patent 4421656.

Improvement in the rheology has been proposed by Halliburton by adding silicone oils to a "clay-free" oil based mud. The silicone oils could be added to the oil-based drilling fluid at levels between 0.1 and 400 pounds per barrel before or while it is circulating through a wellbore as the wellbore is being drilled, swept, cemented, and/or cased, to facilitate the drilling operation. The addition of the silicone-based fluid could afford real-time monitoring and rapid adjustment of the fluid to accommodate changes during drilling operations by adding silicone oils with different viscosities and thereby changing the invert drilling-fluid viscosity as needed [70].

The use of a silane that cross-links at the water–oil interface and stabilizes water droplets in a W/O emulsion has been disclosed by Schlumberger in U.S. Patent 6156805 [75]. The water droplet will be encapsulated in the oil-external phase, thus stabilizing the emulsion.

In an attempt to reduce toxicity and increase thermal stability of OBMs, MI Swaco disclosed silicone based fluids, such as dimethylsiloxane polymers, as the oil phase of an OBM [76,77].

Higher downhole well temperatures are the result of the increment of the average depth of wells. New wells are being discovered every day in remote and deeper zones, onshore and offshore. Silicone oils are stable fluids at temperatures higher than 300°F, which make them attractive for deepwater drilling operations, where temperatures could reach 500°F.

Logging is one of the most important actions taken while drilling, particularly for deviated wells. The measurements while drilling or MWD have the purpose of providing real-time data of the well while drilling operations are in place, such as formation characteristics of rocks, pressures, porosity, and fluids [78].

The MWD tools help during well interventions, reservoir evaluation, sampling, and pipe recovery. An electrical circuit is created and the resistance and other electrical properties of the circuit may be measured while the logging tool is retracted from the well. The resulting data is a measure of the electrical properties of the drilled formations versus the depth of the well.

The nonconductive nature of OBMs makes it difficult to obtain measurements while drilling. The normal resistivity and self-potential measurements cannot be performed [78]. The drilling bit and streams must be pulled out of the well for measurements, increasing the nonproductive time and the possibility of wellbore instability.

The nonconductivity disadvantage has been addressed in different invention disclosures. Patel et al. [78] have proposed to solve the conductivity problem and the loss of conductivity (when the fluid is exposed to air) using a different approach. They have disclosed the use of oil mixtures, including silicone oils, as the continuous phase in combination with quaternary amine salts as the dispersed phase, in order to create an OBM able to conduct electricity while a gas, carbon dioxide buffer, is pumped through the fluid. These authors disclosed a method of electronically logging subterranean wells using a conductive double emulsion fluid. A double emulsion or a multiple emulsion is an emulsion within an emulsion, stabilized by emulsifiers at the inner and the outer interface [79]. Multiple emulsions, present in the OBMs, are oil-in-water-in-oil (O/W/O) emulsions (see Figure 11.5).

Multiple emulsions are usually unstable with a fast coalescence. However, certain types of emulsifiers offer a higher stability and create smaller droplet sizes (such as synthetic block copolymers based on silicone backbones with polyethylene oxide side-chains), which serve also as a steric stabilizer for both interfaces [79].

Patel et al. [80] disclosed the composition of the double emulsion OBMs as

- A miscible combination of an oleaginous fluid (diesel, mineral oil, synthetic oil, vegetable oil, silicone oil, or combinations thereof)
- A hydrophilic emulsifier capable of forming a microemulsion (HLB > 9)
- A hydrophobic emulsifier capable of forming an invert emulsion (4 < HLB < 8)
- An electrolytic salt (quaternary amine salt)
- They also suggested that fluid may additionally contain a polar organic solvent (such as glycols) and a carbon dioxide buffer

In this particular disclosure, a microemulsion is the continuous phase of an inverted emulsion. The electrolytic salt or brine is present in a concentration sufficient to

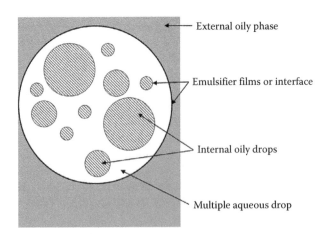

FIGURE 11.5   Schematic representation of an O/W/O double emulsion droplet.

permit the electrical logging of the subterranean well. The double emulsion fluid also contains other components (viscosifiers, weighting agents, loss circulation materials, etc.), in order to reach drilling fluid properties.

11.2.4 PACKER FLUIDS

Gas and oil well drilling operations are followed by well completion operations. During completion, it is standard practice to set a packer between the tubing and the casing above the productive interval, and fill the annulus space with a packer fluid [81] to protect the casing [82]. Annular pressure buildup is generated by thermal expansion of the fluids trapped in casing annuli, commonly between the top of the cement and the wellhead or seal assembly [83].

One of the most common causes for the collapse or burst of casing and tubing strings is heat transfer (See Figure 11.6). During production, the near wellbore-zone temperature may increase to more than 200°F [84]. Fluids in a confined space can only expand until the available space has been filled. When this happens, confined fluids can rapidly exert very high forces on the confinement vessel [84].

The packer fluid reduces the pressure differential between the inside of the tubing and the annulus, and between the outside of the casing and the annulus (Figure 11.7). The fluid will control the formation pressure, as well as isolate and protect the casing from corrosion or failure [81].

Hydrocarbon and mineral and synthetic oils have been suggested and used to control trapped annular pressure [86]. In order to enhance the oil-based fluid, other fluids, such as silicone oils, were added to the packer fluid. These authors disclosed an oil-based packer fluid gelled by inclusion of an organosilane useful for insulating the annular space of an oil well from surrounding permafrost. Carlos [87] proposed the use of a thermally insulating packer fluid consisting of a silicone-based fluid. The proposed fluid has straight chains of PDMS, which are terminated with one or more trimethylsilyl groups.

Quintero et al. [84] disclosed a silicone-based surfactant fluid able to create self-organized structures known as liquid crystals. The structures formed with

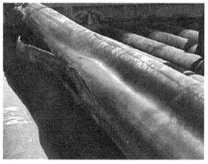

FIGURE 11.6 Casing failure due to annular pressure buildup—SPE 89775-MS. (From Pattillo, P.D. et al., *SPE Drill Completion*, 21(04), 242, 2006.)

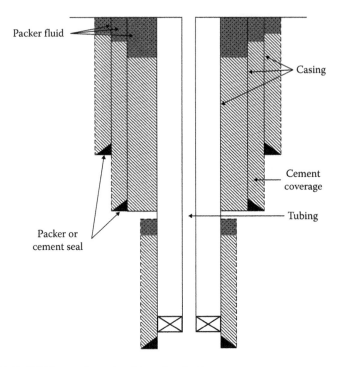

FIGURE 11.7   Wellbore schematic with packer fluid.

silicone-based products can be fine-tuned by controlling the tilt angle of chiral mesogens within layers of the organized system, which control the physical and mechanical properties of the polymeric system and facilitate the stabilization of multiphase-component emulsions.

In general, silicone-based fluids are attractive as packer fluids because they are hydrophobic and Newtonian fluids with very low vapor pressures, pour points, and surface tensions. Conversely, silicone-based fluids also exhibit high flash points and are resistant to oxidation. They are miscible with a wide range of hydrocarbon fluids and other silicone-based products [87]. They have a high dielectric strength and function to reduce molecular vibration by damping [88].

The compressibility and thermal coefficient of expansion properties of silicone fluids make them useful in this application. Table 11.1 summarizes thermal expansion (β) and compressibility (κ) coefficients for several liquids. Silicone oils have small β and κ values compared to *n*-octane and acetone.

11.2.5   Restriction of Water Production in Gas Wells and Surface Modification

Unwanted water production in gas wells leads to liquid loading and eventually stops the gas production. Gelling polymers are the main chemicals used to shut off water-bearing channels or fractures within the formation and significantly reduce the relative

TABLE 11.1

Thermal Expansion Coefficient and Isothermal Compressibility Coefficient for Several Fluids

Product	Temperature, °C	β, K^{-1}	κ, bar^{-1}	Reference
Silicone Oil DMS-10	15–25	0.9×10^{-3}		[89]
Silicone Oil 10 cSt	15–150	1.06×10^{-3}		[90]
Silicone Oil 10 cSt	25–150	1.04×10^{-3}		[90]
Silicone Oil 20 cSt	25–150	1×10^{-3}	1.5×10^{-4}	[91]
Water	20–80	0.45×10^{-3}	0.364×10^{-4}	[91]
Water	—	0.21×10^{-3}		[92]
Gasoline	—	0.9×10^{-3}		[90]
n-Octane	—	1.14×10^{-3}		[92]
Acetone	20	1.487×10^{-3}	0.62×10^{-4}	[93]

permeability to water near the wellbore area. Others chemicals have been used for this purpose including silicones that alter permeability by modifying rock wettability.

Lakatos et al. [11,12] summarized an extensive laboratory study on the use of silicone compounds for restriction of water production in gas wells—in this case, the function of the silicone was to hydrophobe rock surfaces. The first paper [12] presents a broad view of the use of silicone materials for restriction of water production in gas wells. The laboratory characterization involved reactive silanes, a "siloxane," a silicone resin, and a silicone oil. The paper concludes that rock surfaces and pore structure can be modified using siloxanes. The second paper [11] reports on the use of silicone microemulsions for restriction of water production in gas wells. According to their study, silicone microemulsions induced a permeability modification that resulted in a large reduction of water mobility. This method was tested in a Hungarian gas field; results showed that gas production tripled for a half year.

US6206102 [94] discloses the use of organosilicone compounds as the dispersed phase to adsorb to the rock formation, hydrophobing the rock surface, and thereby decreasing water permeability and increasing gas permeability.

3M has a family of patents in which they claim the use of fluorosurfactants [95–97], fluoropolymers, and fluorosilanes [13,98] to improve oil- and gas-well productivity. All of these patents seek to apply a low surface energy coating to rock and/or proppant surfaces. The silane is useful because it promotes the formation of chemical bonds between the coating and the substrate.

US 2010/0276142 [13] references the problem of light-weight and porous proppant particles and discloses the use of a fluorosilane to prevent liquid penetration into the particles. This patent application discloses that the presence of the fluorinated siloxane improves fracture conductivity as measured using API RP 61 test for short-term conductivity testing proposed by the American Petroleum Institute (API). In addition, this patent application teaches that the treatment described is useful in facilitating removal of fracturing fluids that have been injected into the formation (called flowback).

In another related strategy to increase near-wellbore conductivity, Shaari [99] injected organosilane during an acidizing operation to stabilize formation fines and fix them in place. The rationale presented in the paper is that the silane hydrolyzes and then condenses onto available surfaces forming an effectively nonwetting surface coating. Fines that are coated become hydrophobic and agglomerate rather than dispersing into the aqueous phase and this tends to keep them in place in the formation rather than flowing into the near-wellbore region and blocking conductivity.

In US7552771 [100], Halliburton disclosed a method to enhance gas production comprising the combination of a relative permeability modifier (RPM) together with a water drainage rate enhancing agent (WDREA). Disclosed WDREAs include zwitterionic or nonionic surfactants, certain cationic polyorganosiloxanes, and a solvent/surfactant blend that forms a microemulsion.

In US8053395 [14], Halliburton disclosed treatment fluids for increasing gas production from subterranean formations comprising certain cationic polyorganosiloxanes. This patent states that a cationic polyorganosiloxane functions by adsorbing onto rock surfaces and increasing the water contact angle (hence hydrophobing the rock surface). This is supposed to reduce capillary pressure and increase water drainage rates from the formation.

In US7723274 [101], Trican disclosed an aqueous slurry composition in which a slurry of proppant particles is stabilized by hydrophobing the particle surface. The slurry is said to be rendered sufficiently stable so as to be pumped into a formation as a fracturing fluid without use of viscosifying polymer. The patent describes the treatment as rendering the particle surface "extremely hydrophobic," defined to mean a water contact angle greater than $90°$. The materials used to treat the particle surface include organosilanes and fluorosilanes and cationic and betaine functional polysiloxanes. In their US 2010/0267593 [102], the additional feature of adding small amounts of hydrocarbon or silicone oil to enhance the "hydrophobic aggregation" is disclosed.

The flow of gas in tight, low-pressure gas wells can be partially blocked by the water saturation buildup near the hydraulic fracture face if the drawdown pressure does not exceed the capillary pressure. To increase the productivity, the water saturation may be reduced by alteration of the near-wellbore wettability from water-wet to intermediate-wet. Many surfactants have been identified which change the wettability of carbonate and sandstone rocks from water-wet to intermediate-wet in water–air–rock systems [103,104]. Among fluorosilanes, as the number of fluoro groups increases, rocks become less water-wet [105]. One day of aging period and 1 wt.% concentration appears to be sufficient for altering wettability.

11.2.6 Foam Assisted Lift

Gas wells often produce hydrocarbons, water, or mixture of both. Accumulated liquids can create hydrostatic head pressure that obstructs removal of fluids from the wellbore, which leads to reservoir pressure decline and production drop. For the production process to occur, liquids must be removed from the wellbore and transferred to the surface. Foam agents have been applied with success to convert liquid load into foam to unload the wellbore. This method is called foam assisted lift (FAL).

If the liquid load contains mostly hydrocarbons, then fluorocarbon surfactants are used as a foamants. However, if the liquid load contains significant amount of water and hydrocarbon (50% or higher), foams are unstable and cannot be used for FAL [106]. Stability is achieved if the aqueous pseudoemulsion film between the hydrocarbon drop and the gas phase is stable, then the oil drop cannot enter the aqueous surface and will not be detrimental to the foam, achieving high liquid loading from the wellbore [107].

Koczo et al. [108] studied silicone polyether surfactants (SiPE-1 and SiPE-6) and trisiloxane superspreader (TS8EOMe) as aqueous foamants for FAL in the presence of high amounts of hexane or octane/decane/dodecane mixture (model oils). They found that TS8OMe forms unstable pseudoemulsion and therefore do not form stable aqueous foams with these model oils. Whereas, SiPE-1 and SiPE-6 form stable foams in presence of these model oils up to 50°C, therefore could be used as efficient candidates for FAL.

11.2.7   OTHER APPLICATIONS

Although not strictly part of the exploration and production of oil and gas, silicones have been discussed with regard to many aspects of the manufacture and use of hydrocarbon fuels, for example, to remove water haze from distillate fuel [109].

11.2.7.1   Marine Oil Spills

Silicone surfactants have been used to treat crude oil spills. Oil spills happen when petroleum is released into the environment, most notably marine oil spills. The crude oil residues cause severe environmental damage in aquatic environments, affecting flora and wildlife. Hydrocarbon spills occur daily in harbors and marinas, but large marine spills are the ones that capture public attention [110]. Spill response strategies have been designed to minimize environmental impacts. Several techniques have been developed and utilized for removing the hydrocarbon from water, and the usefulness of these methods depend on the size of the spill. The response strategy must be evaluated for operational limitations, potential effectiveness, environmental impact, etc. Common methods include mechanical containment, controlled burning, shoreline treatment, and spill-treating agents [110,111].

The treating agent could be an oil dispersant, emulsion breaker, coating agent, oil-gelling agent, or a "herder" [112–114]. Most oil-soluble surfactants will "herd" oil. These surfactants should be less water-soluble and only slowly oil-soluble and float on water. Oil herders are used in *in situ* burning applications, coating, and dispersants. As a dispersant agent, they will mix with oil. Then, it will move to the oil–water interface. Once at the interface, the effect of the dispersant will be to increase the number of small droplets formed by wave action as well as to keep these small droplets from coalescing. Because the droplets formed are so much smaller than the original crude oil slick, they can be biodegraded by other organisms in the surrounding waters [114].

Buist et al. [115] and Nedwed et al. [111] considered the use of the orgnosilicone superwetter surfactants as oil herders. Silicone-based surfactants could enhance the herding of oil slicks on water due to the development of low surface tensions. These surfactants reduce surface tension to 20 mN/m and produce a spreading pressure in the 50 mN/m range. Silicone-based surfactants performed better than hydrocarbon-based surfactants at the same conditions.

According to Buist et al. [115], the flow rate application, pressure, and atomizing nozzle types are critical to produce the right droplet size distribution of the herder product. The herder must be applied with a drop size large enough to minimize drift (close to 0.3 mm) yet small enough to limit penetration into the water (close to 1 mm) [111,115]. They also found that the silicone surfactant film survived more than 45 minutes in a calm sea, which is more than enough time for *in situ* burning operations without fire-resistant booms [111].

Pelletier and Siron [110] used silicones as part of a herding treatment, which contained Brij 76 (surfactant) and a mixture of silicone-based products (trichlorosilane, octadodecyltrichlorosilane, trimethoxysilane, and silicone fluid) dispersed in petroleum ether. The treatment was sprayed over and around the slick. A strong herding effect was observed and the slick reduced to a third of the original size. A white coating was formed over the slick avoiding dispersion after mild stirring, and no debris or free droplets were observed.

11.2.7.2 Gas Hydrates

Most natural gas is produced from the formation along with water. If the water vapor cools below certain temperatures, hydrates can form [116]. Gas hydrates are solid crystalline compounds formed by the physical combination of gases and water under pressure [117]. They are crystals of water around a host gas molecule [118]. The gas hydrates generate considerable operational and safety concerns in subsea pipelines and process equipment [117]. They pose a significant threat to flow assurance [118].

When gas emerges from the wellhead, it is usually at high temperature and pressure. As the gas cools and water condenses, hydrate formation becomes a real danger [116], especially for wells drilled and completed in deep water or in the artic [118].

Hydrates may pack solidly in gathering lines and equipment, blocking the flow of gas [116]. They are relatively immobile and impermeable [119]. Heating the gas and injecting chemical inhibitors are two ways of preventing formation of gas hydrates [116]. Chemical inhibitors are products added to the production stream that inhibit the production of the hydrates by lowering the freezing point of water vapor so that it does not condense as easily. Some products promote the dissociation of gas hydrate, while others prevent its formation.

Kawamura et al. [120,121] proposed the use of silicone oils as an inhibitor for gas hydrate formation (preventer). A laboratory-made methane hydrate pellet was dissociated in silicone oil at different pressures and temperatures. At temperatures between $-15°C$ and $-10°C$, the dissociation time of the hydrate takes longer in silicone oils than water. Therefore, at these conditions, silicone oils prevent dissociation and could be used for slurry transportation and storage systems. Temperatures higher than $2°C$ showed very fast dissociation rates, where silicone oils behave as a promoter.

11.2.8 PROBLEMS

Silicone polymers and copolymers are sometimes said to cause poisoning of cracker catalysts in refineries processing crude oil containing silicone residues. Pape and Haensel provide a useful discussion of this issue [9,122]. Levels as high as 13 ppm in the crude oil have not shown any unusual problems in refineries using North Slope crude oil [9].

11.2.9 ENVIRONMENTAL FATE

Environmental fate of chemicals used in well treatment has become a very important issue, especially for hydraulic fracturing. It is well known that silicones do not biodegrade, but chemically degrade in the environment to form CO_2 and silicic acid [123]. The rate of chemical degradation depends on where in the environment the silicone is found—volatile silicones in the atmosphere degrade differently than do high molecular weight silicones in wastewater sludge, which degrade differently than do water-soluble silicones such as the silicone polyethers mentioned earlier. This is too large of a subject to be dealt with adequately here, but is thoroughly treated by Chandra [123] to which the reader is referred.

REFERENCES

1. Hill, R.M., *Silicone Surfactants.* Surfactant Science Series. Vol. 86. 1999, New York: Marcel Dekker. 360pp.
2. Baney, R.H. and X. Cao, Polysilsesquioxanes, in *Silicon-Containing Polymers*, R.G. Jones, W. Ando, and J. Chojnowski, eds. 2000, Dordrecht, the Netherlands: Kluwer Academic Publishers. pp. 157–184.
3. Pape, P.G. and E.P. Plueddemann, Methods for improving the performance of silane coupling agents. *Journal of Adhesion Science and Technology*, 1991. **5**(10): 831–842.
4. Plueddemann, E.P., Chemistry of silane coupling agents, in *Silylated Surfaces*, D.E. Leyden and W.T. Collins, eds. 1980, New York: Gordon & Breach Science Publishers. pp. 31–53.
5. Brook, M.A., *Silicon in Organic, Organometallic, and Polymer Chemistry.* 2000, New York: John Wiley & Sons, Inc.
6. Owen, M.J., Siloxane surface-activity. *Advances in Chemistry Series*, 1990. **224**: 705–739.
7. Hill, R.M., Dynamics of surfactant enhanced spreading. *European Coatings Journal*, 1998. **7–8**: 550–553.
8. Economides, M.J. and T. Martin, *Modern Fracturing: Enhancing Natural Gas Production.* 2007, Houston, TX: ET Publishing.
9. Pape, P.G., Silicones: Unique chemicals for petroleum processing. *Journal of Petroleum Technology*, 1983. **35**(7): 1197–1204.
10. Daniel-David, D., A. Le Follotec, I. Pezron, C. Dalmazzone, C. Noik, L. Barre, and L. Komunjer, Destabilisation of water-in-crucle oil emulsions by silicone copolymer demulsifiers. *Oil and Gas Science and Technology*, 2008. **63**(1): 165–173.
11. Lakatos, I., J. Toth, J. Lakatos-Szabo, B. Kosztin, G. Palasthy, and H. Woltje, Application of silicone microemulsion for restriction of water production in gas wells. In *Proceedings of the 2002 13th SPE European Petroleum Conference.* October 29–31, 2002, Aberdeen, U.K.: Society of Petroleum Engineers (SPE).

12. Lakatos, I., J. Toth, K. Bauer, J. Lakatos-Szabo, B. Kosztin, G. Palasthy, and H. Woltje, Comparative study of different silicone compounds as candidates for restriction of water production in gas wells. In *Proceedings of the 2003 SPE International Symposium on Oilfield Chemistry*. February 5–7, 2003, Houston, TX: Society of Petroleum Engineers.

13. Skildum, J.D., J.R. Baran, W.W. Fan, and M.P. Shinbach, US 2010/0276142, Method of treating proppants and fractures in-situ with fluorinated silane, 3M Innovative Properties Co., 2010.

14. Reddy, B.R., L.S. Eoff, D.L. Zhang, E.D. Dalrymple, and P.S. Brown, US8053395, Compositions for increasing gas production from a subterranean formation, Halliburton Energy Services, Inc., 2011.

15. Owen, M.J., Defoamers, in *Kirk-Othmer Encyclopedia of Chemical Technology*, A. Seidel and M. Bickford, eds. 2000, Chapter 20, New York: John Wiley & Sons, Inc. pp. 269–283.

16. Fink, J.K., *Oil Field Chemicals*. 2003, Burlington, MA: Elsevier Science.

17. Kelland, M.A., *Production Chemicals for the Oil and Gas Industry*. 2009, Boca Raton, FL: Taylor & Francis.

18. Lucassen, J. and M. Van Den Tempel, Dynamic measurements of dilational properties of a liquid interface. *Chemical Engineering Science*, 1972. **27**(6): 1283–1291.

19. McBain, J. and J.V. Robinson, Surface properties of oils, DTIC Document No. NACA-TN-1844. 1949, Stanford, CA: Stanford University.

20. DeVries, A.J., Foam stability. A fundamental investigation of the factors controlling the stability of foams. *Rubber Chemistry and Technology*, 1958. **31**(5): 1142–1205.

21. Derjaguin, B.V. and A.S. Titievskaya, Static and kinetic stability of free films and froths. In *Proceedings of the Second International Congress on Surface Activity*. 1957, London, U.K.: Butterworth's Scientific Publications.

22. Fransen, G., R. Carbajal, and I.J. Guevara, Foam detection in process units. In *Latin American and Caribbean Petroleum Engineering Conference*. 2009, Richardson, TX: Society of Petroleum Engineers.

23. Schramm, L.L., *Foams: Fundamentals and Applications in the Petroleum Industry*. Vol. 242. 1994, Washington, DC: An American Chemical Society Publication.

24. Brandrup, J., E. Immergut, and E.A. Grulke, *Polymer Handbook*, 2nd edn. 1975, New York: Wiley.

25. Bauget, F., D. Langevin, and R. Lenormand, Dynamic surface properties of asphaltenes and resins at the oil–air interface. *Journal of Colloid and Interface Science*, 2001. **239**(2): 501–508.

26. Poindexter, M.K., N.N. Zaki, P.K. Kilpatrick, S.C. Marsh, and D.H. Emmons, Factors contributing to petroleum foaming. 1. Crude oil systems. *Energy and Fuels*, 2002. **16**(3): 700–710.

27. Callaghan, I., A. McKechnie, J. Ray, and J. Wainwright, Identification of crude oil components responsible for foaming. *Society of Petroleum Engineers Journal*, 1985. **25**(02): 171–175.

28. Langevin, D., Influence of interfacial rheology on foam and emulsion properties. *Advances in Colloid and Interface Science*, 2000. **88**(1): 209–222.

29. Scheludko, A. and E. Manev, Critical thickness of rupture of chlorbenzene and aniline films. *Transactions of the Faraday Society*, 1968. **64**: 1123–1134.

30. Callaghan, I., C. Gould, R. Hamilton, and E. Neustadter, The relationship between the dilatational rheology and crude oil foam stability. I. Preliminary studies. *Colloids and Surfaces*, 1983. **8**(1): 17–28.

31. Binks, B.P., A. Rocher, and M. Kirkland, Oil foams stabilised solely by particles. *Soft Matter*, 2011. **7**(5): 1800–1808.

32. Sheng, J., B. Maini, R. Hayes, and W. Tortike, Experimental study of foamy oil stability. PETSOC-97-04-02. *Journal of Canadian Petroleum Technology*, 1997. **36**(04). Doi: 10.2118-97-04-02.

33. Sperling, L.H., *Introduction to Physical Polymer Science*, 3rd edn. 2001, New York: John Wiley & Sons.

34. Prud'homme, R.K., *Foams: Theory, Measurements, Applications*. Vol. 57. 1995, Boca Raton, FL: CRC Press.

35. Shearer, L. and W. Akers, Foam stability. *The Journal of Physical Chemistry*, 1958. **62**(10): 1264–1268.

36. Callaghan, I.C., S. Hickman, F. Lawrence and P. Melton. *Spec. Publ. R. Soc. Chem.* 59, 48 (1987).

37. Keil, J.W., US4537677, Oil emulsions of fluorosilicone fluids, Dow Corning Corporation, 1985.

38. Evans, E.R., US4329528, Method of defoaming crude hydrocarbon stocks with fluorosilicone compounds, General Electric Company, 1982.

39. Rome, C. and T. Hueston, *Silicone in the Oil and Gas Industry*, 2002, Dow Corning Corp. Ref No. 26-1139.01, Midland, MI.

40. Wylde, J.J., Successful field application of novel, nonsilicone antifoam chemistries for high-foaming heavy-oil storage tanks in Northern Alberta. *SPE Production and Operations*, 2010, **25**(01), 25–30.

41. Bikerman, J.J., *Foams*. Vol. 10. 2013, New York: Springer Science and Business Media.

42. Callaghan, I.C. and E.L. Neustadter, Foaming of crude oils: A study of non-aqueous foam stability. *Chemistry and Industry (London)*, 1981. **2**: 53–57.

43. Rocker, J., A. Mahmoudkhani, L. Bava, and B. Wilson, Low environmental impact nonsilicone defoamers for use in oil/gas/water separators. In *SPE Eastern Regional Meeting*. 2011, Colombus, OH: Society of Petroleum Engineers.

44. Cowan, K. and L. Eoff, Surfactants: Additives to improve the performance properties of cements. In *SPE International Symposium on Oilfield Chemistry*. 1993, Richardson, TX: Society of Petroleum Engineers.

45. Bava, L. and B. Wilson, Evaluation of defoamer chemistries for deepwater drilling and cementing applications. In *SPE Deepwater Drilling and Completions Conference*. 2014, Richardson, TX: Society of Petroleum Engineers.

46. Wilson, B., A. Mahmoudkhani, L. Levy, and L. Bava, New generation of "Green" defoamers for challenging drilling and cementing applications. In *SPE Production and Operations Symposium*. 2013, Richardson, TX: Society of Petroleum Engineers.

47. Roddy, C.W., J. Chatterji, B.J. King, and D.C. Brenneis, US8030253, Foamed cement compositions comprising oil-swellable particles, Halliburton Energy Services, Inc., 2011.

48. Baret, J.F. and B. Boussouira, EP0315243 A1, Oil-well cement slurries with good fluid-loss control, Pumptech N.V., 1989.

49. Bosma, M., E. Cornelissen, P. Reijrink, G. Mulder, and A. De Wit, Development of a novel silicone rubber/cement plugging agent for cost effective thru' tubing well abandonment. In *IADC/SPE Drilling Conference*. 1998, Richardson, TX: Society of Petroleum Engineers.

50. Pietrangeli, G.A., L. Quintero, T.A. Jones, and Q. Darugar, Treatment of water in heavy crude oil emulsions with innovative microemulsion fluids. In *SPE Heavy and Extra Heavy Oil Conference—Latin America*. 2014, Medellin, CO: Society of Petroleum Engineers.

51. Dalmazzone, C., C. Noik, and L. Komunjer, Mechanism of crude-oil/water interface destabilization by silicone demulsifiers. *SPE Journal*, 2005. **10**(1): 44–53.

52. Phukan, M., A. Saxena, M.K. Dubey, A. Palumbo, and K. Koczo, US8779012, Biodegradable polyorganosiloxane demulsifier composition and method for making the same, Momentive Performance Materials Inc., 2014.

53. Saxena, A., M. Phukan, U. Senthilkumar, I. Procter, S. Gonzalez, K. Koczo, S. Azouani, and V. Kumar, US8507565, Polyorganosiloxane demulsifier compositions and methods of making same, Momentive Performance Material Inc., 2013.

54. Noik, C., C. Dalmazzone, and L. Komunjer, Mechanism of crude oil/water interface destabilization by silicone demulsifiers. In *International Symposium on Oilfield Chemistry*. 2003, Houston, TX: Society of Petroleum Engineers.

55. Goldszal, A. and M. Bourrel, Demulsification of crude oil emulsions: Correlation to microemulsion phase behavior. *Industrial and Engineering Chemistry Research*, 2000. **39**(8): 2746–2751.

56. Salager, J.L., Físicoquimica De Los Sistemas Surfactante-Agua-Aceite. Aplicación a La Recuperación Del Petróleo. *Instituto Mexicano del Petróleo*, 1979. **XI**(3): 59–71.

57. Perino, A., C. Noïk, and C. Dalmazzone, Effect of fumed silica particles on water-in-crude oil emulsion: Emulsion stability, interfacial properties, and contribution of crude oil fractions. *Energy and Fuels*, 2013. **27**(5): 2399–2412.

58. Daaou, M. and D. Bendedouch, Water PH and surfactant addition effects on the stability of an algerian crude oil emulsion. *Journal of Saudi Chemical Society*, 2012. **16**(3): 333–337.

59. Nuraini, M., H. Abdurahman, and A. Kholijah, Effect of chemical breaking agents on water-in-crude oil emulsion. *International Journal of Chemical and Environmental Engineering*, 2011. **2**(4): 250–254.

60. Rondon, M., P. Bouriat, J. Lachaise, and J.-L. Salager, Breaking of water-in-crude oil emulsions. 1. Physicochemical phenomenology of demulsifier action. *Energy and Fuels*, 2006. **20**(4): 1600–1604.

61. Daniel-David, D., I. Pezron, C. Dalmazzone, C. Noik, D. Clausse, and L. Komunjer, Elastic properties of crude oil/water interface in presence of polymeric emulsion breakers. *Colloids and Surfaces A*, 2005. **270**: 257–262.

62. Gotz, K. and R. Gerd, US3677962, Process for breaking petroleum emulsions, Goldschmidt Ag Th, 1972.

63. Langevin, D., Oil-water emulsions, in *Encyclopedia of Surface and Colloid Science*, 2nd edn., P. Somasundaran, ed. Vol. 6. 2006, New York: Taylor & Francis.

64. Daniel-David, D., I. Pezron, D. Clausse, C. Dalmazzone, C. Noik, and L. Komunjer, Interfacial properties of a silicone copolymer demulsifier at the air/water interface. *Physical Chemistry Chemical Physics*, 2004. **6**(7): 1570–1574.

65. Phukan, M., K. Koczo, B. Falk, and A. Palumbo, New silicon copolymers for efficient demulsification. In *SPE Oil and Gas India Conference and Exhibition*. 2010, Mumbai, India: Society of Petroleum Engineers.

66. Dalmazzone, C. and C. Noik, Development of new green demulsifiers for oil production. In *SPE International Symposium on Oilfield Chemistry*. 2001, Houston, TX: Copyright Society of Petroleum Engineers Inc.

67. Le Follotec, A., I. Pezron, C. Noik, C. Dalmazzone, and L. Metlas-Komunjer, Triblock copolymers as destabilizers of water-in-crude oil emulsions. *Colloids and Surfaces A*, 2010. **365**(1–3): 162–170.

68. Salazar, J.M. and C. Torres-Verdín, Quantitative comparison of processes of oil-and water-based mud-filtrate invasion and corresponding effects on borehole resistivity measurements. *Geophysics*, 2008. **74**(1): E57–E73.

69. Caenn, R., H.C. Darley, and G.R. Gray, *Composition and Properties of Drilling and Completion Fluids*. 2011, Houston, TX: Gulf Professional Publishing.

70. Wagle, V.B., S. Savari, and S.D. Kulkarni, US9133385, Method for improving high temperature rheology in drilling fluids, Halliburton Energy Services, Inc., 2015.

71. Amani, M., M. Al-Jubouri, and A. Shadravan, Comparative study of using oil-based mud versus water-based mud in HPHT fields. *Advances in Petroleum Exploration and Development*, 2012. **4**(2): 18–27.

72. Quintero, L., D.E. Clark, T.A. Jones, J.-l. Salager, and A. Forgiarini, US8091645, In situ fluid formation for cleaning oil- or synthetic oil-based mud, Baker Hughes Incorporated, 2012.

73. Romenesko, D.J. and H.M. Schiefer, US4381241, Invert emulsions for well-drilling comprising a polydiorganosiloxane and method therefor, Dow Corning Corporation, 1983.

74. Donatelli, P.A. and J.W. Keil, US4421656, Silicone emulsifier composition, invert emulsions therefrom and method therefor, Dow Corning Corporation, 1983.

75. Smith, P.S. and J.A. Hibbert, US6156805, Stabilizing emulsions, Schlumberger Technology Corporation, 2000.

76. Patel, A.D., US5712228, Silicone based fluids for drilling applications, M-I Drilling Fluids L.L.C., 1998.

77. Patel, A.D., US5707939, Silicone oil-based drilling fluids, M-I Drilling Fluids, 1998.

78. Patel, A.D., R.J. Bell, B. Hoxha, and S. Young, US6405809, Conductive medium for openhold logging and logging while drilling, M-I L.L.C., 2002.

79. Garti, N. and A. Aserin, Double emulsions stabilized by macromolecular surfactants. *Advances in Colloid and Interface Science*, 1996. **65**(0): 37–69.

80. Patel, A.D., R.J. Bell, S. Young, and A. Tehrani, US6793025, Double emulsion based drilling fluids, M-I L.L.C., 2004.

81. Darley, H.C. and G.R. Gray, *Composition and Properties of Drilling and Completion Fluids*. 1988, Houston, TX: Gulf Professional Publishing.

82. Perrin, D., M. Caron, and G. Gaillot, *Well Completion and Servicing: Oil and Gas Field Development Techniques*. 1999, Paris, France: Editions Technip.

83. Bloys, J.B., M.E. Gonzalez, J. Lofton, R.B. Carpenter, S. Azar, D. Wiliams, J.D. McKenzie, J. Cap, R.E. Hermes, and R.G. Bland, Trapped annular pressure mitigation: Trapped annular pressure—A spacer fluid that shrinks (update). In *IADC/SPE Drilling Conference*. 2008, Dallas, TX: Society of Petroleum Engineers.

84. Quintero, L., D.E. Clark, A.E. Cardenas, J.-L. Salager, A. Forgiarini, and A.H. Bahsas, WO2013173233A1, Dendritic surfactants and extended surfactants for drilling fluid formulations, Baker Hughes Incorporated, USA. 2013.

85. Pattillo, P.D., B.W. Cocales, and S.C. Morey, Analysis of an annular pressure buildup failure during drill ahead. *SPE Drilling and Completion*, 2006. **21**(04): 242–247.

86. House, R.F. and F.A. Scearce, US4528104, Oil based packer fluids, NL Industries, 1985.

87. Carlos, W., US8186436, Thermal insulating packer fluid, Warren Carlos, 2012.

88. Webmaster, Silicone solutions, products and technologies. 2016; Available from: www. dowcorning.com.

89. Martinez, I., Properties of liquids. Available from: http://webserver.dmt.upm. es/~isidoro/dat1/eLIQ.pdf. (accessed January 11, 2016).

90. Webmaster, Silicone fluid. 2016. Available from: http://www.shinetsusilicone-global. com/catalog/pdf/kf96_e.pdf. (accessed January 11, 2016).

91. Ramirez, J.C., R. Ogle, A. Carpenter, and D. Morrison, The hazards of thermal expansion. In *Second Latin American Process Safety Conference and Expo*, Sao Paulo, Brazil. 2010.

92. Webmaster, Thermomechanical analysis. 2016. Available from: http://www.engineering toolbox.com/cubical-expansion-coefficients-d_1262.html. (accessed January 11, 2016).

93. Smith, J., H. Van Ness, and M. Abbott, *Introduction to Chemical Engineering Thermodynamics*, 7th edn. Chemical Engineer Series. Vol. 18. 1996, New York: McGraw-Hill Science. pp. 1–817.

94. Pusch, G., R. Meyn, W. Burger, and M. Geck, US6206102, Method for stabilizing the gas flow in water-bearing natural gas deposits or reservoirs, Wacker-Chemie GmbH, 2001.

95. Pope, G.A. and J.R. Baran Jr., US7585817, Compositions and methods for improving the productivity of hydrocarbon producing wells using a non-ionic fluorinated polymeric surfactant, Board of Regents, The University of Texas System and 3M Innovative Properties Company, 2009.

96. Pope, G.A., M.M. Sharma, V. Kumar, and J.R. Baran, US7772162, Use of fluorocarbon surfactants to improve the productivity of gas and gas condensate wells, Board of Regents, The University of Texas System and 3M Innovative Properties Company, 2010.

97. Pope, G.A., J.R. Baran Jr., J. Skildum, V. Bang, and M.M. Sharma, US8043998, Method for treating a fractured formation with a non-ionic fluorinated polymeric surfactant, Board of Regents, The University of Texas System and 3M Innovative Properties Company, 2011.

98. Arco, M.J. and R.J. Dams, US7629298, Sandstone having a modified wettability and a method for modifying the surface energy of sandstone, 3M Innovative Properties Company, 2009.

99. Shaari, N.A.E., A. Swint, and L.J. Kalfayan, Utilizing organosilane with hydraulic fracturing treatments to minimize fines migration into the proppant pack Â" a field application. In *SPE Western Regional and Pacific Section AAPG Joint Meeting*. 2008, Bakersfield, CA: Society of Petroleum Engineers.

100. Eoff, L.S., B.R. Reddy, E.D. Dalrymple, D.M. Everett, M. Gutierrez, and D. Zhang, US7552771, Methods to enhance gas production following a relative-permeability-modifier treatment, Halliburton Energy Services, Inc., 2009.

101. Zhang, K., US7723274, Method for making particulate slurries and particulate slurry compositions, Trican Well Service Ltd., 2010.

102. Zhang, K., US20100267593, Control of particulate entrainment by fluids, Trican Well Service Ltd., 2010.

103. Kim, T.W. and A.R. Kovscek, Wettability alteration of a heavy oil/brine/carbonate system with temperature. *Energy and Fuels*, 2013. **27**(6): 2984–2998.

104. Menezes, J.L., J. Yan, and M.M. Sharma, Mechanism of wettability alteration due to surfactants in oil-based muds. In *Proceedings: 1989 SPE International Symposium on Oilfield Chemistry*. February 8–10, 1989. Houston, TX: Society of Petroleum Engineers of AIME.

105. Adibhatla, B., K.K. Mohanty, P. Berger, and C. Lee, Effect of surfactants on wettability of near-wellbore regions of gas reservoirs. *Journal of Petroleum Science and Engineering*, 2006. **52**(1–4): 227–236.

106. Lea, J.F., H.V. Nickens, and M. Wells, *Gas Well Deliquification*. 2011, Houston, TX: Gulf Professional Publishing.

107. Nikolov, A.D., D.T. Wasan, D.W. Huang, and D.A. Edwards, *The Effect of Oil on Foam Stability: Mechanisms and Implications for Oil Displacement by Foam in Porous Media*. 1986, Richardson, TX: Society of Petroleum Engineers.

108. Koczo, K., O. Tselnik, and B. Falk, Silicon-based foamants for foam assisted lift of aqueous-hydrocarbon mixtures. In *SPE International Symposium on Oilfield Chemistry*. 2011, The Woodlands, TX: Society of Petroleum Engineers.

109. Easton, T. and B. Thomas, US4818251, Removal of water haze from distillate fuel, Dow Corning Ltd., 1989.

110. Pelletier, É. and R. Siron, Silicone-based polymers as oil spill treatment agents. *Environmental Toxicology and Chemistry*, 1999. **18**(5): 813–818.

111. Nedwed, T., G. Canevari, E. Febbo, and T. Coolbaugh, Use of chemical spreading agents to provide a new oil spill response option. In *Abstracts of Papers, 243rd ACS National Meeting*. 2012, Washington, DC: American Chemical Society.

112. Fingas, M., *The Basics of Oil Spill Cleanup*. 2012, Boca Raton, FL: CRC Press.

113. Fingas, M., *Review of Oil Spill Herders*. 2013. Available from: http://www.pwsrcac.org/programs/environmental-monitoring/non-dispersing-oil-spill-response-technologies/. (accessed January 18, 2016).

114. Larson, H., Responding to oil spill disasters. The regulations that govern their response, 2010. Available from: http://www.wise-intern.org/journal/2010/hattielarsonwise2010.pdf (accessed January 22, 2016).

115. Buist, I., S. Potter, and T. Nedwed, Herding agents to thicken oil spills in drift ice for in situ burning: New developments. *International Oil Spill Conference Proceedings*, 2011. **2011**(1): abs230.

116. Van Dyke, K., *Fundamental of Petroleum*. 1997, Austin, TX: The University of Texas Austin: Petroleum Extension Service.

117. Ahmed, T., *Equations of State and PVT Analysis*. 2007, Houston, TX: Gulf Publishing Company. pp. 1–553.

118. Economides, M.J., *Modern Fracturing*. 1998, Houston, TX: Energy Tribune Publishing Inc.

119. Kamath, I.S.K. and S.S. Marsden, *A Wettability Scale for Porous Media*. 1966, Richardson, TX: Society of Petroleum Engineers.

120. Kawamura, T., Y. Sakamoto, M. Ohtake, Y. Yamamoto, T. Komai, H. Haneda, and J.-H. Yoon, Dissociation behavior of pellet-shaped methane hydrate in ethylene glycol and silicone oil. Part 1: Dissociation above ice point. *Industrial and Engineering Chemistry Research*, 2006. **45**(1): 360–364.

121. Kawamura, T., Y. Yamamoto, J.-H. Yoon, Y. Sakamoto, T. Komai, H. Haneda, M. Ohtake, and K. Ohga, Dissociation behavior of methane gas hydrate in silicone oil at the temperature condition of under 0°C. In *Proceedings of the 14th (2004) International Offshore and Polar Engineering Conference*. 2004, Toulon, France: The International Offshore and Polar Engineering Conference. pp. 48–51.

122. Haensel, R., M. Fiedel, M. Ferenz, and J. Venzmer, US 2013/0217930, Use of self-cross-linked siloxanes for the defoaming of liquid hydrocarbons, Evonik Industries, 2013.

123. Chandra, G., *Organosilicon Materials*. Vol. 3. 2013, New York: Springer.

Index

Printed and bound by CPI Group (UK) Ltd, Croydon, CR0 4YY

01/11/2024

01782617-0011